Theorems, Corollaries, Lemmas, and Methods of Proof

PURE AND APPLIED MATHEMATICS

A Wiley-Interscience Series of Texts, Monographs, and Tracts

A complete list of the titles in this series appears at the end of this volume.

Theorems, Corollaries, Lemmas, and Methods of Proof

Richard J. Rossi
Department of Mathematical Sciences
Montana Tech
Butte, MT

WILEY-INTERSCIENCE
A JOHN WILEY & SONS, INC., PUBLICATION

Published by John Wiley & Sons, Inc., Hoboken, New Jersey.
Published simultaneously in Canada.

Library of Congress Cataloging-in-Publication Data:

Rossi, Richard J., 1956-
Theorems, corollaries, lemmas, and methods of proof / Richard J. Rossi.
p. cm.
Includes bibliographical references and index.
ISBN-13 978-0-470-04295-3 (acid-free paper)
ISBN-10 0-470-04295-8 (acid-free paper)
1. Proof theory—Textbooks. 2. Mathematical analysis—Foundations—Textbooks. 3. Logic, Symbolic and mathematical—Textbooks. I. Title

QA9.54.R668 2006
511.3'6—dc22 2006041609

10 9 8 7 6 5 4 3 2 1

To my parents, my wife Debbie, and in memory of Maggie and Bayes

Contents

Preface

I have written this textbook to help students who are studying mathematics make the transition from the calculus courses to the typical advanced core courses found in an undergraduate math program. Specifically, this book has been written to prepare students for rigorous mathematical reasoning of junior/senior-level courses on advanced calculus, real analysis, and modern algebra. Furthermore, in writing this book it is my hope that students taking a course from this textbook will begin to appreciate the beauty of the axiomatic structure of modern mathematics.

The topics chosen for this book were chosen for pedagogical reasons and have been tried, tested, and adjusted over the last 12 years of teaching a course on "methods of proof." In particular, the following topics are presented in this text.

- Chapter 1 provides an introduction to the axiomatic nature of modern mathematics, key terminology, and commonly used symbols.
- Chapter 2 presents an introduction to symbolic logic and is used to help the student understand why the methods of proof by contrapositive and proof by contradiction are valid methods of proof in Chapter 3.
- Chapter 3 discusses the method of forward direct proof, proof by contrapositive, and proof by contradiction. Also included in this chapter are specialized proofs for uniqueness and existence theorems, the methods of mathematical induction, proof by cases, proofs of biconditional theorems, and disproving a conjecture by using a counterexample.
- Chapter 4 provides a gentle introduction to numbers and number theory. Specifically, this chapter includes topics on binary operators, the natural numbers, whole numbers, integers, rational numbers, irrational numbers, real numbers, properties of numbers, divisibility, prime numbers, and recursively defined numbers.
- Chapter 5 introduces the students to real analysis through the study of sequences and convergence, limits of real-valued functions, continuity, and differentiability. This chapter also introduces the students to convergence proofs of the ϵ-N and ϵ-δ forms.
- Chapter 6 introduces the students to sets and set theory, indexed families of sets, countable and uncountable sets, and group theory.

This book is not meant to cover the foundations of mathematics; therefore, topics such as relations, equivalence classes, and functions as relations have not been included. Furthermore, this text is not meant to be a book on discrete mathematics, and thus topics such as combinatorics and graph theory have not been included. The topics and the ordering of their presentation have been chosen for purely pedagogical reasons.

It is also my experience that the order of presentation is appropriate for the nurture and development of the student's confidence and mathematical maturity. These topics also provide the student with the necessary mathematical tools required to succeed in advanced math courses such as advanced calculus, modern algebra, number theory, and real analysis.

Three special features of this book are (1) a basic discussion of the axiomatic nature of modern mathematics, (2) presentation of algorithms for several different types of proofs, and (3) the idea that scratchwork must occur as part of the proof process. In Chapters 1 and 3, the basic structure of modern mathematics is discussed and each of the key components of modern mathematics is defined. In particular, the following terms are defined and examples of each are presented: definition, axiom, conjecture, proof, theorem, corollary, and lemma.

Throughout the text, algorithms are given providing the students with an outline for attacking a particular type of proof. It is my experience that proving a mathematical result is a very difficult skill for an undergraduate math student to master. For this reason, I have provided the students with a clear approach to attack several different types of proof. In particular, algorithms are provided for forward direct proofs, proof by contrapositive, proof by contradiction, mathematical induction, uniqueness proofs, existence proofs, proof by cases, closure proofs, convergence proofs for both sequences and limits of functions, element chasing proofs, and group theoretical proofs. These algorithms are not intended to present proofs in a cookbook fashion, rather, these algorithms are presented as guides for the student to use when faced with the problem of proving a theorem.

Another distinctive feature of this book is the idea of scratchwork. It is important to emphasize to the students that proving a mathematical result is unlike any problem they have encountered in their previous algebra and calculus classes. Furthermore, it is unlikely that the typical sophomore math student will be able to quickly and easily prove most of the problems in this text. Thus, I emphasize that the process of proving a theorem generally involves creative work other than that presented in the proofs included in this text. My goal is to convince the students to do their scratchwork and creative thinking as a first step in their attempts to prove a theorem; once they are satisfied that their scratchwork successfully demonstrates the truth of the theorem, they can then proceed to begin writing their proof up in a clear and concise fashion. Throughout the text there are several theorems whose proof will be preceded by my scratchwork in an attempt to get the student thinking about the thought processes that went into developing the actual proof.

Numerous exercises have been included in each chapter of this text. I believe that the exercises accompanying this text do indeed cover a wide range of topics and levels of difficulty. I believe that the successful completion of these exercises will help the student gain the confidence necessary to be successful in junior/senior-level mathematics courses, which, of course, is the goal of this book.

When teaching from this book I have used it for a one-semester transition course by covering Chapters 1–3, and parts of Chapters 4–6; for a two-semester sequence I cover Chapters 1–4 in the first semester and Chapters 5 and 6 in the second semester. While I have not taught a course from this book on the quarter system, I believe that Chapters 1–3, Sections 4.1 and 4.2 of Chapter 4, and Sections 5.1 and 5.2 of Chapter 5 would make a suitable course to be taught in a single quarter; for a two-quarter sequence, Chapters 1–3 and Sections 4.1 and 4.2 could be covered in the first quarter with the remainder of the book left for the second quarter. However, there are many different ways to teach from this book, and I leave that to the discretion and goals of the particular instructor.

I am grateful to a number of friends, colleagues, and students for their help and motivation during the writing of this book. I am especially indebted to Lloyd Gavin and Dan Brunk, two very inspirational advisors from whom I learned so much and the two people who are indirectly responsible for this book. A great deal of my motivation for writing this book came from long discussions of modern mathematics with two great colleagues, Steve Cherry and Dennis Haley. Special thanks go to Susan Patton, VCAAR at Montana Tech, for supporting a sabbatical to work on this book. Finally, I wish to thank the following individuals who have also contributed in one way or another to this book: Ray Carroll, David Ruppert, Fred Ramsey, Jay Devore, Scott Lewis, Erin Esp, Celeste McGregor, Michelle Johnson, Russ Akers, and Donielle Biers.

Finally, it was my intent to write a book that introduces the students to the philosophy and structure of modern mathematics as well as prepare them for future courses in theoretical mathematics. It is my hope that I have accomplished this task.

R. J. Rossi

Butte, Montana

Chapter 1

Introduction to Modern Mathematics

The field of mathematics was born out of the human necessity for counting items and determining areas and the desire to explain the natural world. The word *mathematics* is derived from the Greek words *mathema*, which means "science, knowledge, or learning," and *mathematikos*, which means "fond of learning." In addition to being responsible for the roots for the term *mathematics*, the ancient Greeks were also the first people to study pure mathematics and to record their logical arguments in proofs. Furthermore, the ancient Greek mathematicians were the first mathematicians to think abstractly about mathematics; the Babylonians and ancient Egyptians, unlike the Greeks, tended to think of mathematics in only practical terms with applications to trade and other universal problems. Thus, the ancient Greek mathematicians are generally credited with providing the foundation for modern mathematics.

The term "modern mathematics" is generally used to refer to the current formal axiomatic system of mathematics that is based on rigorous logical foundations. *Mathworld*, a popular Internet Website maintained by Wolfram Research, describes the field of mathematics as follows:

> *Mathematics is a broad-ranging field of study in which the properties and interactions of idealized objects are examined. Whereas mathematics began merely as a calculational tool for computation and tabulation of quantities, it has blossomed into an extremely rich and diverse set of tools, terminologies, and approaches which range from the purely abstract to the utilitarian.*

Whereas the roots of mathematics are based on counting and the study of geometric shapes, modern mathematics is much more than just the study of numbers and shapes. In particular, modern mathematics is the science of operations on collections of arbitrary objects. Modern mathematics, or axiomatic mathematics, is developed according to the following structure:

$$\begin{aligned} &\text{Axioms} \Longrightarrow \text{definitions} \Longrightarrow \text{conjectures} \Longrightarrow \text{proofs} \\ &\Longrightarrow \text{theorems} \Longrightarrow \text{generalizations and extensions} \Longrightarrow \cdots \end{aligned}$$

It is this formal structure, along with the abstract nature of mathematics, that sets modern mathematics apart from the earlier developments in mathematics.

1.1 Inductive and Deductive Reasoning

For the most part, the development of early mathematics was motivated by the study of the physical world and natural phenomena by physical scientists. In fact, early mathematicians made most of their discoveries from their observations of physical phenomena and everyday occurences. The process of making inferences based on observations is called *inductive reasoning.*

Definition 1.1.1: *Inductive reasoning* is the method of reasoning based on making inferences and conclusions from observations.

Inductive reasoning is often used to extrapolate from a particular set of observations to a more general conclusion or future event. An example of inductive reasoning is given below.

> *Since the sun has come up every day of my life, it follows that the sun will come up tomorrow.*

This statement is based completely on making inferences from past experiences to what is to be expected to occur in the future. Much of the primary focus of the earliest development of mathematics was based on observed results, and did not rely on the formal justification of the mathematical conclusions. One reason why the inductive approach was the common theme in the early development of mathematics was that it was motivated primarily by the study of physical phenomenon (i.e., physics).

Inductive reasoning is also used in the developing mathematical conjectures; however, inductive reasoning can never be relied on as concrete proof of the validity of a conjecture. A classic example of the fallibility of inductive reasoning is due to Pierre de Fermat's (1601-1665) conjecture that $2^{2^n}+1$ is prime for all natural numbers n. While it is true that $2^{2^n}+1$ is prime for $n = 1, 2, 3, 4$, Leonhard Euler (1707-1783) disproved Fermat's conjecture by showing that $2^{2^5}+1$ is not a prime number. While large amounts of empirical data are likely to be used as evidence to support an unproven conjecture, data based reasoning can never provide absolute proof that a conjecture is true.

The writings of the ancient Greek mathematician Thâles (circa 600 B.C.) provide the first documented use of sound logical reasoning in the justification of a mathematical result. Thales is credited with being the first person to write down a set of postulates, a set of mathematical conclusions, and provide a justification of these conclusions with a sequence of sound logical arguments. Thales' writings provide the first known to use of *deductive reasoning.*

Definition 1.1.2: *Deductive reasoning* is the method of reasoning where a conclusion is reached by logical arguments based on a collection of assumptions.

Following Thales, Greek mathematicians such as Pythagoras (569–500 B.C.), Aristotle (384–322 B.C.), and Euclid (325–265 B.C.) used deductive reasoning in justifying their mathematical results. In fact, it was Euclid who is credited for first proving that there are infinitely many primes and a student of Pythagoras who first proved the irrationality of $\sqrt{2}$.

An example of the use of deductive reasoning to prove a mathematical result is given in Examples 1.1.1 and 1.1.2, which follow.

Example 1.1.1: Let $x = 0.\overline{9}$. Then, using deductive reasoning, it can be shown that the conjecture $x = 1$ is true. In fact, there are many different ways to show deductively that $x = 0.\overline{9} = 1$, including the following deductive argument. Let $x = 0.\overline{9}$. Then

$$10x = 9.\overline{9} \tag{1}$$

$$10x - x = 9x = 9 \tag{2}$$

$$9x = 9 \tag{3}$$

$$x = 1 \tag{4}$$

Example 1.1.2: Conjecture: $\left|\int_1^\infty \frac{\cos(x)}{x^2}\,dx\right| < \infty$.

Proof: Since $\frac{\cos(x)}{x^2} < \frac{1}{x^2}$ on $[1, \infty)$ and

$$\left|\int_a^b f(x)\,dx\right| \le \int_a^b |f(x)|\,dx$$

it follows that

$$\left|\int_1^\infty \frac{\cos(x)}{x^2}\,dx\right| \le \int_1^\infty \left|\frac{\cos(x)}{x^2}\right|\,dx < \int_1^\infty \frac{1}{x^2}\,dx = 1 < \infty$$

Thus, it is true that $\left|\int_1^\infty \frac{\cos(x)}{x^2}\,dx\right| < \infty$.

Whereas deductive reasoning is the method mathematicians must use in the justification of a mathematical result, inductive reasoning still plays an important role in modern mathematics. As mentioned before, mathematical conjectures are often based on empirical data and inductive reasoning. However, empirical data can serve as proof of a conjecture only if (1) there are finitely many cases to consider in the conjecture and (2) all the possible cases are considered and the conjecture is shown to be true in each of these cases. However, except for these rare conjectures involving only finitely many cases, no amount of empirical data is sufficient to prove that a more general mathematical conjecture is actually true; mathematical proof comes only from logically sound deductive reasoning. Thus, new contributions to mathematics are justified using only deductive reasoning. Furthermore, deductive reasoning has made it possible for mathematics to become a formalized axiomatic system of the

Axiom–definition–conjecture–theorem–proof–generalization–extension

form.

1.2 Components of Modern Mathematics

The components of the modern axiomatic mathematical system are the axioms, definitions, conjectures, proofs, theorems, corollaries, lemmas, and counterexamples. The basic components on which the mathematical structure is built are the axioms and the definitions.

Definition 1.2.1: An *axiom* or *postulate* is a mathematical statement that is taken to be self-evidently true without proof.

Definition 1.2.2: A mathematical *definition* is a statement that gives precise meaning to a mathematical concept or word.

Mathematical axioms are the building blocks on which an axiomatic system is built. In fact, the validity of any further implications and mathematical conclusions in an axiomatic system will be based on the basic axioms and deductive reasoning. An example of one of the important axioms in axiomatic set theory is the "axiom of choice," given below.

> ***Axiom:*** Let C be a nonempty set, and if A_α is a nonempty set for each α in C, then it is possible to chose an x_α from the set A_α for each $\alpha \in C$.

The axiom of choice is a very important axiom in the foundation on which axiomatic set theory is based. The following two axioms were stated and used throughout Euclid's *Elements*, the first book of axiomatic mathematics.

> ***Axiom:*** Two things that are equal to the same thing are also equal to one another (i.e., "If $a = c$ and $b = c$, then $a = b$").

Axiom: If equals be added to equals, the wholes are equals (i.e., "If $a = b$, then $a + c = b + c$").

One of the most famous axioms is the *parallel-line axiom*, which is also known as the *parallel postulate.*

Parallel-Line Axiom: Given any straight line and a point not on it, there exists one and only one straight line that passes through that point and never intersects the first line, no matter how far the lines are extended.

The parallel-line axiom is equivalent to the Fifth Postulate of Book I of Euclid's *Elements* and is an important axiom of Euclidean geometry. In fact, the foundations of non-Euclidean geometry were developed by mathematicians who did not accept the parallel-line axiom. Euclid's fifth postulate is given below:

Euclid's Fifth Postulate: If a straight line falling on two straight lines makes the interior angles on the same side less than two right angles, the two straight lines, if produced indefinitely, meet on that side on which are the angles less than the two right angles.

Along with the axioms, the other basic building block in an axiomatic mathematical system are the definitions. Now, unlike the dictionary definition of a word, mathematical definitions are designed to have one and only one interpretation. Specifically, a mathematical definition is a precise statement that is used to give explicit conditions for the mathematical term being defined. Furthermore, a mathematical definition is designed to prevent two different mathematicians from using the same word to represent different mathematical ideas. For example, two mathematicians discussing the continuity of a function are basing their discussion on the following definition:

Definition: A real-valued function $f(x)$ is said to be *continuous* at a point x_0 in the domain of f if and only if

$$\lim_{x \to x_0^-} f(x) = \lim_{x \to x_0^+} f(x) = f(x_0).$$

While there are alternative definitions of the continuity of a function, they are all equivalent to this definition of continuity. On the other hand, consider what might happen with two people discussing the paint color white. Clearly, there can be variations in the actual color of the white paint due to the shade of white or the paint company that produced the paint. In fact, it is not unusual for a person to buy a can of white paint, paint a room, and then be unsatisfied with the resulting shade of white. Therefore, to ensure the consistency of mathematics, it is important that mathematical definitions be clear, precise, and uniformly understood within the mathematical community.

Now, an axiomatic mathematical system begins with explicitly stated axioms and definitions, and from these initial ideas new mathematical results are added using deductive reasoning. Furthermore, the addition of new mathematical results follows from studying and making hypotheses concerning the implications of the axioms and definitions. When the truth of a hypothesized result is not yet known, the result is called a *conjecture.*

Definition 1.2.3: A *conjecture* is any mathematical statement that has not yet been proved or disproved.

Whereas the truths of many mathematical conjectures remain unknown today, one of the most famous and heavily studied conjectures, *Fermat's Last Theorem,* was more recently proved by Andrew Wiles of Princeton and his former student Richard Taylor. Fermat's Last Theorem is stated below.

> **Fermat's Last Theorem:** The equation $x^n + y^n = z^n$ has solutions in positive integers x, y, z and n only when $n = 2$; but there are no solutions for $n > 2$.

Fermat's Last Theorem had been studied intensively for over 300 years before Wiles and Taylor finally proved this result in 1995. While Wiles' proof of Fermat's Last Theorem was an incredible accomplishment, even more importantly, the 300 years of study on this particular problem has led to many important and useful mathematical results. An interesting book detailing the history of Fermat's Last Theorem and Wiles' work is *Fermat's Enigma* by Simon Singh (1997).

Three of the most famous unproven mathematical conjectures are listed below:

> **Goldbach's Conjecture:** Every even integer greater than 2 can be expressed as the sum of two prime numbers.

> **The Odd Perfect Number Conjecture:** There do not exist any perfect odd numbers.

> **The Twin Prime Conjecture:** There are an infinite number of twin primes.

Now, once a conjecture has been shown to be true with a mathematical *proof,* the conjecture can now be called a *theorem*; on the other hand, when a conjecture is shown not to be true, it is no longer of much interest in the mathematical world and is discarded.

Definition 1.2.4: A *proof* of a mathematical result is a sequence of rigorous mathematical arguments that are presented in a clear and concise fashion, and which convincingly demonstrates the truth of the given result.

Definition 1.2.5: A *theorem* is any mathematical statement that can be shown to be true using accepted logical and mathematical arguments.

Note that inductive reasoning is often used in the development of a conjecture; however, the proof of a conjecture or theorem is always based on deductive reasoning. In Examples 1.1.1 and 1.1.2, a mathematical conjecture was considered and then proved; since the conjectures in these two theorems have been proved, these conjectures can now be called *theorems*. The conjectures in Examples 1.1.1 and 1.1.2 are stated below in the form of theorems along with their respective proofs.

Theorem: Let $x = 0.\overline{9}$. Then, $x = 1$.

Proof: Let $x = 0.\overline{9}$. Then

$$10x = 9.\overline{9} \tag{1}$$

$$10x - x = 9x = 9 \tag{2}$$

$$9x = 9 \tag{3}$$

$$x = 1 \tag{4}$$

■

Theorem: $\left| \int_1^\infty \frac{\cos(x)}{x^2}\, dx \right| < \infty.$

Proof: Since $\frac{\cos(x)}{x^2} < \frac{1}{x^2}$ on $[1, \infty)$ and

$$\left| \int_a^b f(x)\, dx \right| \le \int_a^b |f(x)|\, dx$$

it follows that

$$\left| \int_1^\infty \frac{\cos(x)}{x^2}\, dx \right| \le \int_1^\infty \left| \frac{\cos(x)}{x^2} \right| dx < \int_1^\infty \frac{1}{x^2}\, dx = 1 < \infty$$

Thus, it is true that $\left|\int_1^\infty \frac{\cos(x)}{x^2}\,dx\right| < \infty$.

■

It is important to note that a mathematical proof is very different from an empirical proof, which is often used in the sciences, or proof beyond a reasonable doubt, which is used in our legal system. For example, at one time scientists believed beyond a reasonable doubt, on the basis of empirical data, that the earth was the center of the universe; however, it is now understood that the sun is the center of the universe. A mathematical proof must represent absolute truth, so that a theorem is absolutely true regardless of any and all empirical data. For example, the fact that Euclid proved using deductive reasoning that there are infinitely many prime numbers is irrefutable (i.e., absolutely true). Furthermore, a mathematical proof provides a sequence of rigorous logical arguments where each step of the proof and the connection between steps is completely justified using mathematical and/or logical arguments.

Now, the proof of a theorem may be extremely long and complicated, or it may be very short and easily understood. A theorem that has a complicated or long proof is often referred to as a "deep theorem." A proof that takes a novel or unusual approach is often called an "elegant proof." A common feature in the Mathematical Association of America (MAA) publication *Mathematics Magazine* is "Proofs without Words," where a mathematical result is proved without using any words; a proof without words usually involves only formulas and/or graphical representation of the proof and should be self-explanatory.

Similarly, the proof of a theorem might be described as complicated or deep when the proof is long or difficult to follow because of its complexity. For example, the apparent simplicity of Fermat's Last Theorem is betrayed by the length and complexity of its proof. In fact, the proof of Fermat's Last Theorem, due to Wiles and Taylor, is long and very difficult for most mathematicians to follow. Examples of some very important mathematical theorems are listed below. Note that the theorems in this list contain results that are based on only addition and multiplication.

The Pythagorean Theorem: The sum of the squares of the lengths of the legs of a right triangle is equal to the square of the length of the hypotenuse.

Fermat's Last Theorem: $x^n + y^n = z^n$ has no nonzero integer solutions for x, y, and z when $n > 2$.

The Fundamental Theorem of Arithmetic: Every natural number greater than 1 either is prime or can be uniquely factored as a product of primes.

Among these three theorems, the Pythagorean theorem is one of the oldest and widely used theorems in mathematics, Fermat's Last Theorem is most likely the most famous mathematical theorem, and the Fundamental Theorem of Arithmetic shows that the prime numbers are the atoms from which the natural numbers are formed.

Often, great contributions are made to mathematical knowledge in the study of a difficult problem as occurred in the pursuit to prove Fermat's Last Theorem. Hence, the importance of an individual theorem is based not only on its utility but also on its complexity or the difficulty of its proof. Moreover, a theorem may be referred to as a "revolutionary" theorem when its impact on mathematics is dramatic or far-ranging. Often a revolutionary theorem will be important in opening up new directions in mathematical research. An example of a revolutionary and very important mathematical theorem is due to Kurt Gödel (1906–1978), who proved the following theorem in 1931 (Gödel 1931):

> **Gödel's Incompleteness Theorem:** In any consistent formalization of mathematics that is sufficiently strong to axiomatize the natural numbers — that is, sufficiently strong to define the operations that collectively define the natural numbers — one can construct a true statement that can be neither proved nor disproved within that system itself.

Prior to Gödel's Incompleteness Theorem, several influential mathematicians believed that all mathematical truths could be logically derived. In fact, David Hilbert (1862–1943), Bertrand Russell (1872–1970), and Alfred North Whitehead (1861–1947) believed that mathematics could be expressed as an axiomatic system that is free of inconsistencies and is also complete. Specifically, Hilbert, Russell, and Whitehead believed that an axiomatic mathematical system could be constructed where true statements are always true regardless of method of proof (i.e., a consistent system) and that all mathematical truths could be proved from the basic axioms of the system (i.e., a complete system). Gödel's Incompleteness Theorem shows that no mathematical axiomatic system that axiomatizes the natural numbers can be complete and hence, all of the mathematical truths cannot be proved from the basic axioms.

1.3 Commonly Used Mathematical Notation

In the communication of mathematics it is often useful to write mathematical sentences using symbols rather than words. The reason for this is that it makes it easier to read, shortens the communication while conveying all the information, and in essence creates a language of mathematics. Effectively communicating mathematical ideas is like writing an essay; it requires well-composed sentences, paragraphs, and correct mathematical grammar. Often, a mathematical essay is written using a great deal of mathematical shorthand and thus, the reading of modern mathematics will require the knowledge of the symbolic language of mathematics. Over the years, mathematicians have created their own language on the basis of standard mathematical shorthand (i.e., symbols used to shorten mathematical communications). A summary of the standard mathematical notation that used is in this text follows.

Throughout this text several different sets of numbers will be studied. In particular, the collections of numbers that are discussed in this text are the natural numbers, whole numbers, integers, rational numbers, real numbers, and irrational numbers. The notation used in this text to represent each of these sets of numbers is given below.

The Natural Numbers: $\mathbb{N} = \{1, 2, 3, 4, \ldots\}$

The Whole Numbers: $\mathbb{W} = \{0, 1, 2, 3, 4, \ldots\}$

The Integers: $\mathbb{Z} = \{0, \pm 1, \pm 2, \pm 3, \pm 4, \ldots\}$

The Rationals: $\mathbb{Q} = \left\{q : q = \frac{p}{r} \text{ for } p \text{ and } r \neq 0 \text{ integers}\right\}$

The Reals: $\mathbb{R} = \{x : -\infty < x < \infty\}$

The Positive Reals: $\mathbb{R}^+ = \{x : 0 < x < \infty\}$

The Negative Reals: $\mathbb{R}^- = \{x : -\infty < x < 0\}$

The Irrationals: $\mathbb{I} = \{r : r \text{ is a real number but not rational}\}$

The Complex Numbers: $\mathbb{C} = \left\{\xi : \xi = a + bi, \ a, b \in \mathbb{R} \text{ and } i = \sqrt{-1}\right\}$

Note that notation analogous to $\mathbb{R}^+$ and $\mathbb{R}^-$ can also be used with the sets $\mathbb{Z}, \mathbb{Q}$, and $\mathbb{I}$. For example, $\mathbb{Z}^+$ is used to represent the positive integers and $\mathbb{Q}^-$ would be used to denote the negative rational numbers. Also, the following standard notation will be used for the intervals of $\mathbb{R}$:

a. **Open Intervals**: The open interval containing the points lying strictly between two endpoints, say, a and b, is denoted by (a, b). This set of numbers can also be represented by $a < x < b$. Note that either a or b, or both a and b may be ∞. In particular, $\mathbb{R} = (-\infty, \infty)$, $\mathbb{R}^+ = (0, \infty)$, and $\mathbb{R}^- = (-\infty, 0)$.

b. **Closed Intervals**: The closed interval containing the points lying between and including two finite endpoints, say, a and b, is denoted by $[a, b]$. This set of numbers can also be represented by $a \leq x \leq b$. Note that a closed interval must have finite endpoints.

c. **Half-Open Intervals**: The half-open/half-closed intervals contain all the points lying strictly between the endpoints a and b and either a or b, but not both a and b. The half-open intervals are denoted by $(a, b]$ and $[a, b)$. These sets of numbers may also be represented by $a < x \leq b$ and $a \leq x < b$, respectively.

Example 1.3.1: Write out, using interval notation, the following sets of real numbers:

a. $-1 < x < 10$

b. $0 \leq x < \infty$

c. $0 < x \leq 10$

d. $-3 \leq x \leq -1.5$

Solutions:

a. $(-1, 10)$

b. $[0, \infty)$

c. $(0, 10]$

d. $[-3, -1.5]$

It is very important that mathematical results be presented using precise and consistent notation. Even the earliest mathematicians began developing and using a symbolic language in their presentations of mathematics. Moreover, in the twentieth century mathematicians began to standardize the symbols and notation used in modern mathematics. Some commonly used mathematical shorthand (i.e., symbols and notation) that is universal within the field of mathematics is given below:

a. $:=$ is often used for "defined to be." For example, the notation $:=$ might be used in defining the set $\mathbb{Z}_E$ that contains the even integers as follows:

$$\mathbb{Z}_E := \{x \in \mathbb{Z} : \ x = 2z \text{ for some integer } z.\}$$

b. s.t. or $\ni$: is often used for "such that." For example, the statement "there exists $x > 0$ such that $f(x) = 0$" could be written as "there exists $x > 0 \ \ni: \ f(x) = 0$.

c. $\in$ is often used for "is an element of" or "is a member of." For example, the statement "x is an element of $\mathbb{R}$" could be written as "$x \in \mathbb{R}$."

d. $x!$ is used to denote "x factorial," where, for a positive integer x, $x! = x(x-1)(x-2)\cdots 3 \cdot 2 \cdot 1$. For example, $6! = 6 \cdot 5 \cdot 4 \cdot 3 \cdot 2 \cdot 1 = 120$.

e. $\sum$ is mathematical shorthand for summation. For example

$$\sum_{n=3}^{7} np^{n-1} = 3p^2 + 4p^3 + 5p^4 + 6p^5 + 7p^6$$

f. Π is mathematical shorthand for product. For example

$$\prod_{i=1}^{10} x^i(1-x)^{10-i} = x(1-x)^9 \times x^2(1-x)^8 \cdots x^9(1-x) \times x^{10}$$

g. $\forall$ is the mathematical shorthand for "for all" or "for every" or "for each." The symbol $\forall$ is referred to as the *universal quantifier*. The symbol $\forall$ was first used by Gerhard Gentzen (1909–1945) in 1934 according to "Earliest Uses of Some of the Words of Mathematics." by Jeff Miller (2006). An example of the usage of $\forall$ is

$$\forall\ i \in \{1, 2, \ldots, 10\},\ g(i) \geq 10$$

This statement is the mathematical shorthand for "for all i ranging from 1 to 10, $g(i) \geq 10$."

h. $\exists$ is the mathematical shorthand for "there exists" or "there is at least one." The symbol $\exists$ is referred to as the *existential quantifier*. The symbol $\exists$ was first used by Giuseppe Peano (1858–1932) in 1895 according to "Earliest Uses of Some of the Words of Mathematics." by Jeff Miller (2006). An example of the usage of $\exists$ is

$$\exists\ i \in \{1, 2, \ldots, 10\} \ni: g(i) \geq 10$$

This statement is the mathematical shorthand for "there exists a value of i between 1 and 10 such that $g(i) \geq 10$."

i. The symbol ∞ is used to represent the concept of the unbounded quantity "infinity." Jon Wallis (1616–1703) is credited for first using the symbol ∞ to represent infinity in 1655 according to *The Penguin Dictionary of Mathematics*, third edition (Nelson 2003). For example, $x \in (0, \infty)$ is the mathematical shorthand for the statement "x is an element of the open interval ranging from 0 to infinity.

j. The symbol $\Longrightarrow$ or $\rightarrow$ is the mathematical shorthand often used for "implies"; for example, $0 < x < y \Longrightarrow 0 < \frac{1}{y} < \frac{1}{x}$.

k. The symbol $\Longleftrightarrow$ or $\leftrightarrow$ is the mathematical shorthand for often used "if and only if"; for example, $ab = 0 \Longleftrightarrow a = 0$ or $b = 0$.

Example 1.3.2: Write the following sentences using as much mathematical notation as possible:

a. If there exists a real number x such that $e^x = 10$, then $x = \ln(10)$.

b. Let a and b be real numbers. The product of a and b is zero if and only if a is zero or b is zero.

Solutions:

a. $\exists\, x \in \mathbb{R} \ni: \ e^x = 10 \Longrightarrow x = \ln(10)$.

b. Let $a, b \in \mathbb{R}$. $ab = 0 \Longleftrightarrow a = 0$ or $b = 0$.

Example 1.3.3: Using common English sentences, write out each of the statements below.

a. $\exists\, x \in (0, \infty) \ni: \ e^x = 3$.

b. $\forall\, x \in \mathbb{Z}, \exists\, y \in \mathbb{Z} \ni: \ x + y = 0$.

c. $\forall\, \epsilon > 0, \ \exists\, N \in \mathbb{N} \ \ni: \left|\frac{1}{n}\right| < \epsilon, \forall\, n \geq N$.

Solutions:

a. There exists a nonnegative real number x such that e^x is equal to 3.

b. For every integer x, there exists an integer y such that $x + y = 0$.

c. For every ϵ greater than 0, there exists a natural number N such that $\left|\frac{1}{n}\right| < \epsilon$ whenever n is greater than or equal to N.

In honor of ancient Greek mathematicians, who laid the foundation for modern mathematics, Greek letters are commonly used in mathematical expressions. For example, π is the universal symbol used to express a constant that denotes the ratio of the circumference of a circle to the diameter of that circle; the uppercase version of π is Π and is used as mathematical shorthand to represent products. The Greek alphabet is listed in Table 1.3.1.

TABLE 1.3.1 The Greek Alphabet

Name	Uppercase Symbol	Lowercase	Name	Uppercase Symbol	Lowercase
Alpha	A	α	Nu	N	ν
Beta	B	β	Xi	Ξ	ξ
Gamma	Γ	γ	Omicron	O	o
Delta	Δ	δ	Pi	Π	π
Epsilon	E	ϵ	Rho	P	ρ
Zeta	Z	ζ	Sigma	Σ	σ
Eta	H	η	Tau	T	τ
Theta	Θ	θ	Upsilon	Y	υ
Iota	I	ι	Phi	Φ	ϕ
Kappa	K	κ	Chi	X	χ
Lambda	Λ	λ	Psi	Ψ	ψ
Mu	M	μ	Omega	Ω	ω

Example 1.3.4: The Greek letters π, γ, Γ, β, ψ and ζ are commonly used in mathematics as follows:

$$\pi = \text{ratio of circumference of every circle to its diameter} \approx 3.14159$$

$$\gamma = \lim_{n \to \infty} \left(\sum_{i=1}^{n} \frac{1}{k} - \ln(n) \right) \approx 0.577 \qquad \text{(Euler–Mascheroni constant)}$$

$$\pi(x) = \text{number of prime numbers} \leq x$$

$$\Gamma(k) = \int_0^\infty x^k e^{-x}\, dx \qquad \text{(gamma function)}$$

$$\beta(j, k) = \frac{\Gamma(j)\Gamma(k)}{\Gamma(k + j)} \qquad \text{(beta function)}$$

$$\psi(x) = \frac{d}{dx}\left[\ln(\Gamma(x)\right] = \frac{\Gamma'(x)}{\Gamma(x)} \qquad \text{(digamma function)}$$

$$\zeta(s) = \sum_{n=1}^{\infty} \frac{1}{n^s} \qquad \text{(Riemann–Zeta function)}$$

EXERCISES

1.1 Briefly explain how inductive and deductive reasoning differ.

1.2 Briefly explain how a theorem differs from a

a. Corollary
b. Conjecture
c. Axiom
d. Lemma
e. Definition

1.3 Using the local library or the Internet, identify and summarize two well-known and unproven mathematical conjectures.

1.4 Briefly explain how an axiom and a definition differ.

1.5 Using the Internet, find short biographies of the following Greek mathematicians:

a. Thales
b. Euclid
c. Pythagoras

1.6 Using the Internet, find short biographies of the following mathematicians:

a. Gauss
b. Gödel
c. Cantor
d. Hilbert
e. Riemann
f. Fermat
g. Russell
h. Euler
i. Fibonacci

1.7 Using the local library or the Internet, find statements of the following mathematical results:

a. Zorn's lemma

b. The parallel postulate

c. The Central Limit Theorem

d. Fatou's lemma

e. Heine–Borel theorem

1.8 Using the local library or the Internet, find three different definitions of "continuous function."

1.9 Write out each of the following sentences using as much mathematical notation as possible:

a. For every epsilon greater than 0, there exists a delta greater than 0 such that $|f(x) - f(y)|$ is less than epsilon whenever $|x - y|$ is less than delta.

b. If n is a natural number and n is an odd number, then there exists an integer k such that $n = 2k + 1$.

c. If n is a natural number and n is an even number, then there exists an integer k such that $n = 2k$.

d. The sum of the first n natural numbers is $\frac{n(n+1)}{2}$.

e. If x is strictly between 0 and 1, then x^n is less than 1 for all n in the natural numbers.

1.10 Translate each of the following mathematical sentences into English:

a. $\exists\, \alpha \in \mathbb{R} \ni: \beta_\alpha \geq 2$.

b. If $\forall\, n \in \mathbb{N},\ F_{n+2} = F_{n+1} + F_n$, then $\sum_{i=1}^{n} F_n = F_{n+2} - 1$.

c. $\forall\, \psi \in \mathbb{R}^+,\ \exists\, \rho > 0 \ni:\ \rho\psi < 1$ and $\frac{\psi}{\rho} > 1$.

d. If $n \in \mathbb{N}$, then $\prod_{i=1}^{n} i = n!$.

Chapter 2

An Introduction to Symbolic Logic

Mathematical logic and symbolic logic form the foundation on which all of modern mathematics is built. George Boole (1815–1864) is primarily responsible for uniting mathematics and logic. Boole's publication of *The Mathematical Analysis of Logic, Being an Essay towards a Calculus of Deductive Reasoning* (Boole 1847) is considered by many to be the starting point for the axiomatic formalization of mathematics. Prior to Boole's work, deductive reasoning had been utilized by many, if not most, mathematicians. The list of the most influential mathematicians who are credited with using deductive reasoning and strong mathematical logic includes Aristotle, Isaac Newton (1642–1727), Gottfried Leibniz (1646–1716), Leonhard Euler, and Karl Friedrich Gauss (1777–1855). Gottlob Frege (1848–1925) extended and strengthened Boole's work with his development of the predicate calculus. In fact, Frege is generally credited with introducing modern symbolic logic to the field of mathematics. A very good discussion of mathematical logic can be found in *Mathematics of the 19th Century* edited by Kolmogorov and Yushkevich (2001).

The foundation of modern mathematics (i.e., axiomatic deductive mathematics) is based entirely on deductive reasoning and mathematical logic. Therefore, an important prelude to the discussion of theorems, corollaries, lemmas, and their proofs is an introduction to mathematical logic. Mathematical logic is an area of mathematics/philosophy that studies the truth of mathematical statements along with the implications of these statements. Furthermore, every mathematical result, including axioms, definitions, theorems, and proofs, must be composed of well-defined and logically correct statements. Also, the communication of mathematics requires the use of correct mathematical grammar and sound logical arguments.

2.1 Statements and Propositional Functions

Mathematics is communicated through the use of mathematical statements and sentences that are used in the axioms, definitions, theorems, proofs, and discussions of mathematical results. The definition of a *logical statement* is given below.

Definition 2.1.1: A *statement* or *proposition* is a declarative sentence that is either true or false.

Example 2.1.1: Determine which of the following mathematical sentences are statements:

a. Is 2 a solution to $x^2 - 4 = 0$?
b. $x^2 - 4x + 4 = 0$.
c. $x^2 - 4x + 4 \geq 0$ for every real number x.
d. $f(x)$ is a continuous function.
e. $|x|$ is an everywhere differentiable function.

Solutions:

a. No, this sentence is not a statement since this is an interrogative sentence.
b. No, this sentence is not a statement since the value of x needs to be known in order to determine the truth of this sentence.
c. Yes, this sentence is a statement since $x^2 - 4x + 4 = (x - 2)^2$, which is greater than or equal to 0 for every real number, so this sentence is true no matter what the value of x is.
d. No, this sentence is not a statement since the explicit form of the function $f(x)$ must be known in order to determine the truth of this sentence.
e. Yes, this sentence is a statement since $|x|$ is not differentiable at $x = 0$, and hence this sentence is false.

Note that a statement is a declarative sentence and not an imperative, interrogative, or exclamatory sentence. For example, "$\sqrt{2}$ is an irrational number" is a statement; however, "Is $\sqrt{2}$ an irrational number?" is not. Furthermore, as seen in Example 2.1.1, some mathematical sentences are not logical statements. In particular, when a mathematical sentence involves a variable, then the truth of the mathematical sentence will depend on the values of the variables. For example, the truth or falsity of the mathematical sentence in Example 2.1.1 part (b) depends on the value of x; if $x = 2$, then this sentence is true, and it is false for any other value of x.

Definition 2.1.2: A declarative sentence $P(x)$ involving a variable x that takes on values in a set Δ is said to be a *propositional function* if and only if $P(x)$ has a well-defined truth value for each value of x in Δ. The set Δ is called the *domain* of the propositional function $P(x)$.

Note that a propositional function is a declarative sentence containing a variable whose truth depends on the specific value of the variable. Thus, the truth or falsity of a propositional function cannot be determined without knowing a specific value of the unknown variable. For example, the declarative sentence "$x^2 - 4 = 0$" is a propositional function that is true when $x = \pm 2$ and false otherwise. However, it is possible for a declarative sentence involving a variable to be true (or false) for all values of the variable, and hence such a sentence is actually a statement and not a propositional function. For example, the sentence "$x^2 + 4 = 0$ has no real solutions" is a declarative sentence that is always true and therefore is a statement, not a propositional function. Furthermore, note that by assigning a value, say, $x = x_0$, to the variable x in a propositional function $P(x)$, $P(x_0)$ becomes a statement.

Example 2.1.2: Determine which of the following sentences are statements and which are propositional functions:

a. $x = 2$ is a solution to $f(x^2 - 4) = 0$.

b. $e^x > 0$ for every real number x.

c. $f(2) = 7$.

d. $f(x)$ is a differentiable function.

e. $\lim_{x \to 1} f(x) = 0$.

Solutions:

a. This sentence is a propositional function since the explicit form of the function $f(x)$ is not specified.

b. This sentence is a statement since the value of e^x is greater than 0 for every real number x.

c. This sentence is a propositional function since the explicit form of the function $f(x)$ is not specified.

d. This sentence is a propositional function since the explicit form of the function $f(x)$ is not specified.

e. This sentence is a propositional function since the explicit form of the function $f(x)$ is not specified.

A mathematical theorem can be composed of both statements and propositional functions. For example, the theorem "$\sqrt{2}$ is an irrational number." is simply a statement. However, the theorem stated below is based on the two propositional functions "$f(x)$ is differentiable at $x = c$" and "$f(x)$ is continuous at $x = c$."

Theorem: If $f(x)$ is differentiable at $x = c$, then $f(x)$ is continuous at $x = c$.

2.2 Combining Statements

In the previous section, simple statements concerning a single object were discussed; however in order to state definitions, theorems, and proofs, more complicated statements are usually needed. In particular, it will often be necessary to combine simple logical statements to form a compound statement that is based on more than one object and the use of the logical operators OR, AND, and NOT. The definitions of these compound statements are given below.

Definition 2.2.1: Let P and Q be statements. The statement "P AND Q" is called the *conjunction* or *meet* of the statements P and Q and is denoted symbolically by $P \wedge Q$.

Definition 2.2.2: The statement "P OR Q" is called the *disjunction* or *join* of the statements P and Q and is denoted symbolically by $P \vee Q$.

Definition 2.2.3: The statement "NOT P" is called the *negation* of the statement P and is denoted symbolically by $\neg P$.

Now, given two statements P and Q, a new statement can be created using any combination of $\wedge$, $\vee$, and/or $\neg$ to combine the statements P and Q. It is important to note that (1) the statement $P \wedge Q$ is true only when both of the statements P and Q are true and (2) the statement $P \vee Q$ is true when either of the statements P and Q is true. That is, since $P \wedge Q$ requires both P and Q to be true in order for $P \wedge Q$ to be true, while $P \vee Q$ requires only that at least one of the statements P and Q is true in order for $P \vee Q$ to be true, it follows that $P \wedge Q$ is a more restrictive statement than is $P \vee Q$. Also, the statement $\neg P$ will be true only when the statement P is false.

Example 2.2.1: Consider the statement S:="$\ln(x)$ is continuous at $x = 3$ and 13 is a prime number." The statement S consists of the two statements P:="$\ln(x)$ is continuous at $x = 3$" and Q:="13 is a prime number." Furthermore, S is a true statement because both of the statements P and Q are true.

Example 2.2.2: Consider the statement S:="$(-1)^2 > 0$ and 3 is a root of $x^2 - 3x + 3 = 0$. The statement S is composed of the two statements P:="$(-1)^2 > 0$" and Q:="3 is a root of $x^2 - 3x + 3 = 0$." Furthermore, S is a true statement since the statement P is true.

Example 2.2.3: Consider the statement P:="4 divides 17," then the negation of P is $\neg P$="4 does not divide 17." In this case, P is a false statement meaning that $\neg P$ is true.

Example 2.2.4: Let P:="Every odd number is a prime number," Q:="4 divides every even number," R:="3 is larger than e," and S:="3 is smaller than π." Express in sentence form and then determine the truth of each of the following compound statements:

a. $P \wedge Q$

b. $\neg P \vee \neg Q$

c. $\neg P \wedge \neg Q$

d. $R \wedge S$

Solutions:

a. The statement $P \wedge Q$ can be written as "Every odd number is a prime number and 4 divides every even number." This statement is false since neither of two these statements is true.

b. The statement $P \vee \neg Q$ can be written as "Every odd number is a prime number or 4 does not divide every even number." This statement is true since the statement $\neg Q$ is true.

c. The statement $\neg P \wedge \neg Q$ can be written as "Every odd number is not a prime number and 4 does not divide every even number." This statement is true since both of these statements are true.

d. The statement $R \wedge S$ can be written as "3 is larger than e and 3 is smaller than π." This statement is true since both of these statements are true.

2.3 Truth Tables

Since a statement must be either true or false, there are two states of nature for any statement P that can be summarized in a truth table as shown below:

P
T
F

Also, since $\neg P$ and P always have the opposite truth values, it follows that when P is true, $\neg P$ is false and vice versa. Thus, the possible states of nature for a statement P and its negation ($\neg P$) can be summarized in the following truth table:

P and $\neg P$

P	$\neg P$
T	F
F	T

Now, when a compound statement S is created from more than one statement, the truth table for S must reflect all possible states of nature for each of the statements used to build S. For example, when the statement of interest is a compound statement based on the two statements P and Q, then there are four possible states of nature, tabulated as follows:

P	Q
T	T
T	F
F	T
F	F

If the statement of interest is a compound statement based on the three statements P, Q, and R, then there are eight different states of nature that must be accounted for in the truth table. The eight possible states of nature for the statements P, Q, and R are given below:

P	Q	R
T	T	T
T	T	F
T	F	T
T	F	F
F	T	T
F	T	F
F	F	T
F	F	F

In general, when a compound statement is based on n different statements, there will be 2^n possible states of nature to account for in the truth table. For example, if a compound statement is built from 10 different statements, then there will be $2^{10} = 1024$ states of nature in the truth table.

The truth tables for the compound statements "P AND Q" and "P OR Q" are given below:

$P \wedge Q$

P	Q	$P \wedge Q$
T	T	T
T	F	F
F	T	F
F	F	F

$P \vee Q$

P	Q	$P \vee Q$
T	T	T
T	F	T
F	T	T
F	F	F

Note that the statement "P AND Q" is true only when both of the statements P and Q are true, and the statement "P OR Q" is false only when both of the statements P and Q are false.

When forming complex truth tables it is often useful to build the truth table according to the following sequence of steps. First, list the statements and their possible states of nature. Next, create intermediate columns that lead to the statement of interest. Finally, the last column to be added to the table should be the column showing the truth of the statement of interest. This approach is illustrated in the following example.

Example 2.3.1: Let P, Q, and R be statements. Write out the truth table for $P \vee (Q \wedge \neg R)$.

Solution: The first step in the solution is to list the eight possible states of nature for the statements P, Q, and R. Next, a column for the statement $Q \wedge \neg R$ will be added to the table, and then its truth will be determined. Finally, add a column for the statement of interest $P \vee (Q \wedge \neg R)$ to the table, and then determine the truth values for this statement. The resulting truth table for $P \vee (Q \wedge \neg R)$ is shown below:

$$P \vee (Q \wedge \neg R)$$

P	Q	R	$Q \wedge \neg R$	$P \vee (Q \wedge \neg R)$
T	T	T	F	T
T	T	F	T	T
T	F	T	F	T
T	F	F	F	T
F	T	T	F	F
F	T	F	T	T
F	F	T	F	F
F	F	F	F	F

Two special types of statement that are often encountered are (1) statements that are always true and (2) statements that are always false. A statement that is always true is called a *tautology*, and a statement that is always false is called a *contradiction*.

Definition 2.3.1: A statement that is true for all of its states of nature is called a *tautology*, and a statement that is false for all of its states of nature is called a *contradiction*.

Since at least one of the statements P and $\neg P$ must be true, the statement "P OR $\neg P$" ($P \vee \neg P$) is an example of a statement that is a tautology. Also, since it is impossible for both P and $\neg P$ to be true, it follows that "P AND $\neg P$" ($P \wedge \neg P$) is always false and hence, the statement "P AND $\neg P$" is a contradiction.

Now, one approach that can be used to determine whether a compound statement is either a tautology, contradiction, or neither is to simply create the truth table for the statement. This approach is illustrated in the following example.

Example 2.3.2: Show that

a. $P \vee \neg P$ is a tautology.

b. $P \wedge \neg P$ is a contradiction.

c. $\neg(P \wedge Q) \vee (P \vee Q)$ is a tautology.

Solutions:

a. The truth table for $P \vee \neg P$ is given below:

$P \vee \neg P$

P	$\neg P$	$P \vee \neg P$
T	F	T
F	T	T

b. The truth table for $P \wedge \neg P$ is given below:

$P \wedge \neg P$

P	$\neg P$	$P \wedge \neg P$
T	F	F
F	T	F

c. The solution to part (c) is left as an exercise.

The following theorem provides a method for creating tautologies and contradictions. In particular, this theorem shows that the disjunction of any statement with a tautology is also a tautology and that the conjunction of any statement with a contradiction is a contradiction.

Theorem 2.3.1: Let P be a statement, T a tautology, and C a contradiction. Then

(i) $P \vee T$ is a tautology.

(ii) $P \wedge C$ is a contradiction.

Proof: Let P be a statement, T a tautology, and C a contradiction.

Proof of part (i): The truth table for $P \vee T$ is given below:

$P \vee T$

P	T	$P \vee T$
T	T	T
F	T	T

Proof of part (ii): The proof of part (ii) is left as an exercise.

■

Now, it turns out that tautologies play an important role in mathematics since a mathematical definition is a tautology. Contradictions also play a key role in mathematics. In fact, finding a contradiction will be the critical step in a very useful method of proof called "proof by contradiction" or "reductio ad absurdum."

In mathematics there is often more than one way to represent a quantity or state a mathematical result. For example, one might write $-7 < x - y < 7$ or equivalently, $|x - y| < 7$ to represent all the pairs (x, y) whose difference is less than 7. Clearly, these two expressions have the same meaning. Similarly, there may often be more than one way to represent a compound logical statement such that each representation has the same meaning. When two compound statements, say, X and Y, are based on the same set of statements and have the same truth tables, then X and Y have the same logical meaning and are said to be *logically equivalent.*

Definition 2.3.2: Two statements X and Y are said to be *logically equivalent* when they have identical truth tables. When two statements X and Y are logically equivalent, this will be denoted by $X \Leftrightarrow Y$.

Example 2.3.3: Let P and Q be statements. Show that the statements $\neg P \vee \neg Q$ and $\neg(P \wedge Q)$ have identical truth tables, and hence they are logically equivalent.

Solution: The truth table for both $\neg P \vee \neg Q$ and $\neg(P \wedge Q)$ is given below:

P	Q	$\neg P \vee \neg Q$	$\neg(P \wedge Q)$
T	T	F	F
T	F	T	T
F	T	T	T
F	F	T	T

Clearly, the truth tables for these two statements are the same and thus these two statements are logically equivalent and have the same logical implications. Now, if two compound statements X and Y are based on the same set of statements and are also logically equivalent, then the statement X may be substituted for the statement Y. Substitution often can be used to simplify a compound statement or help in determining the truth of a compound statement. The next several theorems provide some useful logical equivalences.

Theorem 2.3.2: Let P be a statement, then $\neg(\neg P)$ is logically equivalent to P.

Proof: Let P be a statement. To prove this theorem, it will be shown that the truth tables for the statements $\neg(\neg P)$ and P are exactly the same.

P	$\neg P$	$\neg(\neg P)$
T	F	T
F	T	F

Since the truth tables for P and $\neg(\neg P)$ are identical, it follows that P and $\neg(\neg P)$ are logically equivalent statements.

■

Theorem 2.3.3: Let P and Q be statements. Then

(i) $P \wedge Q \Leftrightarrow Q \wedge P$.

(ii) $P \vee Q \Leftrightarrow Q \vee P$.

Proof: Let P and Q be statements.

Proof of part (i): Consider the truth tables for $P \wedge Q$ and $Q \wedge P$:

P	Q	$P \wedge Q$	$Q \wedge P$
T	T	T	T
T	F	F	F
F	T	F	F
F	F	F	F

Since the truth tables for $P \wedge Q$ and $Q \wedge P$ are identical, it follows that $P \wedge Q$ and $Q \wedge P$ are logically equivalent statements.

Proof of part (ii): The proof of part (ii) is left as an exercise.

∎

Theorem 2.3.3 shows that the order in which the statements are combined in a conjunction or disjunction does not matter; that is, $\wedge$ and $\vee$ are reflexive or commutative operations. The next theorem provides a very useful logical equivalence for the negations of the disjunction and conjunction statements.

Theorem 2.3.4 (DeMorgan's Laws): Let P and Q be statements. Then

(i) $\neg(P \vee Q)$ is logically equivalent to $\neg P \wedge \neg Q$.

(ii) $\neg(P \wedge Q)$ is logically equivalent to $\neg P \vee \neg Q$.

Proof: Let P and Q be statements.

Proof of (i): The truth table for $\neg(P \vee Q)$ and $\neg P \wedge \neg Q$ is given below:

P	Q	$P \vee Q$	$\neg(P \vee Q)$	$\neg P \wedge \neg Q$
T	T	T	F	F
T	F	T	F	F
F	T	T	F	F
F	F	F	T	T

Since the last two truth columns are identical, it follows that $\neg(P \vee Q)$ is logically equivalent to $\neg P \wedge \neg Q$.

Proof of part (ii): The proof of part (ii) is left as an exercise.

∎

The following theorem shows that $\vee$ can be distributed over $\wedge$ and vice versa. In fact, the distributive laws for distributing $\vee$ over $\wedge$ and $\wedge$ over $\vee$ are analogous to the arithmetic law for distributing multiplication over addition (i.e., $a(b + c) = ab + ac$).

Theorem 2.3.5 (The Distributive Laws): Let P, Q, and R be statements. Then

(i) $P \vee (Q \wedge R)$ is logically equivalent to $(P \vee Q) \wedge (P \vee R)$.

(ii) $P \wedge (Q \vee R)$ is logically equivalent to $(P \wedge Q) \vee (P \wedge R)$.

Proof: Let P, Q, and R be statements.

Proof of part (i): The truth table for the statements $P \vee (Q \wedge R)$ and $(P \vee Q) \wedge (P \vee R)$ is given below:

P	Q	R	$P \vee (Q \wedge R)$	$(P \vee Q) \wedge (P \vee R)$
T	T	T	T	T
T	T	F	T	T
T	F	T	T	T
T	F	F	T	T
F	T	T	T	T
F	T	F	F	F
F	F	T	F	F
F	F	F	F	F

Since the last two truth columns are identical, it follows that $P \vee (Q \wedge R)$ is logically equivalent to $(P \vee Q) \wedge (P \vee R)$.

Proof of part (ii): The proof of part (ii) is left as an exercise.

∎

The previous theorems are often used to simplify or rewrite a particular statement in a logically equivalent form that will be easier to work with. In particular, the substitution of logically equivalent statements is often used to simplify complex statements, making them easier to comprehend. Example 2.3.4 illustrates how DeMorgan's laws and the Distributive law can be used.

Example 2.3.3: Let P, Q, and R be statements. Find a logically equivalent form that simplifies each of the statements below:

a. $(P \vee Q) \wedge (P \vee \neg R)$

b. $\neg(P \vee Q) \wedge R$

Solutions: Let P, Q, and R be statements:

a. $(P \vee Q) \wedge (P \vee \neg R)$

$$(P \vee Q) \wedge (P \vee \neg R) \underbrace{\Leftrightarrow P \vee (Q \wedge \neg R)}_{\text{By Theorem 2.3.5}}$$

b. $\neg(P \vee \neg Q) \wedge R$

$$\neg(P \vee \neg Q) \wedge R \underbrace{\Leftrightarrow \neg P \wedge \neg\neg Q \wedge R}_{\text{By Theorem 2.3.4}}$$

$$\underbrace{\Leftrightarrow \neg P \wedge Q \wedge R}_{\text{By Theorem 2.3.2}}$$

The following theorems show that DeMorgan's laws and the Distributive laws can be generalized from dealing with two statements to three statements. Note that this type of generalization is often used to extend a property from two items to three, then from three items to four, and so on.

Theorem 2.3.6: Let P, Q, and R be statements. Then

(i) $\neg(P \vee Q \vee R) \Leftrightarrow \neg P \wedge \neg Q \wedge \neg R$

(ii) $\neg(P \wedge Q \wedge R) \Leftrightarrow \neg P \vee \neg Q \vee \neg R$

Proof: Let P, Q, and R be statements.

Proof of part (i): Define $S = Q \vee R$. Then

$$\neg(P \vee Q \vee R) \Leftrightarrow \neg(P \vee S) \quad \underbrace{\Leftrightarrow \neg P \wedge \neg S}_{\text{By Theorem 2.3.4(i)}}$$

$$\Leftrightarrow \neg P \wedge \neg(Q \vee R) \underbrace{\Leftrightarrow \neg P \wedge \neg Q \wedge \neg R}_{\text{By Theorem 2.3.4(i)}}$$

Proof of part (ii): The proof of part (ii) is left as an exercise.

■

Theorem 2.3.7: Let P, Q, R, and S be statements. Then

(i) $P \vee (Q \wedge R \wedge S) \Leftrightarrow (P \vee Q) \wedge (P \vee R) \wedge (P \vee S)$

(ii) $P \wedge (Q \vee R \vee S) \Leftrightarrow (P \wedge Q) \vee (P \wedge R) \vee (P \wedge S)$

Proof: The proof of Theorem 2.3.7 is left as an exercise.

■

2.4 Conditional Statements

Mathematical sentences, especially in the case of theorems and their proofs, are often stated in the form "statement/propositional function P implies statement/propositional function Q." Statements of this form are called *conditional statements*, and conditional implication is another way of combining two or more statements to create a compound statement.

Definition 2.4.1: Let P and Q be statements. Then the declarative sentence "P implies Q" is called a *conditional statement* and is denoted by $P \to Q$.

In a conditional statement $P \to Q$, the statement P, called the *antecedent*, is said to be a sufficient condition for Q; the statement Q is called the *consequent* and is said to be a necessary condition for P. Other ways of stating $P \to Q$ include "If P, then Q," "If P is true, then Q is true," "P only if Q," "Q if P," "P is a sufficient condition for Q," "Q is necessary for P," "Q assuming P," "Q whenever P", and "Q given P." The truth table for $P \to Q$ is given below:

$P \to Q$

P	Q	$P \to Q$
T	T	T
T	F	F
F	T	T
F	F	T

Note that "P implies Q" is more often than not stated as "If P is true, then Q is true" or simply "If P, then Q." Furthermore, from the truth table for $P \to Q$ it can be seen that when the statement P is false, the conditional statement is always true; that is, if P is not true, a truth value still needs to be assigned for $P \to Q$ in order for $P \to Q$ to be a statement. Thus, it seems reasonable to assign the conditional statement $P \to Q$ the truth value true since the P is sufficient to conclude Q; however, this does not mean that Q follows only from P. For example, consider the conditional statement "If today is Thursday, then Math 222 meets." If the true state of nature is Wednesday, this does not preclude Math 222 from meeting and thus cannot negate this conditional statement.

The following theorem shows that $P \to Q$ is logically equivalent to the disjunction of the statements $\neg P$ and Q.

Theorem 2.4.1: If P and Q are statements, then $P \to Q \Leftrightarrow \neg P \vee Q$.

Proof:

P	Q	$P \to Q$	$\neg P \vee Q$
T	T	T	T
T	F	F	F
F	T	T	T
F	F	T	T

■

Corollary to Theorem 2.4.1: $\neg(P \to Q) \Leftrightarrow P \wedge \neg Q$.

Proof: This result follows directly from Theorem 2.4.1 and DeMorgan's laws:

$$\neg(P \to Q) \Leftrightarrow \neg(\neg P \vee Q) \Leftrightarrow \neg\neg P \wedge \neg Q \Leftrightarrow P \wedge \neg Q$$

■

Note that since $P \to Q$ is logically equivalent to $\neg P \vee Q$ which is logically equivalent to $\neg Q \to \neg P$. Thus, it follows that $P \to Q$ is logically equivalent to $\neg Q \to \neg P$, also.

2.4.1 Converse and Contrapositive Statements

Two special conditional statements that are related to $P \rightarrow Q$ are the *converse* and *contrapositive* to $P \rightarrow Q$. The definitions of the converse and contrapositive statements to $P \rightarrow Q$ are given below.

Definition 2.4.2: Let P and Q be statements. The *converse* of the statement $P \rightarrow Q$ is $Q \rightarrow P$, and the *contrapositive* of the statement $P \rightarrow Q$ is $\neg Q \rightarrow \neg P$.

Students often confuse the two statements $P \rightarrow Q$ and $Q \rightarrow P$. In fact, it is not uncommon for a student to think that the statements $P \rightarrow Q$ and $Q \rightarrow P$ are logically equivalent. The following truth table shows that these two statements are not logically equivalent:

P	Q	$P \rightarrow Q$	$Q \rightarrow P$
T	T	T	T
T	F	F	T
F	T	T	F
F	F	T	T

The truth table for the contrapositive and converse statements of $P \rightarrow Q$ are given below:

P	Q	$P \rightarrow Q$	$\neg Q \rightarrow \neg P$	$Q \rightarrow P$
T	T	T	T	T
T	F	F	F	T
F	T	T	T	F
F	F	T	T	T

Note that the statements $P \rightarrow Q$ and $Q \rightarrow P$ are not logically equivalent whereas the statements $P \rightarrow Q$ and $\neg Q \rightarrow \neg P$ are logically equivalent.

Theorem 2.4.2: Let P and Q be statements. Then, $P \rightarrow Q$ and $\neg Q \rightarrow \neg P$ are logically equivalent.

Proof: Let P and Q be statements:

P	Q	$P \rightarrow Q$	$\neg Q \rightarrow \neg P$
T	T	T	T
T	F	F	F
F	T	T	T
F	F	T	T

Since the last two truth columns are identical, it follows that $P \rightarrow Q$ and $\neg Q \rightarrow \neg P$ are logically equivalent.

■

Thus, a conditional statement and its contrapositive are logically equivalent; however, the converse of a conditional statement is not logically equivalent to the statement. In other words, $P \rightarrow Q \Leftrightarrow \neg Q \rightarrow \neg P$, but $P \rightarrow Q \not\Leftrightarrow Q \rightarrow P$. Also, in Chapter 3 it will be shown that it is possible to prove a theorem of the form "If P, then Q" by proving the contrapositive theorem "If not Q, then not P".

Example 2.4.1: Write out the converse and contrapositive of

a. If $x > 3$, then $f(x) < 0$.

b. If $f'(x) > 0$ on $[a, b]$, then $f(x)$ is increasing on $[a, b]$.

c. If $ab = 0$, then $a = 0$ or $b = 0$.

d. If $f(x)$ is concave upward on an interval $[a, b]$, then $f''(x) > 0$ on the interval $[a, b]$.

Solutions: Recall that the converse of the statement $P \rightarrow Q$ is $Q \rightarrow P$ and the contrapositive is $\neg Q \rightarrow \neg P$.

a. If $x > 3$, then $f(x) < 0$.
   Converse: If $f(x) < 0$, then $x > 3$.
   Contrapositive: If $f(x) \geq 0$, then $x \leq 3$.

b. If $f'(x) > 0$ on $[a, b]$, then $f(x)$ is increasing on $[a, b]$.
   Converse: If $f(x)$ is increasing on $[a, b]$, then $f'(x) > 0$ on $[a, b]$.
   Contrapositive: If $f(x)$ is not increasing on $[a, b]$, then $f'(x) \not> 0$ on $[a, b]$.

c. If $ab = 0$, then $a = 0$ or $b = 0$.
   Converse: If $a = 0$ or $b = 0$, then $ab = 0$.
   Contrapositive: If $a \neq 0$ and $b \neq 0$, then $ab \neq 0$.

d. If $f(x)$ is concave upward on an interval $[a, b]$, then $f''(x) > 0$ on the interval $[a, b]$.
   Converse: If $f''(x) > 0$ on the interval $[a, b]$, then $f(x)$ is concave upward on an interval $[a, b]$.
   Contrapositive: If $f''(x) \not> 0$ on the interval $[a, b]$, then $f(x)$ is not concave upward on an interval $[a, b]$.

2.4.2 Biconditional Statements

Another compound statement that is related to the conditional statement $P \rightarrow Q$ is the conjunction of the statements "If P, then Q" and "If Q, then P" or $(P \rightarrow Q) \wedge (Q \rightarrow P)$. The compound conditional statement $(P \rightarrow Q) \wedge (Q \rightarrow P)$ called a *biconditional* statement and is defined below.

Definition 2.4.3: Let P and Q be statements. The *biconditional statement* "P if and only if Q" is the statement $(P \rightarrow Q) \wedge (Q \rightarrow P)$ and is denoted by $P \leftrightarrow Q$.

Note that the statement $P \leftrightarrow Q$ is true only when both of the statements $P \rightarrow Q$ and $Q \rightarrow P$ are true. That is, $P \leftrightarrow Q$ is true only when both the conditional statement $P \rightarrow Q$ and its converse are true. Also, $P \leftrightarrow Q$ is logically equivalent to $Q \leftrightarrow P$. To see this, consider the following truth table for $P \leftrightarrow Q$ and $Q \leftrightarrow P$:

P	Q	$P \rightarrow Q$	$Q \rightarrow P$	$P \leftrightarrow Q$	$Q \leftrightarrow P$
T	T	T	T	T	T
T	F	F	T	F	F
F	T	T	F	F	F
F	F	T	T	T	T

Now, recall that definitions are tautologies and, moreover, will often be stated as a biconditional statement. For example, consider the definition of the continuity of a function f at a point $x = c$ given below.

> ***Definition:*** A function $f(x)$ is said to be *continuous* at the point $x = c$ if and only if c is in the domain of $f(x)$ and $\lim_{x \to c} f(x) = f(c)$.

Since this is a biconditional statement, it follows that $f(x)$ is continuous at $x = c$ if c is in the domain of the function f and $\lim_{x \to c} f(x) = f(c)$ $(P \to Q)$, and if c is in the domain of the function f and $\lim_{x \to c} f(x) = f(c)$, then $f(x)$ is continuous at $x = c$ $(Q \to P)$.

Biconditional statements are also often used in the statement of a theorem. An example of a biconditional theorem is given below.

> **Theorem:** Let a and b be real numbers. Then, $ab = 0$ if and only if $a = 0$ or $b = 0$.

2.5 Propositional Functions and Quantifiers

Recall that a statement is any declarative sentence, and a propositional function or predicate is any declarative sentence involving a variable for which the declarative sentence has a well-defined truth value for each specific value of the variable. Also, recall that given a specific value for the variable in a propositional function, the propositional function evaluated at this value is a statement. For example, consider the propositional function "$x^2 - 4 = 0$," which is false for $x = 3$ and true for $x = -2$. Moreover, "$x^2 - 4 = 0$" is false for any values of x other than $x = -2$ or $x = 2$.

Definition 2.5.1: In a propositional function, a *variable* is any term whose value is not explicitly stated, implied, or understood and whose value is needed in order to determine the truth or falsity of the proposition. The set of possible values of a variable is called the *domain* of the propositional function and is denoted by Δ.

For example, the sentence "$\sqrt{2}$ is an irrational number" is a statement since it is a declarative sentence that is true. On the other hand, the sentence "$f(x)$ is a rational function" is a propositional function, with variable $f(x)$, since the truth of this sentence cannot be determined without knowledge of $f(x)$. Moreover, most mathematical sentences, results, theorems, and proofs will consist of a combination of statements and propositional functions. For example, in the following mathematical sentence S:="x is an integer" the term x is a variable making S a propositional function. An explicit value of x is needed to determine the truth of this sentence. On the other hand, the sentence "If x is a real number, then $x^2 - 2x + 2 > 0$" is true for every real number x and hence is a statement and not a propositional function.

Example 2.5.1: In each of the propositional functions given below, determine the variable(s):

a. $2^x - 1$ is a prime number.

b. $2x$ is even and $2x + 1$ is odd.

c. $f'(x) < 0$ on the interval I.

d. $|-x| = |x|$.

e. If $f(x)$ is differentiable at $x = x_0$, then $f(x)$ is continuous at $x = x_0$.

Solutions:

a. In the propositional function "$2^x - 1$ is a prime number," x is the variable.

b. In the propositional function "$2x$ is even an $2x + 1$ is odd," x is the variable.

c. In the propositional function "For the function $f(x)$, $f'(x) < 0$ on the interval I," $f(x)$ and I are variables.

d. In the propositional function "$|-x| = x$," x is the variable.

e. In the propositional function "If $f(x)$ is differentiable at $x = x_0$, then $f(x)$ is continuous at $x = x_0$," $f(x)$ and x_0 are the variables.

For notational purposes, let $P(x)$ be a propositional function that depends on a variable x. Often a mathematical sentence will involve a statement such as *for every* x in the set Δ, $P(x)$ is true or *there exists an* x *in the set* Δ *such that* $P(x)$ is true. The operator "for every" is known as the *universal quantifier*, and the operator "there exists" is known as the *existential quantifier*. The universal and existential quantifiers can be used with either a collection of statements or propositional functions in creating new statements or propositional functions.

Definition 2.5.2: The quantifying clause "for every" is called the *universal quantifier* and is denoted by $\forall$.

Definition 2.5.3: The quantifying clause "there exists" is called the *existential quantifier* and is denoted by $\exists$.

Note that the universal quantifier $\forall$ can be used to represent each of the following equivalent quantifying clauses *for every*, *for each*, and *for all*; analogously, the existential quantifier $\exists$ can be used to represent the following equivalent quantifying clauses *there exists*, *there is at least one*, and *there is some*.

Example 2.5.2: Consider the declarative sentences S:="$x^2 + 4 \neq 0$ for all real numbers x" and T:="$x^2 - 4 = 0$ for some real number x." Write out these sentences using the universal and existential quantifiers.

Solution: S can be written as $\forall\ x \in \mathbb{R},\ x^2 + 4 \neq 0$, and T can be written as $\exists\ x \in \mathbb{R}\ \ni:\ x^2 - 4 = 0$. Furthermore, since S and T are declarative statements that are true, it follows that they are not propositional functions. Thus, the value of x is not important, except for being in $\mathbb{R}$, in determining the truth of these declarative sentences. Hence, x is not a free variable in either S or T.

Now, let $P(x)$ be a propositional function that depends on a variable x and let $\Delta = \{x_1, x_2, x_3, \ldots, x_n\}$ be the set of possible values of x. Then, the propositional function $\forall\ x \in \Delta,\ P(x)$ is logically equivalent to

$$P(x_1) \wedge P(x_2) \wedge P(x_3) \cdots \wedge P(x_n)$$

since both of these propositional functions will be true if and only if every one of the propositional functions, $P(x_1)$, $P(x_2)$, $\ldots$, $P(x_n)$ is true. Analogously, $\exists\ x \in \Delta,\ P(x)$ is logically equivalent to

$$P(x_1) \vee P(x_2) \vee P(x_3) \cdots \vee P(x_n)$$

since both of these propositional functions will be true if and only if at least one of the propositional functions $P(x_1)$, $P(x_2)$, $\ldots$, $P(x_n)$ is true.

Example 2.5.3: Write the following statements using the existential and universal quantifiers where appropriate:

a. For every positive real number x, $x^3 - 2x^2 + x > 0$.

b. There exists a real number x such that $x^2 - 3x = 4$.

c. For every positive real number ϵ, there exists a positive number δ such that $|f(x) - f(a)| < \epsilon$ whenever $|x - a| < \delta$.

d. For every positive real number ϵ, there exists a natural number N such that $|a_n - a| < \epsilon$ whenever $n \geq N$.

Solutions:

a. Using the universal quantifier, this mathematical sentence can be written as $\forall\ x > 0,\ x^3 - 2x^2 + x > 0$.

b. Using the existential quantifier, this mathematical sentence can be written as $\exists\ x \in \mathbb{R}\ \ni:\ x^2 - 3x = 4$.

c. Using the universal and the existential quantifiers, this mathematical sentence can be written as $\forall\ \epsilon > 0,\ \exists\ \delta > 0\ \ \ni:\ \ |f(x) - f(a)| < \epsilon$ whenever $|x - a| < \delta$.

d. Using the universal and the existential quantifiers, this mathematical sentence can be written as $\forall\ \epsilon > 0,\ \exists\ N \in \mathbb{N}\ \ni\colon\ |a_n - a| < \epsilon$ whenever $n \geq N$.

Theorem 2.5.1: Let $\Delta = \{x_1, x_2, \ldots, x_n\}$ and let $P(x)$ be a propositional function on Δ. Then

(i) $\neg\Big[\forall\ x \in \Delta,\ P(x)\Big] \ \Leftrightarrow\ \exists\ x \in \Delta\ \ni\colon \neg P(x)$.

(ii) $\neg\Big[\exists\ x \in \Delta\ \ni\colon\ P(x)\Big] \ \Leftrightarrow\ \forall\ x \in \Delta,\ \neg P(x)$.

Proof: Let $\Delta = \{x_1, x_2, \ldots, x_n\}$, and let $P(x)$ be a propositional function on Δ.

Proof of part (i): Since

$$\forall\ x \in \Delta,\ P(x) \ \Leftrightarrow\ P(x_1) \wedge P(x_2) \wedge P(x_3) \cdots \wedge P(x_n)$$

it follows that

$$\neg\Big[\forall\ x \in \Delta,\ P(x)\Big] \ \Leftrightarrow\ \neg\Big[P(x_1) \wedge P(x_2) \wedge P(x_3) \cdots \wedge P(x_n)\Big]$$

Now, by DeMorgan's law for negating $\wedge$ it follows that

$$\neg\Big[P(x_1) \wedge P(x_2) \wedge P(x_3) \cdots \wedge P(x_n)\Big]$$

$$\Leftrightarrow\ \neg P(x_1) \vee \neg P(x_2) \vee \neg P(x_3) \cdots \vee \neg P(x_n)$$

$$\Leftrightarrow\ \exists\ x\ \in\ \Delta,\ \neg P(x)$$

Thus, $\neg\Big[\forall\ x \in \Delta,\ P(x)\Big] \ \Leftrightarrow\ \exists\ x \in \Delta\ \ni\colon \neg P(x)$.

Proof of part (ii): The proof of part (ii) is left as an exercise.

■

Note that the result of Theorem 2.5.1 is that negating a $\forall$ statement produces a $\exists$ statement and negating an $\exists$ statement produces a $\forall$ statement; the propositional function that is associated with the $\forall$ or the $\exists$ must be negated, also. For example, the negation of $\forall\ x > 0,\ x^3 - 2x^2 + x > 0$ requires the negation of the propositional function $x^3 - 2x^2 - x > 0$, which is $x^3 - 2x^2 - x \leq 0$ (i.e., $\neg(>)$ is $\leq$). Thus,

$$\neg\left[\forall\ x > 0,\ x^3 - 2x^2 + x > 0\right] \Leftrightarrow \exists\ x > 0\ \ni:\ x^3 - 2x^2 - x \leq 0$$

Example 2.5.4: Negate each of the following mathematical sentences:

a. $\forall\ n$ in the natural numbers, $n^2 - n$ is an even number.

b. $\exists\ r \in (-\infty, \infty)\ \ni:\ 71 = 3r + 2$.

c. $\forall\ i \in \{1, 2, 3, \ldots, 10\},\ 2^{-i} \geq 0.0005$

Solutions:

a. $\exists\ n$ in the natural numbers $\ni:\ n^2 - n$ is an odd number.

b. $\forall\ r \in (-\infty, \infty),\ 71 \neq 3r + 2$.

c. $\exists\ i \in \{1, 2, 3, \ldots, 10\}\ \ni:\ 2^{-i} < 0.0005$.

EXERCISES

2.1 Determine which of the sentences below are statements, predicates, or neither. Justify your answers.

a. $e^{-x} > 0$ for every real number x.

b. $f(x)$ has a relative maximum at $x = 0$.

c. Does $f(x)$ have a relative minimum at $x = 0$?

d. $\ln(x) > 0$ for every real positive number x.

e. $f(x)$ is continuous at $x = 0$.

f. π is an odd number or 4 is an odd number.

g. If $\sqrt{2}$ is an irrational number, then so is $2\sqrt{2}$.

h. $3 > 5$ if and only if $4 < 5$.

i. If x is an odd number, then so is $x + 2$.

2.2 For the propositional function "$f(x)$ has a relative maximum at $x = 0$," determine

a. The variable in this propositional function.

b. Whether the propositional function is true when $f(x) = x^2$.

c. Whether the propositional function is true when $f(x) = x^3$.

2.3 Given a compound statement S composed from the three statements P, Q, and R, determine

a. The number of possible states of nature that must be listed in a truth table for the compound statement S.

b. The number of possible states of nature in the truth table with either P or Q being true.

c. The list of possible states of nature that must be considered in determining the truth of the compound statement S.

2.4 Construct truth tables for the following statements:

a. $\neg P \vee Q$.

b. $(\neg P \wedge Q) \vee (P \wedge Q)$.

c. $(P \vee Q) \vee (P \wedge Q)$.

d. $P \vee Q \vee \neg R$.

e. $(P \vee Q) \wedge \neg R$.

f. $(P \wedge Q) \rightarrow R$.

g. $(P \vee Q) \rightarrow R$.

2.5 Which of the following statements are true when P is true and Q is false?

a. $P \vee \neg Q$.

b. $\neg(P \wedge Q)$.

c. $\neg(P \vee Q)$.

d. $(P \vee Q) \wedge (\neg P \vee \neg Q)$.

e. $P \rightarrow Q$.

f. $Q \rightarrow P$.

g. $P \leftrightarrow Q$.

h. $\neg Q \rightarrow P$.

i. $\neg P \rightarrow Q$.

j. $(\neg P \rightarrow \neg Q) \wedge (\neg Q \rightarrow P)$.

2.6 Show that

a. $P \vee Q \;\Leftrightarrow\; Q \vee P$.

b. $(\neg P \wedge Q) \vee (P \wedge Q) \;\Leftrightarrow\; Q$.

c. $(P \vee Q) \vee (\neg P \vee \neg Q) \wedge R \;\Leftrightarrow\; R$.

d. $(P \wedge Q) \vee (P \wedge \neg Q) \vee (\neg P \wedge Q) \vee (\neg P \wedge \neg Q)$ is a tautology.

e. $(P \wedge Q) \wedge (P \wedge \neg Q)$ is a contradiction.

f. $\neg(P \wedge Q) \vee (P \vee Q)$ is a tautology.

g. $P \wedge C$ is a contradiction whenever C is a contradiction.

2.7 Prove that $\neg(P \wedge Q)$ is logically equivalent to $\neg P \vee \neg Q$.

2.8 Prove that

a. $P \wedge (Q \vee R)$ is logically equivalent to $(P \wedge Q) \vee (P \wedge R)$.

b. $\neg(P \wedge Q \wedge R)$ is logically equivalent to $\neg P \vee \neg Q \vee \neg R$.

c. $P \wedge (Q \vee R \vee S)$ is logically equivalent to $(P \wedge Q) \vee (P \wedge R) \vee (P \wedge S)$.

2.9 Let P, Q, and R be statements:

a. If $P \rightarrow Q$ is false, under what conditions will $(P \rightarrow Q) \rightarrow R$ be a true statement?

b. If $P \rightarrow Q$ is true, under what conditions will $(P \rightarrow Q) \rightarrow R$ be a true statement?

c. If R is true, under what conditions will $(P \to Q) \to R$ be a true statement?

b. If R is false, under what conditions will $(P \to Q) \to R$ be a true statement?

2.10 Determine the converse, contrapositive, and negation of each of the following statements:

a. $(P \wedge Q) \to R$.

b. If n is a natural number, then $n(n+1)(n+2)$ is an even number.

c. If $f(x)$ is an odd integrable function, then $\int_{-a}^{a} f(x)\,dx = 0$.

d. If $\mathcal{P}$ is the collection of prime numbers, then $\mathcal{P}$ contains infinitely many elements.

e. If n^2 is divisible by 3, then n is divisible by 3.

f. If $x = \sqrt{5}$, then x is an irrational number.

g. If 3 divides A, 3 divides B, and 3 divides C, then 3 divides $A+B+C$.

h. If $a = 0$ or $b = 0$, then $a(b+c) = 0$ and $b(a+c) = 0$.

2.11 Without using a truth table, show that

a. $P \vee Q \vee R \Leftrightarrow \neg(\neg P \wedge \neg Q \wedge \neg R)$.

b. $(P \wedge Q) \vee (P \wedge \neg Q) \Leftrightarrow P$.

c. $P \vee (Q \wedge R \wedge S \wedge T) \Leftrightarrow (P \vee Q) \wedge (P \vee R) \wedge (P \vee S) \wedge (P \vee T)$.

d. $P \wedge (Q \vee R \vee S \vee T) \Leftrightarrow (P \wedge Q) \vee (P \wedge R) \vee (P \wedge S) \vee (P \wedge T)$.

2.12 Using the mathematical symbols for the universal and existential quantifiers, rewrite the following sentences:

a. For every natural number n, $\sum_{i=1}^{n} i = \frac{n(n+1)}{2}$.

b. There exists a real number x such that $x^5 - x + 2 = 0$.

c. For every real number x, there exists a real number y such that $e^{-x} - y = 0$.

d. For every $\epsilon > 0$, there exists $N \in \mathbb{N}$ such that $|A(n) - A| < \epsilon$ whenever $n \geq N$.

2.13 Negate each of the sentences in Problem 2.12.

2.14 Let $\{P_n(x)\}$ be a collection of propositional functions. If $P_8(x)$ is false, what can be concluded about the truth of the propositional function

a. $P_n(x), \ \forall\, n \in \mathbb{N}$?
b. $\exists\, n \in \mathbb{N} \ni: \ P_n(x)$?
c. $\exists\, n \in \mathbb{N} \ni: \ \neg P_n(x)$?

2.15 Let $\{P_n(x)\}$ be a collection of propositional functions. If $P_8(x)$ is true, what can be concluded about the truth of the propositional function

a. $P_n(x), \ \forall\, n \in \mathbb{N}$?
b. $\exists\, n \in \mathbb{N} \ni: \ P_n(x)$?
c. $\neg [\exists\, n \in \mathbb{N} \ni: \ P_n(x)]$?

2.16 Let $\Delta = \{x_1, x_2, \ldots, x_n\}$ and suppose that $P(x)$ is a propositional function on Δ. Prove that

$$\neg [\exists x \in \Delta \ni: \ P(x)] \ \Leftrightarrow \ \forall\, x \in \Delta, \neg P(x)$$

Chapter 3
Methods of Proof

Recall that the axiomatic structure of modern mathematics begins with a set of axioms and definitions that are explicitly stated, and then from these axioms and definitions, new mathematics is created through deductive reasoning. In fact, the first step in the development of new mathematics is to study the implications of this original set of axioms and definitions. The axiomatic structure of modern mathematics proceeds according to the following sequence:

$$\text{Axioms} \Longrightarrow \text{definitions} \Longrightarrow \text{conjectures} \Longrightarrow \text{proofs}$$
$$\Longrightarrow \text{theorems} \Longrightarrow \text{generalization and extension} \Longrightarrow \cdots$$

Furthermore, a new result must be proved rigorously before it is accepted as a new contribution to mathematics. The development of new mathematics is often based on trial and error and is a creative process. Given a mathematical conjecture, its proof or disproof will generally require a great deal of insight, imagination, experimentation, and hard work.

The foundations of this modern approach to mathematics were laid by the ancient Greek mathematicians Thales, Pythagoras, and Euclid. However, the formalization of mathematical theory really began in the nineteenth century with the work of Boole and Frege. By the early twentieth century modern mathematics was flourishing with the work of Georg Cantor (1845–1918), Hilbert, Russell, Whitehead, G. H. Hardy (1877–1947), and Gödel. Today, mathematicians continue to develop new areas of mathematics and at the same time extend and generalize the existing areas of mathematics.

3.1 Theorems, Corollaries, and Lemmas

The starting point of a mathematical system is a set of self-evident truths called *axioms*. The definition of an axiom is given below.

Definition 3.1.1: An *axiom* is a mathematical statement that is taken to be self-evidently true without proof.

Thus, an axiom is a mathematical statement that is believed to be so clearly true that it need not be proved. In an axiomatic mathematical system the initial set of axioms is the starting point from which all mathematics will be derived. Thus, it is very important that the axioms on which a mathematical system is based be true. For example, in set theory the following axioms are often used.

Axiom 1 (The Axiom of Existence): There exists a set.

Axiom 2 (The Axiom of Extensionality): Two sets are equal if and only if they have exactly the same elements.

Another example of commonly encountered axioms are the *axioms of probability*, which form the starting point for probability theory:

Axiom 1: For any event A of a sample space $\mathcal{S}$, $P(A) \geq 0$.

Axiom 2: $P(\mathcal{S}) = 1$

Axiom 3: If $\{A_i\}_{i=1}^{\infty}$ is collection of disjoint events of $\mathcal{S}$, then

$$P\left(\bigcup_{i=1}^{\infty} A_i\right) = \sum_{i=1}^{\infty} P(A_i).$$

Now, given a set of axioms, mathematical properties are defined and the implications of these properties are studied; generally conjectures concerning the implications of the axioms and definitions are made and studied. A mathematical conjecture will be composed of logical statements and/or propositional functions. Once a conjecture is proved, it is called a *theorem*.

Definition 3.1.2: A *theorem* is any mathematical statement that can be shown to be true using accepted logical and mathematical arguments.

The root of the word *theorem* is the Greek word *theorema*, which means "something seen." To a mathematician, a theorem is a result that can be seen to be true. Note that a mathematical result is only a conjecture until it is proved.

Definition 3.1.3: A *proof* of a mathematical result is a sequence of rigorous mathematical arguments that are presented in a clear and concise fashion, and which convincingly demonstrates the truth of the given result.

Note that only after a conjecture is proved can it be called a theorem. Furthermore, in the eyes of a mathematician, it is never enough to simply believe that a mathematical result is true, nor is it enough to be convinced that a result is true beyond a reasonable doubt; a mathematician will accept only those mathematical results that are shown to be absolutely true using sound logical arguments.

Now, given a new theorem, the process of building new mathematics continues by investigating the implications of the new theorem, often by generalizing or extending the theorem to a more general result. In most cases, a theorem will be stated in the conditional form "If H is true, then C is true," where H and C are either logical statements or propositional functions. In a theorem stated as "If H, then C," H is the *hypothesis*, and C is the *conclusion.* However, there are some theorems that will not be stated in the "If H, then C" form, but these theorems can often be rewritten in the conditional form. For example, the following theorem, which will be proved in Chapter 4, is not of the form "If H, then C."

Theorem: $\sqrt{2}$ is not a rational number.

However, note that this theorem could also be stated in the $H \rightarrow C$ form by rewriting it as

Theorem: If $x = \sqrt{2}$, then x is not a rational number.

In some cases, a theorem will provide a very general result and cover many special subcases. The specialized theorems dealing with the subcases of a more general result are called *corollaries.*

Definition 3.1.4: A *corollary* is a theorem that can be stated as a special case of a more general theorem.

Note that a corollary is a theorem itself; however, it is really just a special case of a particular theorem. Thus, once the more general theorem has been proved, the proof of a corollary simply involves showing that it is a special case of the previously proved theorem.

Example 3.1.1: The following theorem and corollary are typically seen in a calculus course.

Theorem A: Let $f(x)$ and $g(x)$ be functions. If $\lim_{x \to a} f(x) = L$ and $g(x)$ is a continuous function, then $\lim_{x \to a} g(f(x)) = g(L)$.

Corollary A: If $\lim_{x \to a} f(x) = L$, then $\lim_{x \to a} e^{f(x)} = e^{L}$.

Proof: Since $g(x) = e^x$ is a continuous function, this corollary follows directly from Theorem A.

■

Example 3.1.2: The following theorem and corollary are typically seen in a course on probability theory.

Theorem B: Let X be a random variable, $g(x)$ a real-valued function, and $c > 0$. Then

$$P\{g(X) \geq c\} \leq \frac{E[g(X)]}{c}$$

Corollary B: Let X be a random variable and $c > 0$. Then

$$P\{|X| \geq c\} \leq \frac{E[|X|]}{c}$$

Proof: Let $g(x) = |x|$; then this corollary follows directly from Theorem B.

■

In some cases, while proving a theorem it becomes evident that a special result is needed primarily for proving the theorem. A theorem that is used primarily in the proof of another theorem is called a *lemma*. The definition of a lemma is given below.

Definition 3.1.5: A *lemma* is any provable result that is used primarily as a necessary step in the proof of another theorem.

As was the case with a corollary, a lemma is also a theorem since it is a provable result. Moreover, lemmas generally precede a theorem and are used mainly to remove certain complications that will be encountered in the proof of a theorem.

Example 3.1.3: The following lemma will be very useful in Chapter 5.

Lemma (The Triangle Inequality): If $x, y \in \mathbb{R}$, then $|x + y| \leq |x| + |y|$.

Now, the general form of a theorem, corollary, or lemma is a conditional statement of the form "If H is true, then C is true." Also, the hypothesis H is the condition from which the conclusion C will follow. In many cases, the hypothesis may be a compound statement made up of sub-hypotheses, say $H_1, \ldots, H_n$, that are joined together by AND's or OR's. Similarly, the conclusion may also be a compound statement made up sub-conclusions, say $C_1, \ldots, C_n$, tied together by AND's or OR's.

Example 3.1.4: Determine the hypothesis and conclusion for each of the following theorems. If the hypothesis is compound, identify the subhypotheses.

a. **Theorem:** If $f(x)$ is differentiable at $x = c$, then $f(x)$ is continuous at $x = c$.

b. **Theorem:** If $f'(c) = 0$, $f'(c^-) < 0$, and $f'(c^+) > 0$, then $f(x)$ has a relative maximum at $x = c$.

c. **Theorem:** If $\lim_{x \to c} f(x) = L$, $\lim_{x \to c} g(x) = M$, and $M \neq 0$, then $\lim_{x \to c} \frac{f(x)}{g(x)} = \frac{L}{M}$.

d. **Theorem:** If a is an even integer, then a^2 is an even integer.

Solutions:

a. Hypothesis: $f(x)$ is differentiable at $x = c$
Conclusion: $f(x)$ is continuous at $x = c$.

b. Hypotheses: $f'(c) = 0$ (H_1), $f'(c^-) < 0$, (H_2), and $f'(c^+) > 0$ (H_3).
Conclusion: $f(x)$ has a relative maximum at $x = c$.

c. Hypotheses: $\lim_{x \to c} f(x) = L$ (H_1), $\lim_{x \to c} g(x) = M$ (H_2), and $M \neq 0$ (H_3).
Conclusion: $\lim_{x \to c} \frac{f(x)}{g(x)} = \frac{L}{M}$.

d. Hypothesis: a is an even integer.
Conclusion: a^2 is an even integer.

Finally, it is important to note that not all theorems will be stated in the "If H, then C" form. However, with a little creative thinking most theorems can be restated in the "If H, then C" form. For example, the following two theorems, which will be proved in Chapter 4, are not stated in the form "If H, then C."

Theorem: $\sqrt{2}$ is an irrational number.

Theorem: There are infinitely many prime numbers.

However, these theorems could be restated in the form "If H, then C" as shown below.

Theorem: If $x = \sqrt{2}$, then x is an irrational number.

Theorem: If $\mathcal{P}$ is the set of prime numbers, then $\mathcal{P}$ contains infinitely many prime numbers.

Example 3.1.5: Restate each of the following theorems in the "If H, then C" form:

a. The limit of the sequence of real numbers a_n is unique.

b. The identity element of a group $(\mathcal{G}, \circ)$ is unique.

c. The sum of an irrational number and a rational number is an irrational number.

Solutions:

a. If a is the limit of the sequence of real numbers a_n, then a is unique.

b. If $(\mathcal{G}, \circ)$ is a group, then the identity element is unique.

c. If x is an irrational number and y is a rational number, then $x + y$ is an irrational number.

3.2 The Contrapositive and Converse of a Theorem

Recall, from Chapter 2, that the converse and the contrapositive of the conditional statement $H \rightarrow C$ are $C \rightarrow H$ and $\neg C \rightarrow \neg H$, respectively. Furthermore, recall that a statement and its contrapositive are logically equivalent. Thus, it follows that a theorem of the form "If H is true, then C is true" is logically equivalent to its contrapositive theorem "If C is not true, then H is not true." Therefore, proving the theorem $H \rightarrow C$ automatically proves the theorem $\neg C \rightarrow \neg H$. Hence, each theorem has a dual theorem to which it is logically equivalent, namely, its contrapositive. On the other hand, since a conditional statement and its converse are not logically equivalent, the converse of a theorem is seldom true.

Example 3.2.1: Determine the contrapositive theorem associated with each of the following theorems:

a. **Theorem:** If $f(x)$ is differentiable at $x = c$, then $f(x)$ is continuous at $x = c$.

b. **Theorem:** If a is an odd integer, then a^2 is an odd integer.

c. **Theorem:** If a and b are an odd integers, then $a + b$ is an even integer.

d. **Theorem:** If $x > 0$, then $x + \dfrac{1}{x} \geq 2$.

Solutions:

a. The contrapositive theorem is "If $f(x)$ is not continuous at $x = c$, then $f(x)$ is not differentiable at $x = c$."

b. The contrapositive theorem is "If a^2 is an even integer, then a is an even integer."

c. The contrapositive theorem is "If $a + b$ is an odd integer, then either a or b is an even integer."

d. The contrapositive theorem is "If $x + \dfrac{1}{x} < 2$, then $x \leq 0$."

Now, when a theorem and its converse are both true, then the theorem can be written as a biconditional theorem of the form "H if and only if C"; a biconditional theorem is sometimes referred to as an "if and only if" theorem. An example of a biconditional theorem is given below.

Theorem: Let a and b be real numbers. Then, $ab = 0$ if and only if $a = 0$ or $b = 0$.

Note that this theorem is composed of the following two theorems:

Theorem: Let a and b be real numbers. If $ab = 0$, then $a = 0$ or $b = 0$.

Theorem: Let a and b be real numbers. If $a = 0$ or $b = 0$, then $ab = 0$.

3.3 Methods of Proof and Proving Theorems

Given a theorem "If H, then C," there are many different approaches that could be attempted when trying to prove this theorem. In fact, many theorems can be proved using several different approaches. However, the particular method used to prove a theorem is not nearly as important as is the fact that a valid proof has been found for the theorem. Two of the most commonly used approaches for proving theorems are the *method of direct proof* and the *method of indirect proof*.

3.3.1 Direct Proof

When faced with the problem of trying to prove a conjecture or a theorem of the form "If H, then C," the first, and often most direct, approach to try is the method of direct proof. For most problems, this approach will lead to a valid proof of the theorem in question. This is generally the first approach that is attempted when trying to prove a conjecture. The method of direct proof is described below.

The Method of Direct Proof: Given a theorem of the form $H \rightarrow C$, a direct proof of $H \rightarrow C$ begins with the assumption of the hypotheses of the theorem. From the hypotheses, a sequence of logical statements is constructed that leads to the conclusion of the theorem. When the sequence of logical arguments leading from the hypotheses (H) to the conclusion (C) is valid, then the theorem will have been proved with a direct proof. A diagram of a direct proof usually follows the pattern

$$H \rightarrow C_1 \rightarrow C_2 \rightarrow C_3 \cdots C$$

where H leads to a conclusion C_1, C_1 leads to conclusion C_2, and so on until the desired conclusion C is reached. This method of direct proof is called the *forward direct* approach. Note that a forward direct proof begins by assuming that the hypothesis (H) is true and then proceeds forward with a sequence of logical arguments that leads to the conclusion (C). An algorithm outlining the method of direct proof is outlined below.

Algorithm for a Direct Proof: The following steps outline the typical procedure used in a direct proof of a theorem $H \rightarrow C$:

1. Identify and list all hypotheses of the theorem.
2. Identify and list all results that follow directly from, or are related to, the hypotheses of the theorem.
3. Begin tying the hypotheses to the results listed in step 2. Begin working toward the conclusion. This is the scratchwork phase of the proof.
4. Develop a complete set of logical arguments leading from the hypotheses of the theorem to the conclusion of the theorem.
5. Clean up and rewrite the scratchwork into a clear and concise proof of the theorem. Make sure that each step of the proof makes sense and is clearly justified.
6. Read the proof over carefully and make any necessary corrections.

It is very important to write clear and concise proofs; therefore, steps 5 and 6 in the algorithm for a direct proof are two of the most important steps in writing a good proof. Furthermore, it is extremely important to provide complete and clear justification for each of the steps leading to the conclusion. Finally, a well-written proof is any proof that can be easily followed and comprehended by any person familiar with the mathematical subject matter involved in the theorem.

It also is important to note that a direct proof requires a sequence of logical arguments leading from the hypothesis to the conclusion. Thus, in carrying out steps 3 and 4 of the preceding algorithm, it will often be useful to focus on some aspect of the conclusion in developing a connection between the hypothesis and the conclusion. For example, suppose that the theorem states that "If n is even, then n^2 is even." In this case, in steps 3 and 4 it will be useful to consider n^2, which can then be written as $(n)^2$, which clearly relates the hypothesis to the conclusion.

Example 3.3.1: Suppose that the following results from calculus have been proved.

> **Theorem 1:** Let f and g be real-valued functions. If $\lim_{x \to c} f(x) = L$ and $\lim_{x \to c} g(x) = M$, then $\lim_{x \to c} f(x)g(x) = \lim_{x \to c} f(x) \cdot \lim_{x \to c} g(x) = L \cdot M$.

Theorem 2: Let f be a real-valued function. If $\lim_{x \to c} f(x) = L$ and $L \neq 0$, then $\lim_{x \to c} \frac{1}{f(x)} = \frac{1}{L}$.

Use these two theorems to prove the following theorem using a direct proof:

Theorem: Let f and g be real-valued functions. If $\lim_{x \to c} f(x) = L$, $\lim_{x \to c} g(x) = M$ and $M \neq 0$, then $\lim_{x \to c} \frac{f(x)}{g(x)} = \frac{L}{M}$.

Scratchwork: Following the algorithm for a forward direct proof:

1. **Hypotheses:** $f(x)$ and $g(x)$ are functions with $\lim_{x \to c} f(x) = L$, $\lim_{x \to c} g(x) = M$, and $M \neq 0$.
2. **Related Results:** The hypotheses of the theorem allow for the use of Theorems 1 and 2. Thus, it follows from Theorems 1 and 2 that

$$\lim_{x \to c} f(x)g(x) = LM \quad \text{and} \quad \lim_{x \to c} \frac{1}{g(x)} = \frac{1}{M}$$

3,4. **Working toward C:** Consider $\frac{f(x)}{g(x)} = f(x) \cdot \frac{1}{g(x)}$.

Now, since $\frac{f(x)}{g(x)} = f(x) \cdot \frac{1}{g(x)}$, it follows that

$$\lim_{x \to c} \frac{f(x)}{g(x)} = \lim_{x \to c} f(x) \cdot \frac{1}{g(x)}$$

and by Theorems 1 and 2 it follows that

$$\lim_{x \to c} \frac{f(x)}{g(x)} = \lim_{x \to c} f(x) \cdot \frac{1}{g(x)} = \lim_{x \to c} f(x) \cdot \lim_{x \to c} \frac{1}{g(x)}$$

$$L \cdot \frac{1}{M} = \frac{L}{M}$$

5,6. **Write a Proof:** Write up this scratchwork into a clear and concise proof of the theorem. Proofread your proof!

Proof: Let f and g be real-valued functions with $\lim_{x \to c} f(x) = L$,

$\lim_{x \to c} g(x) = M$, and $M \neq 0$. Then

$$\lim_{x \to c} \frac{f(x)}{g(x)} = \lim_{x \to c} \left[f(x) \cdot \frac{1}{g(x)} \right]$$

$$= \lim_{x \to c} f(x) \cdot \lim_{x \to c} \frac{1}{g(x)} = \underbrace{\lim_{x \to c} f(x) \cdot \lim_{x \to c} \frac{1}{g(x)}}_{\text{By Theorems 1 and 2}}$$

$$= L \cdot \frac{1}{M} = \frac{L}{M}$$

Thus, $\lim_{x \to c} \frac{f(x)}{g(x)} = \frac{L}{M}$ whenever $\lim_{x \to c} f(x) = L$ and $\lim_{x \to c} g(x) = M$ and $M \neq 0$.

■

Note that in this proof, the hypotheses of the theorem allowed for the application of Theorems 1 and 2 because the hypotheses of Theorems 1 and 2 are the exact hypotheses of this particular theorem. This was no coincidence; rather, it is the natural progression in the modern axiomatic mathematical system. Hence, the theorem proven in Example 3.1.1 turns out to be a natural extension of Theorems 1 and 2 stated above.

Example 3.3.2: For the following theorem, let $\mathbb{N} = \{1, 2, 3, \ldots\}$ be the set of natural numbers. Define an *even* number to be any integer that can be written as $2k$ for some integer k, and define an *odd* number to be any integer that can be written as $2k + 1$ for some integer k . Use a direct proof to prove the following theorem.

Theorem: Let n be a natural number. If n is an even number, then n^2 is also an even number.

Scratchwork: Following the algorithm for a forward direct proof:

1. **Hypotheses:** Let n be a natural number and suppose that n is even.
2. **Related Results:** Since n is an even number, there exists an integer k such that $n = 2k$.

3,4. **Working toward C:** Since n is even, it follows that

$$n^2 = (2k)^2 = 4k^2 = 2(2k^2)$$

which is of the form 2 times the integer $2k^2$. Hence, n^2 is also even.

5,6. **Write a Proof:** Write up this scratchwork into a clear and concise proof of the theorem. Proofread your proof!

Proof: Let n be a natural number and suppose that n is an even number.

Then, there exists an integer k such that $n = 2k$. Consider n^2

$$n^2 = (n)^2 = (2k)^2 = 4k^2 = 2\cdot 2k^2 = 2j$$

where $j = 2k^2$, which is an integer. Hence, n^2 can be written in the form $2j$ for some integer j, and therefore n^2 is an even number whenever n is an even number.

■

Example 3.3.3: Use a forward direct proof to prove the following theorem:

Theorem: If $x, y \in \mathbb{R}$, then $x^2 + y^2 \geq |xy|$.

Proof: Let $x, y \in \mathbb{R}$. Consider $(|x| - |y|)^2$:

$$(|x| - |y|)^2 = (|x|)^2 - 2|x|\cdot|y| + (|y|)^2 = x^2 + y^2 - 2|xy|$$

Furthermore, since $(|x| - |y|)^2 \geq 0$, it follows that

$$(|x| - |y|)^2 = x^2 + y^2 - 2|xy| \geq 0$$

and hence

$$x^2 + y^2 \geq 2|xy| \geq |xy|$$

Therefore, $x^2 + y^2 \geq |xy|$, $\forall\, x, y \in \mathbb{R}$.

■

Note that in the proof of the theorem stated in Example 3.3.3, the proof required a creative approach after assuming the hypotheses of the theorem. In particular, after the assumption $x, y \in \mathbb{R}$, the next step in the proof was to consider $(|x| - |y|)^2$. This step is not obvious to most people trying to prove this theorem, and might only be stumbled across by a few people as a result of their scratchwork. Often a great deal of ingenuity and insight is required in order to successfully prove the desired result. Thus, it is important to attack a tough problem from several different approaches and sometimes with a truly creative approach.

Recall that whenever "If H, then C" is a theorem, then the contrapositive to this theorem is also a theorem; that is, every theorem "If H, then C" has a dual theorem "If not C, then not H," which is also true. Furthermore, if a proof of "If C is not true, then H is not true" is found, then this is equivalent to proving "If H is true, then C is true." Thus, a second method of direct proof that can be used to prove "If H is true, then C is true" is to prove the contrapositive of this theorem: "If not C, then not H." Proving a theorem $H \rightarrow C$ by proving its contrapositive is called a *proof by contrapositive*. Note that the method of proof by contrapositive is also another method of direct proof and is often used when a forward direct proof for a theorem cannot be found.

Now, a proof by contrapositive begins with the assumption that conclusion of the theorem is false, rather than beginning with the hypotheses of the theorem. From the negation of the conclusion (i.e., $\neg C$), a direct proof of the negation of the hypotheses ($\neg H$) is desired as follows. A typical proof by contrapositive begins with $\neg C$, which leads to a conclusion C_1, which leads to a conclusion C_2, and so on until it can be concluded that the hypothesis H is false. An algorithm outlining a recipe for the method of proof by contrapositive is given below.

Algorithm for a Proof by Contrapositive: The following steps outline the typical procedure used in a proof by contrapositive of a theorem $H \rightarrow C$:

1. State the contrapositive of the theorem (i.e., $\neg C \rightarrow \neg H$).
2. Identify and list all the results that are related to $\neg C$.
3. Assume that $\neg C$ is true and begin tying $\neg C$ to the results in step 2 and work toward the conclusion, which is $\neg H$. This is the scratchwork phase of the proof.
4. Complete the scratchwork for the proof by developing a sequence of logical arguments based on steps 1 and 2 that lead to $\neg H$.
5. Clean up and rewrite the scratchwork into a clear and concise proof of the theorem. Make sure that each step of the proof makes sense and is clearly justified.
6. Read the proof over carefully and make any necessary corrections.

Note that as was the case in the method of forward direct proof, it is always important to write clear and concise proofs. Furthermore, it is extremely important to carefully proofread a proof. The following example illustrates how the method of proof by contrapositive is used.

Example 3.3.4: Prove the following theorem using a direct proof:

Theorem: Let n be a natural number. If n^2 is even, then n is an even natural number.

Scratchwork 1: Let n be a natural number and suppose that n^2 is an even natural number. Then, $\exists$ an integer k such that $n^2 = 2k$.

Now, taking the square root of n^2 yields $n = \sqrt{2k}$. However, there is not much that can be done with the $\sqrt{2k}$ with regard to relating it to n. Thus, a direct proof appears to be failing, and thus a proof by contrapositive should be considered.

The contrapositive of the original theorem is "Let n be a natural number. If n is an odd integer, then n^2 is an odd integer."

Scratchwork 2: Let n be a natural number and suppose that n is an odd number. Then, $\exists$ an integer k such that $n = 2k+1$. Consider n^2

$$n^2 = (2k+1)^2 = 4k^2 + 4k + 1 = 2(2k^2 + 2k) + 1 = 2j + 1$$

where $j = 2k^2 + 2k$, which is an integer. Thus, n^2 is an odd number whenever n is an odd number.

Proof (by Contrapositive): Let n be a natural number and suppose that n is an odd number. Then, $\exists$ an integer k such that $n = 2k + 1$. Consider n^2

$$n^2 = (2k+1)^2 = 4k^2 + 4k + 1 = 2j + 1$$

where $j = 2k^2 + 2k$, which is an integer.

Thus, n^2 is an odd number whenever n is an odd number, and therefore, by proving the contrapositive of the original theorem, it follows that n is even whenever n^2 is even.

■

Note that in the first set of scratchwork an impasse was encountered in trying to relate n^2 back to n. When a direct proof fails to prove a theorem, it is always reasonable to try to prove the theorem with the method of contrapositive; in some cases, a direct proof for the theorem of interest will be impossible with the mathematical results at hand. Also, when the method of proof by contrapositive is being used to prove a theorem, it is a good idea to clearly designate this at the beginning of the proof. Clearly designating that an alternative method to the forward direct proof is an important part of a proof by contrapositive.

Example 3.3.5: Prove the following theorem using the method of proof by contrapositive.

> **Theorem:** Let x and y be positive real numbers. If $x \neq y$, then $\ln(x) \neq \ln(y)$.
>
> **Scratchwork 1:** First, the contrapositive of this theorem is "Let x and y be positive real numbers. If $\ln(x) = \ln(y)$, then $x = y$."
>
> Let x and y be positive real numbers and suppose $\ln(x) = \ln(y)$. Note that $x = e^{\ln(x)}$ and $y = e^{\ln(y)}$, and since $\ln(x) = \ln(y)$, it follows that $x = e^{\ln(x)} = e^{\ln(y)} = y$.
>
> Thus, $x = y$ whenever $\ln(x) = \ln(y)$ and therefore, proof by contrapositive shows that $\ln(x) \neq \ln(y)$ whenever $x \neq y$.
>
> ***Proof (by Contrapositive):*** Let x and y be positive real numbers and suppose $\ln(x) = \ln(y)$. Note that $x = e^{\ln(x)}$ and $y = e^{\ln(y)}$. Moreover, since $\ln(x) = \ln(y)$, it follows that $x = e^{\ln(x)} = e^{\ln(y)} = y$. Thus, $x = y$ whenever $\ln(x) = \ln(y)$; therefore, proof by contrapositive shows that $\ln(x) \neq \ln(y)$ whenever $x \neq y$.

■

3.3.2 Indirect Proof

In some cases, neither a forward direct proof nor a proof by contrapositive can be found for a particular theorem. After exhausting all the possibilities with these two methods of proof, another method of proof that can tried is the *method of indirect proof* or *reductio ad absurdum*. The method of indirect proof is described below.

The Method of Indirect Proof: An indirect proof of the theorem "If H is true, then C is true" begins by assuming that the hypothesis (H) is true and the conclusion (C) is false. Working from these two statements, a sequence of logical conclusions is followed until an contradiction develops. Recall that a contradiction is a statement that is always false.

Now, working from H and $\neg C$ to a contradiction proves the theorem $H \rightarrow C$ by showing that $H \wedge \neg C$ is always false, and since $H \wedge \neg C$ is logically equivalent to $\neg(H \rightarrow C)$ (by the corollary to Theorem 2.4.1), it follows that the negation of the theorem is also false. Now, if the negation of the theorem is always false, then it must be the case that the theorem is always true. Thus, the method of proof by contradiction proves $H \rightarrow C$ by proving that $\neg(H \rightarrow C)$ can never be true.

The method of proof by contradiction is often a good approach to try when attempts to prove a result with a forward direct proof or a proof by contrapositive fail to prove the theorem. Moreover, the method of proof by contradiction is often the logical method of proof to use when proving a theorem that states that an object does not have a specific property, involves a mathematical inequality, or states that an object A having a property P is the unique object having this property. For example, a proof by contradiction would be a logical choice for proving the following theorem:

Theorem: $\sqrt{2}$ is not a rational number.

Example 3.3.6: Explain why each of the following theorems is a good candidate for a proof by contradiction:

a. The sum of a rational number and an irrational number is not a rational number.
b. The square root of a prime number is an irrational number.
c. $\forall\, x \in (0, \infty),\ x + \dfrac{1}{x} \geq 2$.
d. The set of all prime numbers is an infinite set.
e. The real solution to the equation $x^3 - 1 = 0$ is unique.
f. The set of all real numbers is an uncountable set.

Solutions:

a. This theorem states that the sum of an irrational number and a rational number does not have the property of being rational.
b. This theorem states that the square root of a prime number is not a rational number.
c. This theorem involves an inequality.
d. This theorem states that the set of all prime numbers is not a finite set.
e. This theorem states that there is a unique solution to $x^3 - 1 = 0$.
f. This theorem states that the set of all real numbers is not a countable set.

Algorithm for an Indirect Proof: The following steps outline the typical procedure used in a proof by contradiction of a theorem $H \rightarrow C$:

1. Identify and list all the hypotheses of the theorem.

2. Negate the conclusion of the theorem.
3. Identify and list all the results that are related to hypotheses and the negation of the conclusion.
3. Begin tying the hypotheses to the results in step 2 and begin working toward a contradiction. This is the scratchwork.
4. Complete the scratchwork proof by developing a logical sequence of arguments based on steps 1 and 2 that lead to a contradiction.
5. Clean up and rewrite the scratchwork into a clear and concise proof of the theorem. Make sure that each step of the proof makes sense and is clearly justified.
6. Read the proof over carefully and make any necessary corrections.

Note that in a proof by contradiction, any sequence of logical arguments that leads to a contradiction of a known fact or contradicts the assumed hypotheses of the theorem is enough to prove the desired theorem. Also, when using a proof by contradiction it is important to clearly designate this at the beginning of the proof. The following example illustrates how a proof by contradiction can be used to prove an inequality.

Example 3.3.7: Prove the following result. For $x > 0$, $x + \frac{4}{x} \geq 4$.

Scratchwork for a Proof by Contradiction:

1. Let $x > 0$.
2. The negation of the conclusion is (i.e., $\neg C$) $x + \frac{4}{x} < 4$.
3. At this point, it might be useful to write down some results concerning inequalities. For example, results of the form "If $c > 0$ and $x < y$, then $cx < cy$" and "If $x < y$, then $c+x < c+y$" might be helpful in proving this theorem.
4. The goal is to work from $x > 0$ and $x + \frac{4}{x} < 4$ to a contradiction. Consider the inequality $x + \frac{4}{x} < 4$.

$$x + \frac{4}{x} < 4 \underbrace{\text{ if and only if }}_{\text{since } x > 0} x^2 + 4 < 4x$$

$$\Longleftrightarrow x^2 + 4 - 4x < 0 \Longleftrightarrow x^2 - 4x - 4 = (x-2)^2 < 0$$

But, $\forall\ x \in \mathbb{R}$, $(x-2)^2 \geq 0$, and thus $(x-2)^2 < 0$ contradicts the fact that $(x-2)^2 \geq 0$.

Hence, proof by contradiction shows that $x + \frac{4}{x} \geq 4$ whenever $x > 0$.

5. Now, write up this scratchwork in a well-written proof.

Proof (by Contradiction): Let $x > 0$ and assume that $x + \frac{4}{x} < 4$. Then

$$x + \frac{4}{x} < 4 \iff x^2 + 4 < 4x \iff x^2 - 4x + 4 < 0$$

Now, $x^2 - 4x + 4 = (x-2)^2$, and thus $(x-2)^2 < 0$ contradicts the fact that $(x-2)^2 \geq 0$, $\forall x \in \mathbb{R}$.

■

Example 3.3.8: Prove the following theorem using a proof by contradiction:

Theorem: For $p > 3$ there are no triples of prime numbers of the form $(p, p+2, p+4)$.

Scratchwork: Note that this might be a good candidate for proof by contradiction because it states that there are no triples of primes of a particular form.

Let $p > 3$, and suppose that there exists at least one triple of prime numbers of the form $(p, p+2, p+4)$. Now, since $p, p+2$ and $p+4$ are all prime numbers greater than 3, it follows that they are all odd numbers.

Consider the following examples of triples of odd numbers: (1,3,5), (3,5,7), (5,7,9). It appears that in each triple of odd numbers one of the numbers is a multiple of 3 showing that it is unlikely for a triple of primes of this form to exist.

Now, since p is a prime number, p is not a multiple of 3. Since p is not a multiple of 3, it follows that $\exists\, k \in \mathbb{Z}$ such that (1) $p = 3k+1$ or (2) $p = 3k+2$:

Case 1: Suppose that $p = 3k+1$. Then, $p+2 = (3k+1)+2 = 3k+3 = 3(k+1)$, and hence $p+2$ is a multiple of 3, contradicting the assumption that $p+2$ is a prime number.

Case 2: Now, suppose that $p = 3k+2$. Then, $p+4 = (3k+2)+4 = 3k+6 = 3(k+2)$, and hence $p+4$ is a multiple of 3, contradicting the assumption that $p+4$ is a prime number.

Thus, in both cases a contradiction is arrived at, and therefore there are no triples of prime numbers of the form $(p, p+2, p+4)$ for $p > 3$.

Proof (by Contradiction): Let $p > 3$, and suppose that there exists at least one triple of prime numbers of the form $(p, p+2, p+4)$.

Now, since p is a prime number and $p > 3$, it follows that p is not a multiple of 3. Furthermore, since p is not a multiple of 3, $\exists\, k \in \mathbb{Z}$ such that (1) $p = 3k + 1$ or (2) $p = 3k + 2$:

Case 1: Suppose that $p = 3k + 1$. Then

$$p + 2 = (3k + 1) + 2 = 3k + 3 = 3(k + 1)$$

and hence $p + 2$ is a multiple of 3, contradicting the assumption that $p + 2$ is a prime number. Therefore, there do not exist any triples of prime numbers of the form $(p, p + 2, p + 4)$ when $p = 3k + 1$.

Case 2: Suppose that $p = 3k + 2$. Then

$$p + 4 = (3k + 2) + 4 = 3k + 6 = 3(k + 2)$$

and hence $p + 4$ is a multiple of 3, contradicting the assumption that $p + 4$ is a prime number. Therefore, there do not exist any triples of prime numbers of the form $(p, p + 2, p + 4)$ when $p = 3k + 2$.

Hence, in either case a contradiction is arrived at, and therefore there are no triples of prime numbers of the form $(p, p+2, p+4)$ for $p > 3$.

■

3.4 Specialized Methods of Proof

Besides the methods of forward direct proof, proof by contrapositive, and proof by contradiction there are also several specialized methods of proof. Moreover, each of these methods deals with a specialized form of the theorem under consideration. In particular, the specialized methods of proof that will be discussed in this section are *proof by mathematical induction*, *uniqueness proofs*, *existence proofs*, and *proof by cases*.

3.4.1 Mathematical Induction

Proof by mathematical induction is a special type of direct proof that can often be used with theorems of the nature "The statement $\mathcal{P}_n$ holds for every natural number n"; that is, a good candidate for proof by mathematical induction is any theorem that involves a statement indexed by n (i.e., $\mathcal{P}_n$) and holds $\forall n \in \mathbb{N}$. For example, the following two theorems would be good candidates to be proved using mathematical induction since both theorems contain results that are indexed by n and hold $\forall n \in \mathbb{N}$.

Theorem: $\displaystyle\sum_{i=1}^{n} i = \frac{n(n+1)}{2}, \ \forall n \in \mathbb{N}.$

Theorem: $(n+1)! \geq 2^n, \ \forall n \in \mathbb{N}.$

Using an induction proof does have one major advantage over the other methods of proof discussed previously; namely, every induction proof involves exactly the same two steps, an initial step and an induction step. Of course, this does not mean that a proof by induction will be easy. Also, there are actually two different methods of mathematical induction, weak and strong induction, which will be denoted by I_1 and I_2, respectively; strong induction is also referred to as *complete induction*. Furthermore, each of these induction methods can be used in an attempt to prove a theorem of the form $\mathcal{P}_n$ is true $\forall n \in \mathbb{N}$. The two steps required in every induction proof, for both weak (I_1) and strong (I_2) induction, are given below.

> **Weak Induction (I_1):** A theorem of the form "$\mathcal{P}_n$ holds $\forall n \in \mathbb{N}$" will be proved if it can be shown that the following two conditions are true:
>
> (i) **The Initial Step:** $\mathcal{P}_1$ is true.
>
> (ii) **The Induction Step:** If $\mathcal{P}_k$ is true for an arbitrary but fixed (ABF) value of $k \in \mathbb{N}$, then it follows that $\mathcal{P}_{k+1}$ is also true.

> **Strong Induction (I_2):** A theorem of the form "$\mathcal{P}_n$ holds $\forall n \in \mathbb{N}$ will be proved if it can be shown that the following two conditions are true:
>
> (i) **Initial Step:** $\mathcal{P}_1$ is true.
>
> (ii) **Induction Step:** If $\mathcal{P}_1, \mathcal{P}_2, \ldots, \mathcal{P}_k$ are all true for an arbitrary but fixed (ABF) value of $k \in \mathbb{N}$, then it follows that $\mathcal{P}_{k+1}$ is also true.

Note that these two versions of mathematical induction are logically equivalent. Therefore, proving the theorem "$\mathcal{P}_n$, $\forall n \in \mathbb{N}$" with strong induction is equivalent to proving the theorem "$\mathcal{P}_n$, $\forall n \in \mathbb{N}$" with weak induction and vice versa. Note that the only difference between the two steps outlined in I_1 and I_2 is that the hypothesis for the induction step in I_2 is stronger than the hypothesis in I_1. Thus, strong induction is based on the condition that $\mathcal{P}_1, \mathcal{P}_2, \ldots, \mathcal{P}_k$ are all true for some arbitrary but fixed value of $k \in \mathbb{N}$, and this is a stronger condition that $\mathcal{P}_k$ is true for an arbitrary but fixed value of $k \in \mathbb{N}$, which is used in weak induction. The question of which version of induction to use in a proof often becomes clear in the scratchwork of the induction step. In fact, when strong induction is required, it will usually become obvious in the induction step of the proof. In general, it is standard practice to attempt an induction proof using weak induction first, and then trying strong induction only when it is clear that $\mathcal{P}_1, \mathcal{P}_2, \ldots, \mathcal{P}_k$ are needed.

Now, theorems that are good candidates for a proof by mathematical induction are theorems that are indexed by n and also hold $\forall n \in \mathbb{N}$. In other words, a theorem of the form $\mathcal{P}_n$ holds $\forall n \in \mathbb{N}$ can often be proved using mathematical induction. For example, the following theorem can be proved with mathematical induction:

Theorem: $\dfrac{d}{dx}[x^n] = nx^{n-1}, \ \forall n \in \mathbb{N}.$

An algorithm that can be used for a proof by weak induction is outlined below.

An Algorithm for Weak Induction: The following steps outline the typical procedure used in a proof by weak induction of a theorem $\mathcal{P}_n$, $\forall n \in \mathbb{N}$:

1. Define the statement $\mathcal{P}_n$.
2. Show that $\mathcal{P}_1$ is true.
3. Assume that $\mathcal{P}_k$ is true for an arbitrary but fixed (ABF) value of $k \in \mathbb{N}$, and write down exactly what this means.
4. Write down the statement $\mathcal{P}_{k+1}$. This is the statement that $\mathcal{P}_k$ is supposed to lead to.
5. Began scratchwork by trying to relate $\mathcal{P}_{k+1}$ to $\mathcal{P}_k$. Determine a set of logical arguments showing that $\mathcal{P}_k \rightarrow \mathcal{P}_{k+1}$.
6. Clean up and rewrite the scratchwork into a clear and concise proof of the theorem. Make sure that each step of the proof makes sense and is clearly justified.
7. Read the proof over carefully and make any necessary corrections.

As is the case with any proof that is based on a method of proof other than the method of forward direct proof, it is important to denote the type of proof being used. Hence, when an induction proof is used to prove a theorem it should be clearly noted that this is the method of proof being used. The following two examples illustrate how weak induction can be used to prove a theorem whose conclusion is indexed by n and holds $\forall\, n \in \mathbb{N}$.

Example 3.4.1: Prove the following result using weak induction:

$$\sum_{i=1}^{n} i = \frac{n(n+1)}{2}, \quad \forall n \in \mathbb{N}$$

Solution: This result is a good candidate for a proof by weak induction since it is of the form $\mathcal{P}_n$, $\forall n \in \mathbb{N}$.

Scratchwork:

1. Define $\mathcal{P}_n := \displaystyle\sum_{i=1}^{n} i = \frac{n(n+1)}{2}$.

2. The initial step is to prove that $\mathcal{P}_1$ is true. First, consider $\displaystyle\sum_{i=1}^{1} i$.

   Now, $\displaystyle\sum_{i=1}^{1} i = 1$, and since $\dfrac{1(1+1)}{2} = 1$, it follows that $\mathcal{P}_1$ is true.

3,4,5. The induction step. The goal here is to show that if $\mathcal{P}_k$ is true for some ABF $k \in \mathbb{N}$, then so is $\mathcal{P}_{k+1}$. The key to carrying out the induction step is often to relate the result in $\mathcal{P}_{k+1}$ to the result in $\mathcal{P}_k$, and then work forward from this relationship.

Suppose that $\mathcal{P}_k$ is true for some ABF $k \in \mathbb{N}$. This means that

$$\sum_{i=1}^{k} i = \frac{k(k+1)}{2}$$

Consider $\mathcal{P}_{k+1}$. If $\mathcal{P}_{k+1}$ is true, then it follows that $\displaystyle\sum_{i=1}^{k+1} i$ will be equal to $\dfrac{(k+1)(k+2)}{2}$.

Now, consider $\sum_{i=1}^{k+1} i$:

$$\sum_{i=1}^{k+1} i = \sum_{i=1}^{k} i + (k+1) = \underbrace{\frac{k(k+1)}{2}}_{\text{since } \mathcal{P}_k \text{ is true}} + (k+1)$$

$$= \frac{k^2 + k + 2k + 2}{2} = \frac{k^2 + 3k + 2}{2} = \frac{(k+1)(k+2)}{2}$$

Thus, $\mathcal{P}_{k+1}$ is true whenever $\mathcal{P}_k$ is true and therefore, it follows that $\sum_{i=1}^{n} i = \frac{n(n+1)}{2}$, $\forall n \in \mathbb{N}$.

6. Write up this set of induction arguments in a formal proof.

Proof (by Induction): Define $\mathcal{P}_n := \sum_{i=1}^{n} i = \frac{n(n+1)}{2}$.

First, $\sum_{i=1}^{1} i = 1$ and $\frac{1(1+1)}{2} = 1$. Therefore, $\mathcal{P}_1$ is true.

Now, suppose that $\mathcal{P}_k$ is true for some ABF $k \in \mathbb{N}$. This means that $\sum_{i=1}^{k} i = \frac{k(k+1)}{2}$. If $\mathcal{P}_{k+1}$ is true, then it follows that $\sum_{i=1}^{k+1} i$ will be $\frac{(k+1)(k+2)}{2}$. Consider $\sum_{i=1}^{k+1} i$:

$$\sum_{i=1}^{k+1} i = \sum_{i=1}^{k} i + (k+1) = \underbrace{\frac{k(k+1)}{2}}_{\text{since } \mathcal{P}_k \text{ is true}} + (k+1)$$

$$= \frac{k^2 + k + 2k + 2}{2} = \frac{k^2 + 3k + 2}{2} = \frac{(k+1)(k+2)}{2}$$

Thus, $\mathcal{P}_{k+1}$ is true whenever $\mathcal{P}_k$ is true; therefore, $\sum_{i=1}^{n} i = \frac{n(n+1)}{2}$, for all $n \in \mathbb{N}$.

■

Example 3.4.2: Prove the following result using weak induction. Let $p \in \mathbb{R}$. If $p \neq 1$, then $\sum_{i=0}^{n} p^i = \frac{1-p^{k+1}}{1-p}$, $\forall n \in \mathbb{N}$.

Solution: This is clearly a good candidate for a proof based on mathematical induction since it is of the form $\mathcal{P}_n$, $\forall n \in \mathbb{N}$.

Scratchwork: Let $p \in \mathbb{R}$ and suppose that $p \neq 1$ and define $\mathcal{P}_n := \sum_{i=0}^{n} p^i = \frac{1-p^{n+1}}{1-p}$.

The Initial Step: For $n = 1$, it follows that

$$\sum_{i=0}^{1} p^i = p^0 + p^1 = 1 + p$$

and

$$\frac{1-p^{1+1}}{1-p} = \frac{1-p^2}{1-p} = \frac{(1-p)(1+p)}{1-p} = 1 + p$$

Therefore, $\mathcal{P}_1$ is true.

The Induction Step: Suppose that $\mathcal{P}_k$ is true for some ABF $k \in \mathbb{N}$. This means that $\sum_{i=0}^{k} p^i = \frac{1-p^{k+1}}{1-p}$. Consider, $\mathcal{P}_{k+1}$. If $\mathcal{P}_{k+1}$ is true, then $\sum_{i=0}^{k+1} p^i$ will be $\frac{1-p^{k+2}}{1-p}$. Consider $\sum_{i=0}^{k+1} p^i$:

$$\sum_{i=0}^{k+1} p^i = \sum_{i=0}^{k} p^i + p^{k+1} = \underbrace{\frac{1-p^{k+1}}{1-p}}_{\text{since } \mathcal{P}_k \text{ is true}} + p^{k+1}$$

$$= \frac{1-p^{k+1}+p^{k+1}-p^{k+2}}{1-p} = \frac{1-p^{k+2}}{1-p}$$

Thus, $\mathcal{P}_{k+1}$ is true whenever $\mathcal{P}_k$ is true, and therefore

$$\sum_{i=0}^{n} p^i = \frac{1-p^{k+1}}{1-p}, \quad \forall n \in \mathbb{N}$$

Proof (by Induction): Let $p \in \mathbb{R}$ and suppose that $p \neq 1$. Define

$$\mathcal{P}_n := \sum_{i=0}^{n} p^i = \frac{1 - p^{n+1}}{1 - p}$$

First, $\sum_{i=0}^{1} p^i = p^0 + p^1 = 1 + p$ and

$$\frac{1 - p^{1+1}}{1 - p} = \frac{1 - p^2}{1 - p} = \frac{(1 - p)(1 + p)}{1 - p} = 1 + p$$

Therefore, $\mathcal{P}_1$ is true.

Suppose that $\mathcal{P}_k$ is true for some ABF $k \in \mathbb{N}$. This means that $\sum_{i=0}^{k} p^i = \frac{1 - p^{k+1}}{1 - p}$.

Now, consider $\mathcal{P}_{k+1}$. If $\mathcal{P}_{k+1}$ is true, then it follows that $\sum_{i=0}^{k+1} p^i$ will be $\frac{1 - p^{k+2}}{1 - p}$. Consider $\sum_{i=0}^{k+1} p^i$:

$$\sum_{i=0}^{k+1} p^i = \sum_{i=0}^{k} p^i + p^{k+1} = \underbrace{\frac{1 - p^{k+1}}{1 - p}}_{\text{since } \mathcal{P}_k \text{ is true}} + p^{k+1}$$

$$= \frac{1 - p^{k+1} + p^{k+1} - p^{k+2}}{1 - p} = \frac{1 - p^{k+2}}{1 - p}$$

Thus, $\mathcal{P}_{k+1}$ is true whenever $\mathcal{P}_k$ is true. Thus, whenever $p \neq 1$, it follows that

$$\sum_{i=0}^{n} p^i = \frac{1 - p^{n+1}}{1 - p}, \quad \forall n \in \mathbb{N}$$

■

In the development of an axiomatic mathematical system, often a result will first be proved for the special case involving only two objects, followed by proving that the result can be extended to a more general result dealing with n objects. Often, mathematical induction is used to prove the more general result that holds for n objects. The following example illustrates how mathematical induction can be used to extend a result concerning two objects to a result about an arbitrary natural number of objects.

Example 3.4.3: Recall, from Chapter 2, DeMorgan's law for negating the conjunction of two statements stating that

$$\neg(P \wedge Q) \iff \neg P \vee \neg Q$$

Using mathematical induction, DeMorgan's law for conjunction can be extended to any collection of n statements, say, $P_1, \ldots, P_n$. Thus, induction can be used to prove the following result:

$$\neg\left(\bigwedge_{i=1}^{n+1} P_i\right) = \bigvee_{i=1}^{n+1} \neg P_i, \ \forall\, n \in \mathbb{N}$$

Proof: The proof of this result is left as an exception.

■

Note that weak induction was used to prove each of the results in Examples 3.4.1 and 3.4.2. Although strong induction could also have been used to prove the results in each of these examples, certain mathematical results will require the use of strong rather than weak induction. An algorithm for a typical strong induction proof is given below.

An Algorithm for Strong Induction: The following procedure can be used for a strong induction proof of a theorem $\mathcal{P}_n$, $\forall\, n \in \mathbb{N}$.

1. Define the statement $\mathcal{P}_n$.
2. Show that $\mathcal{P}_1$ is true.
3. Assume that $\mathcal{P}_1, \mathcal{P}_2, \ldots, \mathcal{P}_k$ are all true for an arbitrary but fixed value of $k \in \mathbb{N}$. Explain in writing what this means.
4. Write down the statement $\mathcal{P}_{k+1}$. This is the statement to which the statements $\mathcal{P}_1, \mathcal{P}_2, \ldots, \mathcal{P}_k$ are supposed to lead.
5. Begin the scratchwork by trying to relate $\mathcal{P}_{k+1}$ to $\mathcal{P}_1, \mathcal{P}_2, \ldots, \mathcal{P}_k$. Determine a set of logical arguments showing that $\mathcal{P}_1 \wedge \mathcal{P}_2 \wedge \cdots \wedge \mathcal{P}_k \rightarrow \mathcal{P}_{k+1}$.

6. Clean up and rewrite the scratchwork in a clear and concise proof of the theorem. Make sure each that step of the proof makes sense and is clearly justified.
7. Read the proof over carefully and make any necessary corrections.

The following example illustrates a result that cannot be easily proved with weak induction, but can be easily proved using strong induction.

Example 3.4.4: Let the sequence of real numbers a_n be defined by $a_1 = 0$, $a_2 = 1$, and $a_{n+2} = 3a_{n+1} - 2a_n$, for $n \in \mathbb{N}$. Prove that $a_{n+2} = 2^{n+1} - 1$, $\forall\, n \in \mathbb{N}$.

Solution: This result is a good candidate for a proof based on mathematical induction since it is of the form $\mathcal{P}_n$, $\forall n \in \mathbb{N}$. The scratchwork below shows that weak induction fails to prove this result, while strong induction does not fail.

Scratchwork: First try weak induction.

Let $\mathcal{P}_n$ be the propositional function "$a_{n+2} = 2^{n+1} - 1$".

For $n = 1$, it follows that $a_{1+2} = a_3 = 3a_2 - 2a_1 = 3 - 0 = 3$ and $2^{1+1} - 1 = 4 - 1 = 3$. Therefore, $\mathcal{P}_1$ is true.

Suppose that $\mathcal{P}_k$ is true for some ABF $k \in \mathbb{N}$. This means that $a_{k+2} = 2^{k+1} - 1$.

If $\mathcal{P}_{k+1}$ is true, then $a_{(k+1)+2} = a_{k+3}$ will be $2^{k+1+1} - 1$. Now try to show that $\mathcal{P}_k \rightarrow \mathcal{P}_{k+1}$.

Consider a_{k+3}. By definition, $a_{k+3} = 3a_{k+2} - 2a_{k+1}$. Now, relate to a_{k+3} to $\mathcal{P}_k$. Note that $\mathcal{P}_k$ gives information on a_{k+2} but not on a_{k+1}, and hence there is no obvious way to relate a_{k+3} to $\mathcal{P}_k$. At this point it appears that weak induction has failed. Now, an alternative approach is to try strong induction.

Since the initial step is the same in weak and strong induction, only the induction step must be reconsidered. The scratchwork now continues with the strong induction hypothesis.

Suppose that $\mathcal{P}_1, \ldots, \mathcal{P}_k$ are all true for some ABF $k \in \mathbb{N}$. This means that $a_{j+2} = 2^{j+1} - 1$ whenever $j = 1, 2, \ldots, k$.

Consider a_{k+3}. By definition, $a_{k+3} = 3a_{k+2} - 2a_{k+1}$. Now, relate a_{k+3} to $\mathcal{P}_1, \ldots, \mathcal{P}_k$. Note that $\mathcal{P}_k$ gives information on a_{k+2} and

$\mathcal{P}_{k-1}$ provides information on a_{k+1}. Thus, from $\mathcal{P}_k$ and $\mathcal{P}_{k-1}$ it follows that $a_{k+2} = 2^{k+1} - 1$ and $a_{k+1} = 2^k - 1$. Hence

$$a_{k+3} = 3a_{k+2} - 2a_{k+1} = 3(2^{k+1} - 1) - 2(2^k - 1)$$

$$= 3 \cdot 2^{k+1} - 3 - 2^{k+1} + 2 = (3 - 1)2^{k+1} - 1$$

$$= 2^{k+2} - 1$$

Thus $\mathcal{P}_1 \wedge \cdots \wedge \mathcal{P}_k \rightarrow \mathcal{P}_{k+1}$, and therefore $a_{n+2} = 2^{n+1} - 1, \forall\, n \in \mathbb{N}$.

Proof: Let $\mathcal{P}_n$ be the propositional function "$a_{n+2} = 2^{n+1} - 1$".

For $n = 1$, it follows that $a_{1+2} = a_3 = 3a_2 - 2a_1 = 3 - 0 = 3$ and $2^{1+1} - 1 = 4 - 1 = 3$. Therefore, $\mathcal{P}_1$ is true.

Suppose that $\mathcal{P}_1, \ldots, \mathcal{P}_k$ are all true for some ABF $k \in \mathbb{N}$. This means that $a_{j+2} = 2^{j+1} - 1$ whenever $j = 1, 2, \ldots, k$.

If $\mathcal{P}_{k+1}$ is true, then a_{k+3} will be $2^{k+2} - 1$. Consider a_{k+3}. Now, by definition, $a_{k+3} = 3a_{k+2} - 2a_{k+1}$, and from $\mathcal{P}_k$ and $\mathcal{P}_{k-1}$ it follows that $a_{k+2} = 2^{k+1} - 1$ and $a_{k+1} = 2^k - 1$. Thus

$$a_{k+3} = 3a_{k+2} - 2a_{k+1} = 3(2^{k+1} - 1) - 2(2^k - 1)$$

$$= 3 \cdot 2^{k+1} - 3 - 2^{k+1} + 2 = (3 - 1)2^{k+1} - 1$$

$$= 2^{k+2} - 1$$

Thus, $\mathcal{P}_{k+1}$ is true whenever $\mathcal{P}_1, \ldots, \mathcal{P}_k$ are true; therefore, by strong induction, it follows that $a_{n+2} = 2^{n+1} - 1, \forall\, n \in \mathbb{N}$.

■

The previous example illustrated how easy it is to switch from weak to strong induction in the scratchwork phase of a proof. Furthermore, strong induction will often be needed when recursive definitions are being used in a result indexed by n. The following theorem, stated without proof, states that the strong and weak forms of induction are logically equivalent; a nice proof of Theorem 3.4.1 can be found in *Elementary Introduction to Number Theory* by Calvin Long (1972).

Theorem 3.4.1: The methods of weak and strong mathematical induction are logically equivalent methods of proof.

It is important to note that in a proof by mathematical induction, the initial step must lead to the induction step so that $\mathcal{P}_1$ leads to $\mathcal{P}_2$ and so on. If $\mathcal{P}_1$ is true but does not lead to $\mathcal{P}_2$, then the induction algorithms as stated cannot be used to prove the result in question. The following example illustrates a scenario where the initial step does not lead directly to the induction step and hence leads to the erroneous conclusion that in any set of n numbers, all the numbers have the same value.

> **Invalid Induction Proof:** Let $\mathcal{P}_n$ be the statement "If A is any set of n numbers, then all the numbers in A have the same exact value."
>
> First, consider a set A containing only one element. Clearly, by default every element in this set has the same value. Thus, $\mathcal{P}_1$ is true.
>
> Suppose that $\mathcal{P}_k$ is true for some ABF $k \in \mathbb{N}$. This means that in any set of k numbers, all the numbers have exactly the same value.
>
> If $\mathcal{P}_{k+1}$ is true, then it will follow that in any set of $k+1$ elements, all the elements have exactly the same value. Consider a set of $k+1$ elements, say, $\{a_1, a_2, \ldots, a_{k+1}\}$.
>
> Note that $\{a_1, \ldots, a_k\}$ and $\{a_2, \ldots, a_{k+1}\}$ are two sets of k numbers, and thus by $\mathcal{P}_k$, it follows that $a_1 = a_2 = \cdots = a_k$ and that $a_2 = a_3 = \cdots = a_{k+1}$. Hence, $a_1 = a_2 = \cdots = a_{k+1}$ and therefore, $\mathcal{P}_{k+1}$ is true whenever $\mathcal{P}_k$ is true. Therefore, if A is a set of n numbers, then each of the numbers in A has exactly the same value, $\forall n \in \mathbb{N}$.
>
> ■

But clearly, in the set of numbers $\{1, 2\}$ all the numbers do not have the same value. Thus, there must be an error in the set of induction steps outlined above. In fact, the problem that set of arguments is that while the initial step does show that $\mathcal{P}_1$ is true, it does not lead to $\mathcal{P}_2$ or any other statement $\mathcal{P}_k$. Since there is no successor to the statement $\mathcal{P}_1$, the induction argument fails to prove this conjecture.

Finally, it is not required that the mathematical result being proved need be indexed by the entire collection of natural numbers ($\mathbb{N}$) in order to use mathematical induction. In other words, if the theorem states that the result

$\mathcal{P}_n$ holds for only n greater than or equal to some integer m, then mathematical induction can still be used to prove the theorem by rewriting it as $\mathcal{P}_{n+m-1}$ holds for $n \in \mathbb{N}$. For example, consider the following theorem:

> **Theorem:** $2^n > n^2$, whenever n is a natural number greater than or equal to 5.

This theorem can be rewritten as

> **Theorem:** $2^{n+4} > (n+4)^2$, $\forall\, n \in \mathbb{N}$.

Example 3.4.5: Rewrite each of the following results in the form $\mathcal{P}_n$, $\forall\, n \in \mathbb{N}$:

a. $3^n < n!$, for all natural numbers $n \geq 7$.

b. $n\ln(n) > n$, for all natural numbers $n \geq 3$.

c. $\displaystyle\int \frac{1}{x^n}\,dx = \frac{1}{(1-n)x^{n-1}} + C$, for all natural numbers $n \geq 2$.

Solutions:

a. $3^{n+6} < (n+6)!$, $\forall\, n \in \mathbb{N}$.

b. The solution to part (b) is left as an exercise.

c. The solution to part (c) is left as an exercise.

3.4.2 Uniqueness Proofs

A special theorem of the nature "Object A that is an element of the set C is the only (i.e., the unique) object having a property P" is called a *uniqueness theorem*. A uniqueness theorem is very important in that it shows that one and only one object has the special property P. Two examples of uniqueness theorems are given below:

> **Theorem:** In the field of real numbers, the multiplicative identity, $e = 1$, is unique.

> **Theorem:** If $(\mathcal{G}, \circ)$ is a group and $e \in \mathcal{G}$ is an identity element, then e is the unique identity element in $\mathcal{G}$.

A uniqueness theorem can be proved with a proof by contradiction, and the steps used to prove a uniqueness theorem are outlined below.

Algorithm for a Uniqueness Proof: A proof by contradiction can be used to prove the theorem "Object A in the set C having property P is unique" as outlined below.

1. Show that object $A \in C$ has property P.

2. Assume that object A is not the only object in the set C having property P.
3. Let object $B \neq A$ be any other object in the set C that has property P.
4. Using logical arguments, show that $B = A$, contradicting statement 3.
5. Clean up and rewrite the scratchwork into a clear and concise proof of the theorem. Make sure that each step of the proof makes sense and is clearly justified.
6. Read the proof over carefully and make any necessary corrections.

The following three examples illustrate the use of the method of proof by contradiction to prove a uniqueness theorem, using the algorithm presented above.

Example 3.4.6: Prove the following result. If $x = 1$, then x is the unique real-valued solution to the equation $x^3 - 1 = 0$.

> ***Proof (Uniqueness Proof):*** First, $x = 1$ is a real-valued solution to $x^3 - 1 = 0$ since $1^3 - 1 = 0$. Now, suppose that $x = 1$ is not the only real-valued solution to $x^3 - 1 = 0$. In other words, there exists more than one real-valued solution to $x^3 - 1 = 0$. Let $s \neq 1$ be any other solution to $x^3 - 1 = 0$. Then, $s^3 - 1 = 0$ and hence
>
> $$0 = s^3 - 1 \text{ if and only if } s^3 = 1 \text{ if and only if } s = 1$$
>
> However, this contradicts the fact that $s \neq 1$, and therefore, $x = 1$ is the unique solution to the equation $x^3 - 1 = 0$.
>
> ∎

Example 3.4.7: Prove the following result. The matrix $I_2 = \begin{bmatrix} 1 & 0 \\ 0 & 1 \end{bmatrix}$ is the unique 2×2 identity matrix.

> ***Proof (Uniqueness Proof):*** Let $A = \begin{bmatrix} a & b \\ c & d \end{bmatrix}$ be an ABF a 2×2 matrix, and let $I_2 = \begin{bmatrix} 1 & 0 \\ 0 & 1 \end{bmatrix}$. Then, $AI_2 = A$ and $I_2A = A$ and hence I_2 is a 2×2 identity matrix.
>
> Now, suppose that there exists more than one 2×2 identity matrix, and let $E \neq I_2$ be any other 2×2 identity matrix.
>
> Since E and I_2 are identity matrices, it follows that
>
> $$AE = EA = A \quad \text{and} \quad AI_2 = I_2A = A$$

for any 2×2 matrix A.

Thus, $EI_2 = E$ since I_2 is an identity matrix, and similarly, $EI_2 = I_2$ since E is an identity matrix. Hence, $E = EI_2 = I_2$, contradicting the fact that $E \neq I_2$ and therefore, I_2 is the unique 2×2 identity matrix.

■

Example 3.4.8: For $a, b \in \mathbb{R}$, define $a \circ b = a + b - 1$. Prove the following theorem.

Theorem: Let $a, b \in \mathbb{R}$. Then, the solution x to the equation $a \circ x = b$ is unique.

Proof (Uniqueness Proof): Let $a, b \in \mathbb{R}$ and suppose that the solution to $a \circ x = b$ is not unique. Let $s_1 \neq s_2$ be two real numbers that are solutions to the equation $a \circ x = b$.

Then

$$a \circ s_1 = b \text{ if and only if } a + s_1 - 1 = b \text{ if and only if } s_1 = b - a + 1$$

and

$$a \circ s_2 = b \text{ if and only if } a + s_2 - 1 = b \text{ if and only if } s_2 = b - a + 1$$

Thus, $s_1 = b - a + 1 = s_2$, which contradicts $s_1 \neq s_2$, and therefore the solution to $a \circ x = b$ is unique.

■

3.4.3 Existence Proofs

Another special type of theorem is a theorem of the nature "There exists an object A that is an element of the set $\mathcal{C}$ that has property P." A theorem of this nature is called an *existence theorem*. The following theorem provides an example of a existence theorem.

Theorem: If $(\mathcal{G}, \circ)$ is a group and $a, b \in \mathcal{G}$, then there exists $y \in \mathcal{G}$ such that $a \circ y = b$.

Now, the proof of an existence theorem requires showing that there exists an object $A \in \mathcal{C}$ that has the property P. An existence proof may simply involve creating an object A in the set $\mathcal{C}$ having the property P, but in many cases, the proof of an existence theorem will require a great deal of creative thinking and mathematical insight. Hence, the method of proof that is typically used in proving an existence theorem is called "proof by construction." The following algorithm outlines the typical approach used in proving a uniqueness theorem.

Algorithm for an Existence Proof: To prove the theorem "There exists an object A in the set C that has property P"

1. Create or construct an element A that has property P.
2. Show that the element A is in the set C.
3. Clean up and rewrite the scratchwork into a clear and concise proof of the theorem. Make sure that each step of the proof makes sense and is clearly justified.
4. Read the proof over carefully and make any necessary corrections.

Note that while this algorithm has only main two steps, the proof of an existence theorem often requires a great deal of scratchwork and creative thinking. In most cases, the existence proof does not reveal the amount of work involved in creating or finding the object A. While an existence proof might be very short once the object A has been created or found, a short proof does not necessarily indicate the level of difficulty in developing the proof. In many existence proofs, the construction of the element $A \in C$ can be found by focusing on the property P, and this may simply amount to solving an equation related to the property P. Also, it is not unusual that a mathematical trick or unusual approach will need to be used in creating the object A in an existence proof. An example of an existence theorem and its proof are given in the following example.

Example 3.4.9: Prove the following existence theorem:

Theorem: There exist prime numbers of the form $2^p - 1$ where p is a prime number.

> **Scratchwork:** First, the set C is the set of prime numbers, and the property of interest is that $2^p - 1$ is a prime number.
>
> To create a prime of the form $2^p - 1$, first consider the prime number $p = 2$. Now, 2 is a prime number and $2^2 - 1 = 3$, which is a prime number. Thus, $p = 2$ is in C and has the property P.
>
> ***Proof (Existence Proof):*** Note that 2 is a prime number, and $2^2 - 1 = 3$ is a prime number. Thus, there do exist prime numbers of the form $2^p - 1$ where p is a prime number.

■

Note that in this theorem, the value of p was found by considering the first possible prime number. Since $p = 2$ is a prime number for which $2^2 - 1$ is also a prime number, this proved the existence theorem, and the proof is very short.

Example 3.4.10: Prove the following theorem:

> **Theorem:** There exist two irrational numbers whose sum is rational.
>
> **Scratchwork:** In this theorem, C is the set of irrational numbers and P is the property that the sum of two irrational numbers is rational. Consider well-known irrational numbers. For example, $\sqrt{2}, \sqrt{3}, \pi$, and e.
>
> The trick here is to focus on the sum being rational. Now, there are many ways to form a rational number from two irrational numbers, but only one is needed.
>
> Let $x = \sqrt{2}$ and $y = -\sqrt{2}$. Then, x and y are both irrational numbers, and $x + y = \sqrt{2} + (-\sqrt{2}) = \sqrt{2} - \sqrt{2} = 0$. Since 0 is a rational number, the result is as follows.
>
> ***Proof (Existence Proof):*** Let $x = \sqrt{2}$ and $y = -\sqrt{2}$. Then, x and y are both irrational numbers.
>
> Consider $x + y$
>
> $$x + y = \sqrt{2} + (-\sqrt{2}) = \sqrt{2} - \sqrt{2} = 0$$
>
> Thus, since 0 is a rational number, it follows that there exist irrational numbers such that the sum of the two irrational numbers is rational.
>
> ■

Although the two previous existence proofs were very short, it is not always the case that the proof of an existence theorem will be short. The key to any existence proof is the construction of an element A in the set C that has property P; the length of the proof may or may not represent the difficulty of the construction of the element A.

Example 3.4.11: For $a, b \in \mathbb{R}$, define the binary operator $\circ$ as follows: $a \circ b = a + b - 1$. Prove the following theorem:

Theorem: Let $a, b \in \mathbb{R}$. Then, there exists $x \in \mathbb{R}$ such that $a \circ x = b$.

Scratchwork: Here the set C is $\mathbb{R}$ and the property P is that $x \in \mathbb{R}$ satisfies the equation $a \circ x = b$ or equivalently, $a + x - 1 = b$. The proof of this theorem requires that a solution to this equation be constructed.

Let $a, b \in \mathbb{R}$ and consider the equation $a \circ x = b$. Now, solve this equation for x.

$$a \circ x = b \text{ if and only if } a + x - 1 = b \text{ if and only if } x = b - a + 1$$

Double-check this solution by plugging it in for x.

$$a \circ (b - a + 1) = a + b - a + 1 - a = b$$

So, the solution is $x = b - a + 1$ which is in $\mathbb{R}$.

Proof (Existence Proof): Let $a, b \in \mathbb{R}$. Consider the equation $a \circ x = b$:

$$a \circ x = b \text{ if and only if } a + x - 1 = b \text{ if and only if } x = b - a + 1$$

Thus, $x = b - a + 1$ is a real number satisfying $a \circ x = b$.

Therefore, there exists $x \in \mathbb{R}$ such that $a \circ x = b$, $\forall\ a, b \in \mathbb{R}$.

■

3.4.4 Proof by Cases

In many proofs, the path of the logical arguments will lead to a statement involving an either/or statement such as condition S_1 or condition S_2. When this happens, it is often useful to consider separate proofs for each of the cases S_1 and S_2, provided these cases are mutually exclusive and exhaustive. This approach is called a *proof by cases*. The need for a proof by cases may be obvious or may be hidden in the implications of a particular statement. Furthermore, a proof by cases may involve more than simply two cases. Examples of the type of statement that may indicate whether a proof by cases is needed are given below:

a. ... either $x > 0$ or $x \leq 0$.

b. ... either x is a prime number or x is a composite number.

c. ... either n is odd or n is even.

d. ... either x is in the set A, x is in the set B, or x is in the set C.

e. ... the set $\mathcal{P}$ is either empty, finite, or infinite.

An example of a proof by cases can be found in the proof of Theorem 2.3.4, part (i), which states that $\neg(P \vee Q)$ and $\neg P \wedge \neg Q$ are logically equivalent. In the proof of this theorem, the truth tables for these two statements were compared and shown to be identical. Thus, it turns out that proving that two statements are logically equivalent by showing that they have identical truth tables is actually a proof by cases. In other words, in a proof utilizing a truth table, all the different possible states of nature (i.e., different cases) for the base statements must be considered. An alternative proof of Theorem 2.3.4(i) is shown in the following example.

Example 3.4.12: The proof of Theorem 2.3.4(i) is restated as a proof by cases below.

Theorem 2.3.4(i): If P and Q are statements, then $\neg(P \vee Q)$ is logically equivalent to $\neg P \wedge \neg Q$.

Proof: Let P and Q be statements. Then either (1) P is true and Q is true, (2) P is true and Q is false, (3) P is false and Q is true, or (4) P is false and Q is false.

Case 1: Suppose that P is true and Q is true. Then, $(P \vee Q)$ is true and hence $\neg(P \vee Q)$ is false. Furthermore, $\neg P$ and $\neg Q$ are both false, and thus $\neg P \wedge \neg Q$ is false. Therefore, when P and Q are both true, then $\neg(P \vee Q)$ and $\neg P \wedge \neg Q$ are both false.

Case 2: Suppose that P is true and Q is false. Then, $(P \vee Q)$ is true and hence $\neg(P \vee Q)$ is false. Furthermore, $\neg P$ is false and $\neg Q$ is true, and thus $\neg P \wedge \neg Q$ is false. Therefore, when P is true and Q is false, then $\neg(P \vee Q)$ and $\neg P \wedge \neg Q$ are both false.

Case 3: Suppose that P is false and Q is true. Then, $(P \vee Q)$ is true and hence $\neg(P \vee Q)$ is false. Furthermore, $\neg P$ is true and $\neg Q$ is false, and thus $\neg P \wedge \neg Q$ is false. Therefore, when P is false and Q is true, then $\neg(P \vee Q)$ and $\neg P \wedge \neg Q$ are both false.

Case 4: Suppose that P is false and Q is false. Then, $(P \vee Q)$ is false and hence $\neg(P \vee Q)$ is true. Furthermore, both $\neg P$ and $\neg Q$ are true, and thus $\neg P \wedge \neg Q$ is true. Therefore, when both P and Q are false, then $\neg(P \vee Q)$ and $\neg P \wedge \neg Q$ are both true.

Thus, in each of the four cases $\neg(P \vee Q)$ and $\neg P \wedge \neg Q$ have the same truth values. Thus, it follows that $\neg(P \vee Q)$ is logically equivalent to $\neg P \wedge \neg Q$.

■

Note that when using a proof by cases to prove a theorem, a separate proof of the theorem is required for each of the possible individual cases. Also, with regard to the clarity of the proof it is important to denote (1) the possible cases and (2) where the proof of each case begins and ends. The following example illustrates how this might be done.

Proof: Hypotheses

Logical arguments

⋮

Thus, it follows that $x \geq y$ or $x < y$.

Case 1: Suppose that $x \geq y$.

⋮

Hence, for $x \geq y$ it follows that ...

Case 2: Suppose that $x < y$.

⋮

Hence, for $x < y$ it follows that ...

Therefore, in either case it follows that ...

■

Note that the beginnings and endings of each of the two cases in the example above are clearly delineated. Furthermore, note that the last line of the proof states that the result holds for both of the possible cases considered. The following two examples illustrate the typical use of the method of proof by cases.

Example 3.4.13: Prove the following theorem:

Theorem: Let n be a natural number. If n is not a multiple of 3, then n^2 is not a multiple of 3.

Scratchwork: Let n be a natural number that is not a multiple of 3. Since n is not a multiple of 3 (i.e., $n \neq 3k$), it follows that $n = 3k + 1$ or $n = 3k + 2$ for some natural number k. Thus, the two cases that must be considered are $n = 3k + 1$ and $n = 3k + 2$.

Case 1: Suppose that $n = 3k + 1$ for some natural number k. Then

$$n^2 = (3k + 1)^2 = 9k^2 + 6k + 1 = 3(3k^2 + 2k) + 1 = 3j + 1$$

where $j = 3k^2 + 2k$, which is an integer. Thus, n^2 is not a multiple of 3.

Case 2: Suppose that $n = 3k + 2$ for some natural number k. Then

$$n^2 = (3k + 2)^2 = 9k^2 + 12k + 4 = 3(3k^2 + 4k + 1) + 1 = 3j + 1$$

where $j = 3k^2 + 2k$, which is an integer. Thus, n^2 is not a multiple of 3.

Hence, in either case n^2 is not a multiple of 3 and therefore, when a natural number n is not a multiple of 3, it follows that n^2 is not a multiple of 3, either.

Proof: Let n be a natural number that is not a multiple of 3. Now, since n is not a multiple of 3, it follows that $n = 3k + 1$ or $n = 3k + 2$ for some natural number k.

Case 1: Suppose that $n = 3k + 1$ for some natural number k. Then

$$n^2 = (3k + 1)^2 = 9k^2 + 6k + 1 = 3(3k^2 + 2k) + 1 = 3j + 1$$

where $j = 3k^2 + 2k$, which is an integer. Thus, n^2 is not a multiple of 3.

Case 2: Suppose that $n = 3k + 2$ for some natural number k. Then

$$n^2 = (3k + 2)^2 = 9k^2 + 12k + 4 = 3(3k^2 + 4k + 1) + 1 = 3j + 1$$

where $j = 3k^2 + 2k$, which is an integer. Thus, n^2 is not a multiple of 3.

Hence, in either case n^2 is not a multiple of 3 and therefore, when a natural number n is not a multiple of 3, it follows that n^2 is not a multiple of 3, either.

■

Example 3.4.14: Prove the following theorem:

Theorem: Let x and y be real numbers. Then $|xy| = |x| \cdot |y|$.

Proof: Let x and y be real numbers. Then, either $x, y \geq 0$ or $x \geq 0, y < 0$ or $x, y < 0$ or $x < 0, y \geq 0$.

Case 1: Suppose that $x, y \geq 0$. Then, $|x| = x$, $|y| = y$, and $|xy| = xy$. Hence, $|xy| = xy = |x| \cdot |y|$, and thus $|xy| = |x| \cdot |y|$.

Case 2: Suppose that $x \geq 0$ and $y < 0$. Then, $|x| = x$, $|y| = -y$, and $|xy| = -xy$. Hence, $|xy| = -xy = |x| \cdot |y|$, and thus $|xy| = |x| \cdot |y|$.

Case 3: Suppose that $x, y < 0$. Then, $|x| = -x$, $|y| = -y$, and $|xy| = xy$. Hence, $|xy| = xy = (-x)(-y) = |x| \cdot |y|$, and therefore $|xy| = |x| \cdot |y|$.

Case 4: Suppose that $x < 0$ and $y \geq 0$. Then, $|x| = -x$, $|y| = y$, and $|xy| = -xy$. Hence, $|xy| = -xy = |x| \cdot |y|$, and thus $|xy| = |x| \cdot |y|$.

Therefore, in each of the four cases $|xy| = |x| \cdot |y|$, and therefore $|xy| = |x| \cdot |y|$, for any two real numbers x, y.

■

Note that if the same arguments are used for the proofs of two or more cases in a proof, then these cases should be combined into a single case. In fact, since the same arguments worked for each of the cases there was no need to consider the cases separately. In some proofs the arguments for proving one or more cases are very similar, *mutatis mutandis* (Latin for "with the necessary changes"), to the arguments needed to prove another case. In this case, a "without loss of generality" (WLOG) statement is usually issued and the proof of the theorem is shortened by presenting only the proof of one of these similar cases. However, when using a WLOG statement in a proof, it is very important to clearly explain why there will be no loss of generality.

Example 3.4.15: Note that in Example 3.4.14 the proofs for cases 2 and 3 are nearly identical. In fact, except for the role changes of the variables x and y, the proofs are exactly the same. Thus, this is an example of a proof that could be shortened using a WLOG statement. The shorter version of the proof with a WLOG statement is shown below.

Proof: Let x and y be real numbers. Then, either $x, y \geq 0$ or $x \geq 0, y < 0$ or $x, y < 0$ or $x < 0, y \geq 0$.

Case 1: Suppose that $x, y \geq 0$. Then, $|x| = x$, $|y| = y$, and $|xy| = xy$. Hence, $|xy| = xy = |x| \cdot |y|$, and thus $|xy| = |x| \cdot |y|$.

Case 2: By symmetry, the proof for $x \geq 0$ and $y < 0$ and for $x < 0$ and $y \geq 0$ are similar. Thus, WLOG assume that $x \geq 0$ and $y < 0$. Then, $|x| = x$, $|y| = -y$, and $|xy| = -xy$. Hence, $|xy| = -xy = |x| \cdot |y|$, and thus $|xy| = |x| \cdot |y|$.

Case 3: Suppose that $x, y < 0$. Then, $|x| = -x$, $|y| = -y$, and $|xy| = xy$. Hence, $|xy| = xy = (-x)(-y) = |x| \cdot |y|$, and therefore $|xy| = |x| \cdot |y|$.

Thus, in each of the cases $|xy| = |x| \cdot |y|$, and therefore $|xy| = |x| \cdot |y|$ for any two real numbers x and y.

■

3.4.5 Proving Biconditional Theorems

Recall that the converse of a theorem is seldom also true. However, when a theorem and its converse are true, then the theorem may be written in the biconditional form "H if and only if C."; a theorem of the form "H if and only if C" is called a *biconditional theorem*. For notational purposes the phrase "if and only if" is commonly abbreviated by "iff." An example of a biconditional theorem is given below.

Theorem: Let a and b be real numbers. Then, $ab = 0$ if and only if $a = 0$ or $b = 0$.

Note that a biconditional theorem $H \leftrightarrow C$ is actually a theorem that is comprised of the two theorems, namely, $H \rightarrow C$ and $C \rightarrow H$. Thus, to prove a biconditional theorem, both of the theorems $H \rightarrow C$ and $C \rightarrow H$ must be proved. The following steps outline the typical approach used in proving a biconditional theorem.

Algorithm for a Biconditional Proof: To prove the theorem "H if and only if C," perform the following sequence of steps:

1. Prove $H \rightarrow C$, which is often referred to as the *forward proof*.
2. Prove $C \rightarrow H$, which is the converse of the forward theorem.
3. After proving $H \rightarrow C$ and $C \rightarrow H$, conclude $H \leftrightarrow C$.
4. Clean up and rewrite the scratchwork into a clear and concise proof of the theorem. Make sure that each step of the proof makes sense and is clearly justified.
5. Read the proof over carefully and make any necessary corrections.

Thus, the proof of a biconditional theorem will actually consist of two standalone proofs, one for each of the theorems $H \rightarrow C$ and its converse. Furthermore, each of the theorems $H \rightarrow C$ and $C \rightarrow H$ may be proved with any valid method of proof. Also, in the presentation of the proof of a biconditional theorem it is very important to designate where the proofs of $H \rightarrow C$ and $C \rightarrow H$ begin and end. Example 3.4.16 illustrates a biconditional theorem and its proof.

Example 3.4.16: Prove the following biconditional theorem:

Theorem: Let x be a real number. Then, $x^2 - 4x + 4 = 0$ if and only if $x = 2$.

> ***Proof:*** Let x be a real number.
>
> First, assume that $x^2 - 4x + 4 = 0$. Then, since $x^2 - 4x + 4 = (x-2)^2$, it follows that if $0 = x^2 - 4x + 4 = (x-2)^2$, then $x = 2$.
>
> Conversely, assume that $x = 2$. Then, $x^2 - 4x + 4 = 2^2 - 4(2) + 4 = 0$.
>
> Hence, $x^2 - 4x + 4 = 0$ if and only if $x = 2$.

■

Example 3.4.17: Prove the following biconditional theorem.

Theorem: Let $m, n \in \mathbb{N}$. Then, $m + n$ is odd if and only if exactly one of m and n is odd.

> ***Proof:*** Let $m, n \in \mathbb{N}$.
>
> First, assume that $m + n$ is odd and it is not the case that exactly one of m and n is odd. Then either m and n are both odd or m and n are both even.
>
> **Case 1:** Suppose that both m and n are odd. Then, there exist $k, j \in \mathbb{Z}$ such that $n = 2k + 1$ and $m = 2j + 1$. Consider $m + n$
>
> $$m + n = (2k + 1) + (2j + 1) = 2j + 2k + 2 = 2(j + k + 1) = 2l$$
>
> where $l = j + k + 1$, which is an integer. Hence, $m + n$ is even, contradicting $m + n$ is odd. Therefore, it is not possible for $m + n$ to be odd when both m and n are odd.
>
> **Case 2:** Suppose that both m and n are even. Then, $\exists\ k, j \in \mathbb{Z}$ such that $n = 2k$ and $m = 2j$. Consider $m + n$
>
> $$m + n = 2k + 2j = 2(j + k) = 2l$$
>
> where $l = j + k$, which is an integer. Hence, $m + n$ is even, contradicting $m + n$ is odd. Therefore, it is not possible for $m + n$ to be odd when both m and n are even.

Thus, if $m + n$ is odd, then exactly one of m and n must be even.

Conversely, suppose that exactly one of m and n is even. Without loss of generality, suppose that n is odd and m is even. Then, there exist $k, j \in \mathbb{Z}$ such that $n = 2k + 1$ and $m = 2j$. Consider $m + n$

$$m + n = (2k + 1) + 2j = 2j + 2k + 1 = 2(j + k) + 1 = 2l + 1$$

where $l = j + k$, which is an integer.

Therefore, $m + n$ is odd whenever exactly one of m and n is odd.

■

3.4.6 Disproving a Conjecture

In the axiomatic development of mathematics, results are hypothesized and conjectures are made, and then an attempt is made to prove the conjectured results. Once a conjecture is proved it is called a *theorem* and is added to the collection of mathematical results. However, when a conjecture cannot be proved, even after substantial effort has been put forth, then it is often reasonable to begin doubting the truth of the conjecture. In this case, an attempt may be made to disprove the conjecture. Now, disproving a mathematical conjecture requires only a single example to be found showing that the conjecture is false. Thus, if an example can be found where the hypothesis of the conjecture is true but the conclusion is false, then conjecture will be disproved. An example used to disprove a conjecture is called a *counterexample.*

Definition 3.4.1: A *counterexample* to the conjecture $H \to C$ is a specific example where the hypothesis H is true, but the conclusion C is false.

Note that only a single counterexample is needed to disprove the conjecture $H \to C$; however, finding a counterexample can be extremely difficult and often requires a great deal of creativity and insight. Furthermore, while a counterexample may disprove the conjecture in general, it is often the case that the conjecture will be true for a few, or even many, specific examples. Moreover, while examples may be used to support the truth of a conjecture, it must be kept in mind that even a massive amount of examples will never constitute a proof of the conjecture. For example, consider the following conjecture:

Conjecture: $\sqrt{x^2 + y^2} = x + y, \ \forall \ x, y \in \mathbb{R}$.

This conjecture is true for the special cases of $x = y = 0$, $x = 1, y = 0$, and $x = 0, y = 1$. However, this conjecture is disproved by the counterexample

$x = 2, y = 1$ since $\sqrt{2^2 + 1^2} = \sqrt{5} \neq 1 + 2 = 3$. Thus, this conjecture is not true $\forall\ x, y \in \mathbb{R}$, even though it is true for special cases of $x = y = 0$, $x = 1, y = 0$, and $x = 0, y = 1$.

Now, when a conjecture $H \rightarrow C$ has been disproved it can be regarded as a theorem of the form "There exist counterexamples to $H \rightarrow C$," which is an existence theorem that be proved by simply constructing a counterexample to $H \rightarrow C$.

Example 3.4.18: It is conjectured that $n^2 + n + 41$ is a prime number for every natural number n. Prove that this result is true or find a counterexample to disprove this result.

Solution: Consider a few trial cases. Clearly, this result holds for the values $n = 1, 2, 3, 4, 5, 6$. Furthermore, since this conjecture has the form $\mathcal{P}_n$, $\forall n \in \mathbb{N}$, if it were true it would be a good candidate for a proof by mathematical induction. Consider the following attempt to prove the conjecture with weak induction.

> **Induction Scratchwork:** Let $\mathcal{P}_n$ be the propositional function "$n^2 + n + 41$ is a prime number."
>
> For $n = 1$, $n^2 + n + 41 = 1 + 1 + 41 = 43$, which is prime. Thus, $\mathcal{P}_1$ is true.
>
> Now, suppose that $\mathcal{P}_k$ is true for some ABF $k \in \mathbb{N}$. This means that $k^2 + k + 41$ is prime.
>
> If $\mathcal{P}_{k+1}$ is true, then $(k + 1)^2 + (k + 1) + 1$ will be prime. Consider, $(k + 1)^2 + (k + 1) + 1$:
>
> $$(k + 1)^2 + (k + 1) + 1 = k^2 + 2k + 1 + k + 1 + 1$$
>
> $$= \underbrace{(k^2 + k + 1)}_{\text{prime since } \mathcal{P}_k \text{ is true}} + (2k + 2)$$
>
> At this point, it is difficult to see whether $\mathcal{P}_{k+1}$ is true.
>
> The induction proof has bogged down, and thus it might be time to consider an attempt to disprove the conjecture. Now, to disprove this conjecture, a value of n must be found so that $n^2 + n + 41$ is not a prime number. One approach is to simply begin computing $n^2 + n + 41$ for different values of n. In this case, $n^2 + n + 41$ is true for $n = 1, 2, \ldots, 39$ which appears to be strong evidence supporting this conjecture. However, for $n = 40$ the conjecture is false since $40^2 + 40 + 41 = 1681 = 41^2$, which is not a prime number. Thus,

$n = 40$ is a counterexample to the conjecture $n^2 + n + 41$ is a prime number for all $n \in \mathbb{N}$.

■

Example 3.4.19: Find a counterexample that disproves each of the following mathematical conjectures:

a. $(x + y)^2 = x^2 + y^2$, $\forall\, x, y \in \mathbb{R}$

b. $\sqrt{x + y} = \sqrt{x} + \sqrt{y}$, $\forall\, x, y \in \mathbb{R}$

c. $\sin(x + y) = \sin(x) + \sin(y)$, $\forall\, x, y \in \mathbb{R}$

Solutions:

1. $(1 + 1)^2 = 4 \neq 1^2 + 1^2 = 2$. Thus, $x = 1, y = 1$ is a counterexample that disproves $(x + y)^2 = x^2 + y^2$, $\forall\, x, y \in \mathbb{R}$.
2. $\sqrt{1 + 1} = \sqrt{2} \neq \sqrt{1} + \sqrt{1} = 2$. Thus, $x = 1, y = 1$ is a counterexample that disproves $\sqrt{x + y} = \sqrt{x} + \sqrt{y}$, $\forall\, x, y \in \mathbb{R}$.
3. $\sin(\pi/2 + \pi/2) = 0 \neq \sin(\pi/2) + \sin(\pi/2) = 1 + 1 = 2$. Thus, the values $x = \pi/2$ and $y = \pi/2$ provide a counterexample that disproving $\sin(x + y) = \sin(x) + \sin(y)$, for all $x, y \in \mathbb{R}$.

3.5 Some Final Notes on Proving Theorems

In concluding this chapter, it should be noted that choosing the particular method of proof that is best suited for proving a theorem is often a difficult task. In fact, the particular method of proof that is used to prove a theorem will often depend on the actual structure of the theorem itself. Some theorems are easily proved with a direct proof, some with a indirect proof, some with proof by contradiction, and some theorems require induction or another specialized method of proof. Furthermore, there will often be more than one method that can be used to prove a theorem. So, when faced with the problem of proving a theorem, it is important to keep the following facts in mind:

1. No method of proof begins with the assumption that the conclusion (C) is true, and no method of proof begins with the assumption that the hypothesis (H) is false. Thus, a valid proof must begin with either the assumption that H is true (forward direct proof), the assumption that H is true and C is false (indirect proof), or the assumption that the conclusion (C) is false (proof by contrapositive).
2. A proof might have to be broken into distinct cases (i.e., $x > 0$, $x < 0$, or $x = 0$).

3. Mathematical induction is often useful in proving a mathematical result of the nature "$\mathcal{P}_n, \ \forall \ n \in \mathbb{N}$."
4. There are specialized methods of proof for uniqueness theorems and existence theorems.
5. If the theorem is a biconditional theorem $H \to C$, then both of the theorems $H \to C$ and $C \to H$ must be proved.
6. If a conjecture cannot be proved, it might be false. Disproving a conjecture can be done by finding a single counterexample.

Example 3.5.1: For each of the theorems in this example, a potential start for a proof has been given. Determine which method of proof, if any, is being used to prove each of the theorems below.

a. **Theorem:** If $f(x)$ is a differentiable function on $[a, b,]$, then $f(x)$ is continuous on $[a, b]$.

   Proof: Let $f(x)$ be a differentiable function on $[a, b]$.

b. **Theorem:** $\sqrt{5}$ is an irrational number.

   Proof: Assume that $\sqrt{5}$ is a rational number.

c. **Theorem:** Let A and B be sets. If $A \subset B$, then $A \cap B = A$.

   Proof: Let A and B be sets, and suppose that $A \cap B = A$.

d. **Theorem:** $7^n - 2^n$ is divisible by 5, $\forall \ n \in \mathbb{N}$.

   Proof: For $n = 1$, $7^1 - 2^1 = 5$ which is divisible by 5. Thus, this result is true for $n = 1$.

e. **Theorem:** Let n be a natural number. If n^2 is odd, then n is odd.

   Proof: Let $n \in \mathbb{N}$ and suppose that n is even.

Solutions:

a. This proof begins with the hypothesis (H) and thus could be a direct proof.

b. This proof begins with negation of the conclusion ($\neg C$) and thus could be a proof by contradiction. It could also be a proof by contrapositive.

c. This is not a proof because it begins by assuming that the conclusion (C) is true.

d. This proof begins by looking at the case where $n = 1$ and thus could be a proof by mathematical induction.

e. This proof begins with negation of the conclusion $(\neg C)$, but not the hypothesis (H), and therefore it could be a proof by contrapositive.

Example 3.5.2: Determine which method of proof is best suited to each of the following theorems:

a. **Theorem:** If n^3 is an even number, then n is an even number.

b. **Theorem:** $\sqrt{17}$ is an irrational number.

c. **Theorem:** There exists an element e in $\mathcal{C}$ such that $a \circ e = e \circ a = a$, $\forall\, a \in \mathcal{C}$.

d. **Theorem:** $a^n - b^n$ is divisible by $a - b$, $\forall\, n \in \mathbb{N}$.

e. **Theorem:** If $(\mathcal{G}, \circ)$ is a group and $a \in \mathcal{G}$, then a^{-1} is unique.

Solutions: The solutions to Example 3.5.2 are left as exercises.

EXERCISES

3.1 Determine the hypothesis and conclusion for each of the following theorems:

a. **Theorem:** If $x, y \in \mathbb{R}$, then $|xy| = |x| \cdot |y|$.

b. **Theorem:** If $\lim_{x \to a} f(x) = F$ and $\lim_{x \to a} g(x) = G$, then $\lim_{x \to a} [f(x) + g(x)] = F + G$.

c. **Theorem:** If $f(x)$ is differentiable on $[a, b]$ and $f(a) = f(b)$, then there exists $\psi \in [a, b]$ such that $f'(\psi) = 0$.

d. **Theorem:** If p and $p + 2$ are prime numbers, then $p + 1$ is divisible by 6.

e. **Theorem:** If $\lim_{n \to \infty} a_n \neq 0$, then $\sum_{n=1}^{\infty} a_n$ diverges.

f. **Theorem:** If $\lim_{n \to \infty} a_n = 0$ and $\lim_{n \to \infty} \left| \frac{a_{n+1}}{a_n} \right| < 1$, then $\sum_{n=1}^{\infty} a_n$ converges.

g. **Theorem:** If $(\mathcal{G}, \circ)$ is a group, then the identity element in $\mathcal{G}$ under $\circ$ is unique.

3.2 Determine the contrapositive of each of the theorems in Exercise 3.1.

3.3 **Theorem:** Let $a, b, c, d \in \mathbb{Z}$. If a is divisible by d, b is divisible by d, and c is divisible by d, then $a + b + c$ is divisible by d.

a. Determine the hypothesis and conclusion of this theorem.

b. Determine the converse of this theorem.

c. Determine the contrapositive of this theorem.

d. Is the converse of this theorem true?

e. Is the contrapositive of this theorem true?

3.4 Prove each of the following theorems by the method of forward direct proof:

a. **Theorem:** Let $n, m \in \mathbb{N}$. If n and m are even, then $n + m$ is even.

b. **Theorem:** Let $n, m \in \mathbb{N}$. If n and m are odd, then $n + m$ is even.

c. **Theorem:** Let $n, m \in \mathbb{N}$. If n and m are odd, then nm is odd.

d. **Theorem:** Let $n, m \in \mathbb{N}$. If n is odd and m is even, then $n + m$ is odd.

3.5 Prove each of the following theorems by the method of contrapositive:

a. **Theorem:** Let $n \in \mathbb{N}$. If n^2 is odd, then n is odd.

b. **Theorem:** Let $n \in \mathbb{N}$. If $n^3 + n^2 + n + 2$ is odd, then n is odd.

c. **Theorem:** Let $m, n \in \mathbb{N}$. If mn is odd, then m is odd and n is odd.

d. **Theorem:** Let $a, b, c \in \mathbb{N}$. If $b + c$ is not divisible by a, then neither b nor c is divisible by a.

e. **Theorem:** Let $n \in \mathbb{N}$. If $n^2 - 1$ is not divisible by 4, then n is even.

f. **Theorem:** Let $x \in \mathbb{R}$. If $x^2 + 2x < 0$, then $x < 0$.

3.6 Prove the following theorems by the method of contradiction:

a. **Theorem:** If $x > 0$, then $x + 1 over x \geq 2$.

b. **Theorem:** If $x > 0$, then $\dfrac{x}{3} + \dfrac{3}{x} \geq 2$.

c. **Theorem:** There is no largest natural number.

d. **Theorem:** Let $n \in \mathbb{N}$. If n^2 is odd, then n is odd.

e. **Theorem:** Let $n \in \mathbb{N}$. If n^2 is divisible by 3, then n is divisible by 3.

f. **Theorem:** The equation $x^3 + x + 1$ has no rational roots.

3.7 A real-valued function $f(x)$ is said to be odd if $f(-x) = -f(x), \forall x \in \mathbb{R}$; a real-valued function $f(x)$ is said to be even if $f(-x) = f(x), \forall x \in \mathbb{R}$. Prove the following theorems:

a. **Theorem:** If $f(x)$ and $g(x)$ are real-valued even functions, then $f + g$ is an even function.

b. **Theorem:** If $f(x)$ and $g(x)$ are real-valued odd functions, then $f(x) + g(x)$ is an odd function.

c. **Theorem:** If $f(x)$ and $g(x)$ are real-valued odd functions, then $f(x)g(x)$ is an even function.

3.8 Prove the following theorems by using the method of proof by cases:

a. **Theorem:** If n is a natural number, then $n^2 + n + 1$ is an odd number.

b. **Theorem:** Let $n \in \mathbb{N}$. If n^2 is divisible by 5, then n is divisible by 5.

c. **Theorem:** If $x \in \mathbb{R}$, then $|ax| = |a| \cdot |x|, \ \forall\, a \in \mathbb{R}$.

d. **Theorem:** If $x \in \mathbb{R}$, then $|1 + x| \leq 1 + |x|$.

e. **Theorem:** If $f(x) = |x|$, then $f(x)$ is differentiable for all $x \neq 0$.

f. **Theorem:** If P is a statement, then $\neg(\neg P) \iff P$.

3.9 For each $n \in \mathbb{N}$, let $\mathcal{P}_n$ denote the statement "$n^2 + 5n + 1$ is an even integer."

a. Prove that $\mathcal{P}_{n+1}$ is true whenever $\mathcal{P}_n$ is true.

b. Is $\mathcal{P}_n$ true for every $n \in \mathbb{N}$?

3.10 Prove the following results using mathematical induction:

a. $2^{2n-1} + 3^{2n-1}$ is divisible by 5 for every natural number n.

b. $\displaystyle\sum_{i=1}^{n} \frac{1}{i(i+1)} = \frac{n}{n+1}$ for every natural number n.

c. Let $a, b \in \mathbb{R}$. Then, $a^n - b^n$ is divisible by $a - b$, $\forall\, n \in \mathbb{N}$.

d. $(n+1)^3 - (n+1)$ is divisible by 3 for every natural number n.

e. $\displaystyle\sum_{i=1}^{n} i^2 = \frac{n(n+1)(2n+1)}{6}$ for every natural number n.

f. $\left(1 + \frac{1}{2}\right)^n \geq 1 + \frac{n}{2}$, $\forall\, n \in \mathbb{N}$.

g. $2^{2n-1} + 3^{2n-1}$ is divisible by 5 for every natural number n.

h. $\displaystyle\sum_{j=1}^{n} j^3 = \left(\sum_{j=1}^{n} j\right)^2$ for every natural number n.

i. Let P_i be a statement for $i \in \mathbb{N}$. Then

$$\neg\left(\bigvee_{i=1}^{n+1} P_i\right) \iff \bigwedge_{i=1}^{n+1} \neg P_i, \ \forall\, n \in \mathbb{N}.$$

j. Let P_i be a statement for $i \in \mathbb{N}$. Then

$$\neg\left(\bigwedge_{i=1}^{n+1} P_i\right) \iff \bigvee_{i=1}^{n+1} \neg P_i, \ \forall\, n \in \mathbb{N}.$$

k. $\displaystyle\sum_{j=1}^{n} \frac{1}{\sqrt{j}} \geq \sqrt{n}$, $\forall\, n \in \mathbb{N}$.

l. $3^{2n-1} + 1$ is divisible by 4 for every natural number n.

m. $8^n - 1$ is divisible by 7 for every natural number n.

n. $\displaystyle\binom{2n}{n} < \frac{4^n}{\sqrt{3n+1}}$, $\forall\, n \in \mathbb{N}$.

o. If $x \in \mathbb{R}$, then $(1+x)^n \geq 1+nx$ for $n \in \mathbb{N}$.

p. If $p \neq -1$, then $\sum_{i=1}^{n} (-p)^i = \frac{1-(-p)^{n+1}}{1+p}$ for every natural number n.

q. $(n+3)! \geq 2^{n+3}$, $\forall\, n \in \mathbb{N}$.

r. If $a \in \mathbb{N}$ is an odd number, then a^{n+1} is odd, $\forall\, n \in \mathbb{N}$.

s. $5^n - 3^n$ is divisible by 2 for every natural number n.

t. $2^{2n-1} + 4^{2n-1}$ is divisible by 6 for every natural number n.

u. $(2n+1) + (2n+3) + (2n+5) + \cdots + (4n-1) = 3n^2$, $\forall n \in \mathbb{N}$.

3.11 Prove the following results using strong induction:

a. Let $y_0 = 1, y_1 = 2$ and $y_{n+1} = \frac{3y_n + 4y_{n-1}}{12}$ for $n \in \mathbb{N}$. Then, $y_{n+1} \leq 1$, $\forall\, n \in \mathbb{N}$.

b. Let $y_0 = 1, y_1 = 1$ and $y_{n+1} = y_n + y_{n-1}$ for $n \in \mathbb{N}$. If $n \in \mathbb{N}$, then, $y_{n+1} \leq 2^n$.

c. Let $y_0 = 1, y_1 = 1$ and $y_{n+1} = y_n + y_{n-1}$ for $n \in \mathbb{N}$. Then, y_{3n} is even, $\forall\, n \in \mathbb{N}$.

3.12 Prove each of the following biconditional theorems:

a. Let $n \in \mathbb{N}$. Then, n is odd if and only if $n^2 - 1$ is even.

b. Let $n \in \mathbb{N}$. Then, $n^2 - 1$ is divisible by 3 if and only if n is not divisible by 3.

c. Let $m, n \in \mathbb{N}$. Then, $n + n$ is even if and only if m and n are both even or m and n are both odd.

d. Let $n, m \in \mathbb{N}$. Then, mn is odd if and only if m and n are both odd.

e. Let $x \in \mathbb{R}$. Then, $|x| = x$ if and only if $x \geq 0$.

3.13 Prove each of the following existence theorems:

a. If p is a prime number, then there exist prime numbers of the form $3^p + 16$.

b. There exists a natural number such that $2^{2^n} + 1$ is not a prime number.

c. There exists a 2×3 matrix such that $A + E = A$ for every 2×3 matrix A.

d. If $A = \begin{pmatrix} 2 & 3 \\ 1 & 4 \end{pmatrix}$, then there exists a matrix B such that

$$AB = \begin{pmatrix} 0 & 1 \\ 1 & 0 \end{pmatrix}.$$

e. If $f(x) = 1 - x\cos(x)$, then there exists a solution to $f(x) = 0$ on the interval $[-\pi, \pi]$.

f. If $f(x) = \dfrac{x}{x^2+1}$, then there exists a real-valued function $F(x)$ such that $F'(x) = f(x)$, $\forall\, x \in \mathbb{R}$.

3.14 Prove each of the following uniqueness theorems:

a. Let $x \in \mathbb{R}$. The solution to the equation $7x - 3 = 0$ is unique.

b. If A is a nonsingular $n \times n$ matrix, then A^{-1} is unique.

c. There exists a unique 2×3 matrix such that $A + E = A$ for every 2×3 matrix A.

d. If $f(x) = \sin(x)\cos(x)$ and $F(0) = 1$, then $F(x) = \dfrac{1}{2}\sin^2(x) + 1$ is the unique antiderivative of $f(x)$.

e. If $f(x) = 1 - x\cos(x)$, then there exists a unique solution to $f(x) = 0$ on the interval $[-\pi, \pi]$.

3.15 Find a counterexample for each of the following conjectures:

a. **Conjecture:** If p is a prime number, then $2^p + 1$ is a prime number.

b. **Conjecture:** Let $a, b, c \in \mathbb{N}$. If $b + c$ is divisible by a, then b and c are both divisible by a.

c. **Conjecture:** Let $a, b, c \in \mathbb{N}$. If bc is divisible by a, then b and c are both divisible by a.

d. **Conjecture:** There is one and only one real solution to the equation $x^3 - 4x^2 + x - 6 = 0$.

e. **Conjecture:** Let $n \in \mathbb{N}$. If n^2 is divisible by 4, then n is divisible by 4.

f. **Conjecture:** $\sqrt{x^2} = x$, $\forall\, x \in \mathbb{R}$.

g. **Conjecture:** If A and B are $n \times n$ matrices, then $AB = BA$.

h. **Conjecture:** If $\lim\limits_{x \to a} g(x) = 0$, $\lim\limits_{x \to a} \dfrac{f(x)}{g(x)}$ does not exist.

i. **Conjecture:** If $f(x)$ is continuous at $x = x_0$, then $f(x)$ is differentiable at $x = x_0$.

j. **Conjecture:** If $\lim\limits_{n \to \infty} a_n = 0$, then $\sum\limits_{n=1}^{\infty} a_n$ converges.

k. **Conjecture:** If a is an irrational number, then a^n is an irrational number for every natural number n.

3.16 Prove each of the following theorems:

a. **Theorem:** Let $a, b \in \mathbb{N}$. If a is divisible by b and b is divisible by a, then $a = b$.

b. **Theorem:** Let $n \in \mathbb{N}$. If n is odd, then there exists $m \in \mathbb{Z}$ such that $n^2 = 8m + 1$.

c. **Theorem:** Let $n \in \mathbb{N}$. If $n^2 - 1$ is not divisible by 8, then n is even.

d. **Theorem:** Let $n, m \in \mathbb{N}$. If n and $n + m$ are both divisible by 3, then m is divisible by 3.

e. **Theorem:** Let $x, y \in \mathbb{R}$ with $y > 0$. Then, $|x| < y$ if and only if $-y < x < y$.

f. **Theorem:** Let $x, y \in \mathbb{R}$. If $|x + y| < |x| + |y|$, then exactly one of x and y is negative.

3.17 Several theorems and potential first lines in a proof of the theorem are given below. In each case determine whether the first line of the proof is consistent with a valid method of proof. If so, determine which method of proof is being used. If not, explain why.

a. **Theorem:** Let $x = \sqrt{3}$. Then x is not a rational number.

Proof: Let $x = \sqrt{3}$ and suppose that x is an irrational number.

b. **Theorem:** Let $x = \sqrt{3}$. Then x is not a rational number.

Proof: Let $x = \sqrt{3}$ and suppose that x is a rational number.

c. **Theorem:** Let $m, n \in \mathbb{N}$. If mn is odd, then m and n are odd.

Proof: Let $m, n \in \mathbb{N}$. Suppose that either n or m is even.

d. **Theorem:** If $n \in \mathbb{N}$, then $n^2 + n + 1$ is odd.

Proof: Let $n \in \mathbb{N}$ and suppose that $n^2 + n + 1$ is even.

e. **Theorem:** Let $n, m \in \mathbb{N}$. If n is even and m is odd, then $n + m$ is odd.

Proof: Let $n, m \in \mathbb{N}$ and suppose that $n + m$ is odd, n is even, and m is odd.

3.18 Write each of the mathematical results in the form "$\mathcal{P}_n, \ \forall\, n \in \mathbb{N}$."

a. If $n \in \mathbb{N}$, then, $2^n < n!$ for $n > 3$.

b. If $n \in \mathbb{N}$, then, $\neg \left(\bigwedge_{i=1}^{n} P_i \right)$ for $n \geq 2$.

c. If $n \in \mathbb{N}$, then, $n^2 \geq n + 1$ for $n \geq 2$.

d. If $n \in \mathbb{N}$, then, $n^2 \leq n!$ for $n \geq 4$.

e. If $n \in \mathbb{N}$, then, $\dfrac{n}{n^2 - 5n + 4} \geq \dfrac{1}{n}$ for $n \geq 5$.

3.19 Determine which method of proof is best suited for proving each of the following theorems and write down the first line for the proof of the theorem:

a. **Theorem:** If p is a prime number, then $\sqrt{p}$ is not a rational number.
b. **Theorem:** Let $n \in \mathbb{N}$. If p is a prime number and n^2 is divisible by p, then n is divisible by p.
c. **Theorem:** Let $n, m \in \mathbb{N}$. Then mn is odd if and only if m and n are both odd.
d. **Theorem:** Let $m, n \in \mathbb{N}$. If n is even and m is odd, then $m + n$ is odd.
e. **Theorem:** The set of prime numbers is not finite.
f. **Theorem:** If A is a nonsinguluar matrix, then A^{-1} is unique.

3.20 Determine a corollary of each of the following theorems:

a. **Theorem:** Let $a, b \in \mathbb{N}$. If a is divisible by b, then a^n is divisible by b for every natural number n.
b. **Theorem:** If $\lim\limits_{x \to a} f(x) = L$, then $\lim\limits_{x \to a} f(x)^n = L^n$, $\forall\, n \in \mathbb{N}$.
c. **Theorem:** If $a \in \mathbb{R}$, then $|ax| = |a| \cdot |x|$, $\forall\, x \in \mathbb{R}$.
d. **Theorem:** Let $a, b \in \mathbb{N}$. If p is a prime number and ab is divisible by p, then either a is divisible by p or b is divisible by p.
e. **Theorem:** If p is a prime number, then $\sqrt{p}$ is an irrational number.

3.21 Determine which method of proof is best suited for proving each of the following theorems:

a. **Theorem:** Let $n \in \mathbb{N}$. If n^2 is an even number, then n is an even number.
b. **Theorem:** $\sqrt{17}$ is an irrational number.
c. **Theorem:** There exists an element e in C such that $a \circ e = e \circ a = a$, $\forall\, a \in C$.
d. **Theorem:** $a^n - b^n$ is divisible by $a - b$, $\forall\, n \in \mathbb{N}$.
e. **Theorem:** If $(\mathcal{G}, \circ)$ is and $a \in \mathcal{G}$, then a^{-1} is unique.

Chapter 4
Introduction to Number Theory

The first evidence of human use of numbers and counting was found in Czechoslovakia in 1937 by archaeologist Karl Absolom. The evidence of counting is found in a tibia bone of a wolf, which carbon dating has shown to be approximately 30,000 years old, and that has two series of 25 and 30 notches carved into it. Moreover, the marks are grouped together in groups of 5s. Thus, it appears that numbers and counting began with the Neanderthal people. However, modern mathematics is concerned with more than just numbers and the counting of numbers. Today's mathematicians study the properties of special classes of numbers with specialized operations applied to these classes. Study of the mathematical properties of numbers and the operations applied to these numbers is known as *number theory*.

4.1 Binary Operators

The most commonly used operations when working with real numbers are the basic arithmetic operations of addition, multiplication, subtraction, and division. Note that each of the basic arithmetic operations operates on two numbers, and from the two initial numbers a third number is created. Operations, or operators, which map two objects in a set to another object, are called *binary operations*. The definition of a *binary operator* is given below.

Definition 4.1.1: A *binary operator* $\circ$ is a rule defined on a set Ω that assigns to the objects $a, b \in \Omega$ an object c and is denoted by $a \circ b = c$.

René Descartes (1596–1650) in his 1637 publication *Geometrie*, states that "arithmetic consists of only four or five operations, namely addition, subtraction, multiplication, division, and the extraction of roots." While addition, multiplication, subtraction, and division are the most commonly used binary operators when working with real numbers, binary operators other than these will be defined and discussed throughout the remainder of this book. Several examples of commonly used binary operators are given in the example below.

Example 4.1.1: Each of the following operators is a binary operator.

a. Addition of two real numbers with $a \circ b = a + b$.

b. Multiplication of two real numbers with $a \circ b = a \times b$.

c. Exponentiation of two real numbers with $a \circ b = a^b$.

d. The conjunction of two statements with $P \circ Q = P \wedge Q$.

e. Scalar multiplication of a scalar with a p vector with

$$k \circ \vec{v} = k \cdot \vec{v} = (kv_1, kv2, \ldots, kv_p)$$

f. The cross-product of two p-vectors with $\vec{u} \circ \vec{v} = \vec{u} \times \vec{v}$.

Note that there nothing in the definition of a binary operator that would require that $a \circ b = b \circ a$. In fact, for many operators it will not be true that $a \circ b$ and $b \circ a$ will have the same value, and it may even be that while $a \circ b$ exists, $b \circ a$ does not. When $a \circ b = b \circ a$ for all a and b, then $\circ$ is called an *Abelian operator* in honor of the outstanding algebraist Niels Henrik Abel (1802–1829).

Definition 4.1.2: A binary operator $\circ$ is said to be *commutative* on a set Ω and is called an *Abelian operator* if and only if $a \circ b = b \circ a$, $\forall a, b \in \Omega$.

Examples of Abelian binary operators include ordinary addition and multiplication of numbers and addition of matrices. Examples of non-Abelian operators include subtraction, division, and exponentiation on the real numbers. For example, exponentiation is not an Abelian operator since $3 \circ 2 = 3^2 = 9$ but $2 \circ 3 = 2^3 = 8$. Example 4.1.3 shows that matrix multiplication is not an Abelian operator.

Algorithm for Proving a Binary Operator is Abelian: Let Ω be a set and $\circ$ a binary operator defined on Ω. To show that $\circ$ is an Abelian operator on Ω:

1. Let a, b be arbitrary but fixed elements in Ω.
2. Compute $a \circ b$ and $b \circ a$.
3. Show that $a \circ b = b \circ a$.
4. Clean up and rewrite the scratchwork in a clear and concise proof of the theorem. Make sure that each step of the proof makes sense and is clearly justified.
5. Read the proof over carefully and make any necessary corrections.

Example 4.1.2: Let $\Omega = \mathbb{R}$ and define $a \circ b = (a - b)^2$. Prove that $\circ$ is an Abelian binary operator on $\mathbb{R}$.

Proof: Let $a, b \in \mathbb{R}$ be ABF. Then

$$a \circ b = (a - b)^2 = (b - a)^2 = b \circ a$$

Thus, $\circ$ is an Abelian operator on $\mathbb{R}$.

Example 4.1.3: Show that when working with matrices, the binary operator multiplication is not an Abelian operator.

Solution: Suppose that matrix multiplication does commute. A counterexample to this conjecture is given by letting $A = \begin{pmatrix} 1 & 4 \\ 3 & -1 \end{pmatrix}$ and $B = \begin{pmatrix} 5 \\ 7 \end{pmatrix}$. Then, $A \circ B = AB = \begin{pmatrix} 33 \\ 8 \end{pmatrix}$ but $B \circ A$ does not exist since a 2×1 matrix cannot be multiplied by a 2×2 matrix. Hence, matrix multiplication is not an Abelian operator.

When a binary operator $\circ$ is Abelian, it can simplify computations with $\circ$. For example, if a solution is desired for the equations $a \circ x = a$ and $x \circ a = a$ and $\circ$ is an Abelian operator, then solving either one of these equations is sufficient since $a \circ x = x \circ a$.

While a binary operator can be applied to only two elements at one time, computations involving more than two elements are not unusual. For example, with three elements a, b, and c it might be necessary to compute $a \circ (b \circ c)$ or $(a \circ b) \circ c$, which for many binary operators do not produce the same result. If $a \circ (b \circ c) = (a \circ b) \circ c$ for all a, b, c, then the binary operator is said to be *associative*. The *associative property* for a binary operator is defined below.

Definition 4.1.3: A binary operator $\circ$ is said to be an *associative* operator on a set Ω if and only if $(a \circ b) \circ c = a \circ (b \circ c)$, $\forall a, b, c \in \Omega$.

Examples of associative binary operators include ordinary addition and multiplication of numbers and also addition and multiplication of matrices. Examples of nonassociative operators include subtraction, division, and exponentiation on the real numbers. For example, subtraction is not an associative operator since $(3 \circ 2) \circ 4 = (3 - 2) \circ 4 = 1 - 4 = -3$ but $3 \circ (2 \circ 4) = 3 \circ (2 - 4) = 3 - (-2) = 5$.

Algorithm for Proving a Binary Operator is Associative: Let Ω be a set and $\circ$ a binary operator defined on Ω. To show that $\circ$ is an associative operator on Ω:

1. Let a, b, c be arbitrary but fixed elements in Ω
2. Compute $(a \circ b) \circ c$ and $a \circ (b \circ c)$.
3. Show that $(a \circ b) \circ c = a \circ (b \circ c)$.
4. Clean up and rewrite the scratchwork in a clear and concise proof of the theorem. Make sure that each step of the proof makes sense and is clearly justified.
5. Read the proof over carefully and make any necessary corrections.

Example 4.1.4: Let $\Omega = \mathbb{R}$ and define $a \circ b = a + b + ab$. Prove that $\circ$ is an associative binary operator on $\mathbb{R}$.

Proof: Let $a, b, c \in \mathbb{R}$ be ABF. Then

$$(a \circ b) \circ c = (a + b + ab) \circ c = a + b + ab + c + (a + b + ab)c$$

$$= a + b + c + ab + ac + bc + abc$$

and

$$a \circ (b \circ c) = a \circ (b + c + bc) = a + b + c + bc + a(b + c + bc)$$

$$= a + b + c + ab + ac + bc + abc$$

Thus, $a \circ b = b \circ a$ for all $a, b \in \mathbb{R}$; therefore, $\circ$ is an associative operator on $\mathbb{R}$.

Throughout this chapter, nine axioms concerning the addition and multiplication of real numbers will be assumed. The first seven of the nine axioms are given below.

Axioms for Addition and Multiplication of Real Numbers: The following axioms will be assumed concerning the arithmetic operators addition and multiplication. If $a, b, c \in \mathbb{R}$, then

A1: $a + b = b + a$ (commutative property of addition)

A2: $a + (b + c) = (a + b) + c$ (associative property of addition)

A3: $a + 0 = a$

A4: $a + {}^{-}a = 0$

A5: $ab = ba$ (commutative property of multiplication)

A6: $a(bc) = (ab)c$ (associative property of multiplication)

A7: $a \times 1 = a$

The mathematical area of study known as "number theory" involves the study of particular subsets of the real numbers and their behavior with respect to the arithmetic binary operators $(+, \times, -, \div)$. In particular, one of the properties that is important when studying the behavior of a class of numbers under a binary operator $\circ$ is *closure*. The definition of a closed set under a binary operator $\circ$ is given below.

Definition 4.1.4: A set Ω is said to *closed* under a binary operator $\circ$ if and only if $a \circ b \in \Omega$ whenever a and b are in Ω.

For example, the real numbers are closed under addition; that is, given any two real numbers a and b it follows that $a+b$ is also a real number. In fact, the real numbers are closed under multiplication, subtraction, and nonzero division, also; however, the real numbers are not closed under exponentiation since $-2 \circ 0.5 = \sqrt{-2}$, which is not a real number. Note that closure will depend on both the set of interest and the binary operator. Furthermore, if a set Ω is closed under the binary operator $\circ$, then there is no way to create an element that is not in Ω by using $\circ$. The last two of the nine axioms concerning the addition and multiplication of real numbers are given below.

A8: $\mathbb{N}, \mathbb{Z}$, and $\mathbb{R}$ are closed under addition.

A9: $\mathbb{N}, \mathbb{Z}$, and $\mathbb{R}$ are closed under multiplication

Example 4.1.5: Show that

a. $\mathbb{N}$ is not closed under subtraction.

b. $\mathbb{N}$ is not closed under nonzero division.

c. $\mathbb{Z}$ is not closed under nonzero division.

d. $\mathbb{R}$ is not closed under exponentiation.

Solutions: A counterexample for showing that

a. $\mathbb{N}$ is not closed under subtraction is to let $a = 3$ and $b = 5$. Then, $3 - 5 = -2 \notin \mathbb{N}$.

b. $\mathbb{N}$ is not closed under nonzero division is to let $a = 3$ and $b = 5$. Then, $3 \div 5 = \frac{3}{5} \notin \mathbb{N}$.

c. $\mathbb{Z}$ is not closed under nonzero division is to let $a = 3$ and $b = 5$. Then, $3 \div 5 = \frac{3}{5} \notin \mathbb{Z}$.

d. $\mathbb{R}$ is not closed under exponentiation is to let $a = -2$ and $b = \frac{1}{2}$. Then, $(-2)^{\frac{1}{2}} = \sqrt{-2} \notin \mathbb{R}$.

An algorithm for proving a set Ω is closed under a binary operator $\circ$ is given below.

Algorithm for Proving Closure: Let Ω be a set and $\circ$ a binary operator defined on Ω. To prove that Ω is closed under $\circ$

1. Let a, b be arbitrary but fixed elements in Ω.

2. Compute $a \circ b$.
3. Show that $a \circ b \in \Omega$.
4. Clean up and rewrite the scratchwork in a clear and concise proof of the theorem. Make sure that each step of the proof makes sense and is clearly justified.
5. Read the proof over carefully and make any necessary corrections.

Example 4.1.6: Let $3\mathbb{Z} = \{z : z = 3k \text{ for some } k \in \mathbb{Z}\}$. Prove that $3\mathbb{Z}$ is closed under addition.

> ***Proof (Closure Proof):*** Let $a, b \in 3\mathbb{Z}$ be ABF. Then, $a = 3k$ for some $k \in \mathbb{Z}$ and $b = 3j$ for some $j \in \mathbb{Z}$. Consider $a + b$
>
> $$a + b = 3k + 3j = 3(k + j) = 3l$$
>
> where $l = k + j$, which is an integer. Thus, $a + b$ is a multiple of 3, and hence, $a + b \in 3\mathbb{Z}$ whenever $a, b \in 3\mathbb{Z}$. Therefore, $3\mathbb{Z}$ is closed under addition.

Example 4.1.7: Let $\Omega = \mathbb{R}$ and for $a, b \in \mathbb{R}$ define $a \circ b = a + b - ab$. Prove that $\mathbb{R}$ is closed under $\circ$.

> ***Proof (Closure Proof:*** Let a, b be arbitrary but fixed elements in $\mathbb{R}$ and consider $a \circ b$:
>
> $$a \circ b = a + b - ab = (a + b) + (-ab)$$
>
> Now, $a + b \in \mathbb{R}$ since $\mathbb{R}$ is closed under addition (A8). Also, both ab and $(-ab)$ are in $\mathbb{R}$ since $\mathbb{R}$ is closed under multiplication (A9). Thus, by A8, $(a + b) + (-ab)$ is in $\mathbb{R}$ since it is the sum of two real numbers.
>
> Therefore, $a \circ b \in \mathbb{R}$ whenever $a, b \in \mathbb{R}$ and hence, $\mathbb{R}$ is closed under $\circ$.

■

Another important issue to consider when studying the behavior of a set Ω under a binary operator $\circ$ is whether there exists an element $e \in \Omega$ such that $a \circ e = e \circ a = a$ for each element $a \in \Omega$. When there is an element $e \in \Omega$ such that $a \circ e = e \circ a$ for every $a \in \Omega$, then the element e is called an *identity element* under $\circ$.

Definition 4.1.5: An element e in Ω is said to be an *identity element* under the binary operator $\circ$ if and only if for every element a in Ω, $a \circ e = e \circ a = a$.

Note that for e to be an identity element, e must satisfy the following three conditions:

1. $e \in \Omega$
2. $a \circ e = a, \ \forall \ a \in \Omega$
3. $e \circ a = a, \ \forall \ a \in \Omega$.

In other words, the identify element e must be in Ω and must also commute with every element in Ω. However, in the special case that the binary operator $\circ$ is Abelian, showing that an element $e \in \Omega$ is an identity element, requires only showing that the first two conditions hold.

Example 4.1.8: Let $\Omega = \mathbb{R}$. Since $a+0 = a = 0+a$ for every real number a, 0 is the additive identity for ordinary addition. Also, since $a \times 1 = a = 1 \times a$ for every real number, 1 is the ordinary multiplicative identity.

Note that the identity element must be an element of Ω, and the identity element will be entirely dependent on the definition of the operator $\circ$. Moreover, proving that an identity element e exists in Ω under $\circ$ will require an *existence proof.* An algorithm for proving the existence of an identity element $e \in \Omega$ under $\circ$ is given below.

Algorithm for Proving the Existence of an Identity: Let Ω be a set and $\circ$ a binary operator defined on Ω. To prove that e is an identity element in Ω

1. Let $a \in \Omega$ be arbitrary but fixed.
2. Compute $a \circ x$ and $x \circ a$.
3. From the equations $a = a \circ x$ and $a = x \circ a$, solve for x.
4. Show that $x \in \Omega$ and that x does not depend on a.
5. Conclude that $e = x$ is an identity element in Ω under $\circ$.
6. Clean up and rewrite the scratchwork in a clear and concise proof of the theorem. Make sure that each step of the proof makes sense and is clearly justified.
7. Read the proof over carefully and make any necessary corrections.

Note that in solving the equations $a = a \circ x$ and $a = x \circ a$, solutions depending on the value of a cannot be the identity; that is, a solution to these equations will be an identity element if and only if the same solution works for every $a \in \Omega$. The following example illustrates how the preceding algorithm is used to find an identify element.

Example 4.1.9: Let $\Omega = \mathbb{R}$ and for $a, b \in \Omega$ define $a \circ b = a + b - ab$. Prove that there exists an identity element in $\mathbb{R}$ under $\circ$.

> ***Proof:*** Let a be an arbitrary but fixed element in $\mathbb{R}$ and consider the equations $a = a \circ x = x \circ a$. Now, $a \circ x = a + x - ax$ and $x \circ a = x + a - xa$. Thus, $a \circ x = x \circ a$ and hence, consider solving $a \circ x = a + x - ax = a$ for x.
>
> $$a + x - ax = a \qquad \text{if and only if} \qquad x(1 - a) = 0$$
>
> Thus, since a is arbitrary, it follows that the only solution to this equation is $x = 0$, which is in $\mathbb{R}$. Therefore, the identity element in $\mathbb{R}$ under $\circ$ is $e = 0$.
>
> ■

Example 4.1.10: Let $\Omega = \mathbb{R}$ and for $a, b \in \Omega$ define $a \circ b = (a + b)^2$. Show that there does not exist an identity element in $\mathbb{R}$ under $\circ$.

Solution: Let $\Omega = \mathbb{R}$ and for $a, b \in \Omega$ be ABF. Clearly, $a \circ b = b \circ a$, so consider the equation $a = a \circ x$. Now, $a = a \circ x = (a + x)^2$ has only solutions $x = -a \pm \sqrt{a}$. Thus, since the only solutions to the equation $a = a \circ x$ depend on the value of a, there is no identity element in $\mathbb{R}$ under $\circ$.

Note that there is nothing in the definition of the identity element that prevents the existence of more than one identity element in Ω under $\circ$. However, the following theorem shows that if a set Ω is closed under the binary operator $\circ$, then the identity element $e \in \Omega$ under $\circ$ is unique.

Theorem 4.1.1: If Ω is closed under the operation $\circ$ and $e \in \Omega$ is an identity element under $\circ$, then the identity element e is unique.

> ***Proof (Uniqueness Proof):*** Let Ω be closed under $\circ$, and let $e \in \Omega$ be an identity element. Furthermore, assume that e is not unique.
>
> Now let $e' \neq e$ be any other identity element in Ω. Then, since e is an identity element, it follows that $e \circ e' = e'$. Similarly, since e' is an identity element it also follows that $e \circ e' = e$.
>
> Thus, $e \circ e' = e'$ and $e \circ e' = e$ and hence, $e \circ e' = e = e'$. Moreover, $e = e'$, which contradicts the assumption that $e \neq e'$, and therefore e is the unique identity element in Ω.
>
> ■

Note that since $\mathbb{R}$ is closed under ordinary addition and multiplication (Axioms A8 and A9), 0 and 1 are the unique additive and multiplicative identity elements in $\mathbb{R}$. However, there is no identity in $\mathbb{R}$ for the arithmetic operator subtraction since there is no real number satisfying $a - e = e - a$, $\forall\, a \in \mathbb{R}$.

Example 4.1.11: Let $\Omega = \mathbb{Z}$ and $a \circ b = a + b - 1$. Prove that there exists an identity element in $\mathbb{Z}$.

> ***Solution (Construction Proof):*** Let $a \in \mathbb{Z}$ be ABF. Note that $x \circ a = x + a - 1 = a \circ x = a + x - 1$ so that $\circ$ is Abelian. Now, consider solving $a \circ e = a$ for e:
>
> $$a = a \circ e \text{ if and only if } a + e - 1 = a \text{ if and only if } e = 1$$
>
> Now, since $1 \in \mathbb{Z}$ and $a \circ 1 = a = 1 \circ a$, $\forall\, a \in \mathbb{Z}$, it follows that $e = 1$ is the identity element for the binary operator $\circ$.

■

Now, once the identity element $e \in \Omega$ has been identified, another logical question to consider with regard to the binary operator $\circ$ is "If $a \in \Omega$ is there an element $a' \in \Omega$ such that $a \circ a' = e$ and $a' \circ a = e$?" If $a \in \Omega$ and there does exist an element $a' \in \Omega$ such that $a \circ a' = e = a' \circ a$, then a' is called an *inverse* of the element a.

Definition 4.1.6: An element $a \in \Omega$ is said to have an *inverse* element $a^{-1} \in \Omega$ under the binary operator $\circ$ if and only if $a \circ a^{-1} = a^{-1} \circ a = e$.

Note that when a set Ω has an identity element e, then e is the identity element for every element in the set. However, every element in Ω will not necessarily have an inverse. For example, there are many matrices that do not have an inverse. In the best of all mathematical structures, a set will be closed under $\circ$; it will include an identity element e; it will include the inverse for every one of its elements. The following example shows that the set of real numbers has all three of these properties under the arithmetic operator addition, but not under the operator multiplication.

Example 4.1.12: Let $\Omega = \mathbb{R}$. By Axioms A8 and A9, $\mathbb{R}$ is closed under addition and multiplication, and it was shown in Example 4.1.8 that 0 and 1 are the additive and multiplicative identities in $\mathbb{R}$.

Now, since $a + -a = 0$ for every real number a and $-a \in \mathbb{R}$, it follows that $-a$ is the additive inverse of a under ordinary addition. Thus, $\mathbb{R}$ contains all the additive inverses of its elements.

Now, if $a \neq 0$, then $a \times 1/a = 1$ and $1/a \in \mathbb{R}$ for every nonzero real number a, thus $1/a$ is the multiplicative inverse of a, provided $a \neq 0$. However, for $a = 0$ there is no element $a^{-1} \in \mathbb{R}$ such that $a \circ a^{-1} = 1$. Thus, $\mathbb{R}$ does not contain all the multiplicative inverses of its elements.

In general, showing that there exists an inverse or identity element in a set Ω under a binary operator $\circ$ requires an existence proof. Furthermore, showing that all the inverses exist in Ω often requires that an explicit representation of the form of the inverse be found. An algorithm for showing that Ω contains all the inverses under $\circ$ is given below.

Algorithm for Proving the Existence of all the Inverses: Let Ω be a set and $\circ$ a binary operator defined on Ω. To show that $a \in \Omega$ has an inverse element under $\circ$ requires an existence proof as outlined below.

0. Determine e.
1. Let $a \in \Omega$ be arbitrary but fixed.
2. Compute $a \circ x$ and $x \circ a$.
3. From the equation $e = a \circ x = x \circ a$ solve for x.
4. Show that $x \in \Omega$.
5. Conclude that $a^{-1} = x$ is the inverse of the element a under $\circ$.
6. Clean up and rewrite the scratchwork in a clear and concise proof of the theorem. Make sure that each step of the proof makes sense and is clearly justified.
7. Read the proof over carefully and make any necessary corrections.

The algorithm above will be used in the following example to show that $\mathbb{R}$ contains all the additive inverses under the binary operator defined by $a \circ b = a + b - 1$.

Example 4.1.13: Let $\Omega = \mathbb{R}$ and define $a \circ b = a + b - 1$ for $a, b \in \mathbb{R}$. Given that the identity element is $e = 1$, prove that Ω contains an inverse for every one of its elements under the binary operator $\circ$.

> ***Proof:*** Let $a \in \mathbb{R}$ be arbitrary but fixed. Since $\circ$ was shown to be an Abelian operator in Example 4.1.11 it follows that $a \circ x = x \circ a$. Consider solving the equation $e = a \circ x$ for x:
>
> $$e = 1 = a \circ x \text{ if and only if } 1 = a + x - 1$$

Solving for x yields $x = 2 - a$, which is in $\mathbb{R}$ since $\mathbb{R}$ is closed under ordinary addition.

Thus, $a^{-1} = 2 - a$ is in $\mathbb{R}$ whenever $a \in \mathbb{R}$ and hence, $\mathbb{R}$ contains all of its inverses under the operator $\circ$.

■

The following theorem shows that if Ω is closed under a binary operator $\circ$ and $\circ$ is an associative operator, then the inverses contained in Ω are unique.

Theorem 4.1.2: Let $\circ$ be an associative binary operator. If Ω is closed under $\circ$ and $a^{-1} \in \Omega$ whenever $a \in \Omega$, then a^{-1} is unique.

Proof: The proof of Theorem 4.1.2 is left as an exercise.

■

Recall that the closure of Ω under $\circ$ also ensured that the identity was unique when it existed in Ω. Thus, if the set Ω is closed under an associative binary operator $\circ$, Theorems 4.1.1 and 4.1.2 show that the inverses and the identity element are unique; however, this is not necessarily the case when Ω is not closed under $\circ$. Furthermore, the following theorem shows that when Ω is closed under $\circ$, the unique inverse of a^{-1} (i.e., $(a^{-1})^{-1}$) is a.

Theorem 4.1.3: If Ω is closed under $\circ$ and $a^{-1} \in \Omega$ whenever $a \in \Omega$, then $\left(a^{-1}\right)^{-1} = a$.

Proof: The proof of Theorem 4.1.3 is left as an exercise.

■

4.2 Commonly Used Number Systems

Several different number systems are commonly encountered in advanced math classes, including the natural numbers, whole numbers, integers, rational numbers, irrational numbers, real numbers, and the complex numbers. In this section, the properties of the arithmetic operators $(+, \times, -, \div)$ on the natural, whole, integer, rational, irrational, and real numbers will be studied;

the complex numbers are generally studied in courses such as abstract algebra, complex analysis, and advanced engineering mathematics, and therefore will not be studied here.

4.2.1 The Natural Numbers

The act of counting is a basic day-to-day operation in the lives of most people. For example, people are always counting something like the days until Christmas, the number of fish they have caught, or the number of people living in a town. The natural numbers are the oldest numbers known to humans dating back at least 30,000 years. Proof that humans were counting objects 30,000 years ago was found by archaeologist Karl Absolom in 1937. Absolom unearthed a wolf bone in Czechoslovakia that had an obvious set of notches carved into it. The bone had 55 notches scratched into it arranged in groups of 5, with a second scratchmark after the first 25 scratchmarks. Clearly, this wolf bone is evidence that early humans were counting some sort of object. Thus, the first set of numbers that a student will encounter in school is the set of counting numbers, which are also called the *natural numbers*.

Definition 4.2.1: The set of *natural numbers* or the *counting numbers* is denoted by $\mathbb{N}$ and consists of the numbers $1, 2, 3, 4, 5, \ldots$.

Several facts concerning properties of the natural numbers are given below. In particular, most of the following facts concern the ordinary arithmetic operators addition, multiplication, subtraction, and division.

Some Facts about $\mathbb{N}$:

a. $\mathbb{N}$ is closed under addition and multiplication (Axioms A8 and A9).
b. $\mathbb{N}$ is not closed under subtraction since $1 - 4 = -3$ and $-3 \notin \mathbb{N}$.
c. $\mathbb{N}$ is not closed under division since $1 \div 4 = 0.25$ and $0.25 \notin \mathbb{N}$.
d. Since 0 is the unique additive identity in $\mathbb{R}$ (Axiom A3) and $0 \notin \mathbb{N}$, the natural numbers do not contain the additive identity.
e. Since 1 is the unique multiplicative identity in $\mathbb{R}$ (Axiom A7) and $1 \in \mathbb{N}$, the natural numbers do contain the multiplicative identity.
f. Since $-a$ is the unique additive inverse of a in $\mathbb{R}$ (Axiom A4) and $-a \notin \mathbb{N}$ for any element $a \in \mathbb{N}$, none of the elements in the natural numbers have additive inverses.
g. Since $1/a$ is the unique multiplicative inverse of a in $\mathbb{R}$ and $1/a \notin \mathbb{N}$ except for $a = 1$, the only natural number having a multiplicative inverse is $a = 1$.
h. $\mathbb{N}$ is made up of the even natural numbers and the odd natural numbers.

Note that $\mathbb{N}$ does not contain the additive identity, nor does it contain any of the additive or multiplicative inverses. The following theorem shows that the sum of the first n natural numbers is $\dfrac{n(n+10)}{2}$. Also, the corollary to this theorem provides a formula for computing the sum of the first n odd natural numbers, as well as the sum of the first n even natural numbers.

Theorem 4.2.1: The sum of the first n natural numbers is

$$\sum_{i=1}^{n} i = \frac{n(n+1)}{2}, \ \forall\, n \in \mathbb{N}.$$

Proof: This theorem was proved in Example 3.4.1 using mathematical induction.

■

Corollary to Theorem 4.2.1: For every natural number n, the sum of the first

a. n odd natural numbers is n^2.
b. n even natural numbers is $n(n+1)$.

Proof: First, note that

(i) The sum of the first n odd numbers is $\sum_{i=1}^{n} (2i-1)$.

(ii) The sum of the first n odd numbers is $\sum_{i=1}^{n} 2i$.

Proof of part (i): Consider $\sum_{i=1}^{n} (2i-1)$:

$$\sum_{i=1}^{n} (2i-1) = \sum_{i=1}^{n} 2i - \sum_{i=1}^{n} 1$$

$$= 2\left(\sum_{i=1}^{n} i\right) - n = \underbrace{2\frac{n(n+1)}{2}}_{\text{By Theorem 4.2.1}} - n$$

$$= n^2 + n - n = n^2$$

Therefore, the sum of the first n odd numbers is n^2 for all $n \in \mathbb{N}$.

Proof of part (ii): Since the sum of the first $2n$ natural numbers is

$$\sum_{i=1}^{2n} i = \frac{2n(2n+1)}{2} = n(2n+1)$$

it follows that

$$\sum_{i=1}^{2n} i = \underbrace{\sum_{i=1}^{n} 2i}_{\text{the evens}} + \underbrace{\sum_{i=1}^{n} (2i-1)}_{\text{the odds}}$$

Therefore

$$\sum_{i=1}^{n} 2i = \sum_{i=1}^{2n} i - \sum_{i=1}^{n} (2i-1)$$

$$= \frac{2n(2n+1)}{2} - n^2 = n^2 + n$$

$$= n(n+1)$$

Therefore, the sum of the first n even numbers is $n(n+1)$ for all $n \in \mathbb{N}$.

■

4.2.2 The Whole Numbers

While people were counting 30,000 years ago, the idea of a number 0 was not a natural result of counting. For example, a family of farmers would say that they had 1, 2, 3, 4, or n sheep, but if they had no sheep, then they were likely to say "We have no sheep" rather than saying they had 0 sheep. The number 0 was not necessary before the concepts of commerce and credit were in common use. In fact, neither the ancient Egyptians, Greeks, nor Romans used the number 0. The first people known to have used 0 were the Babylonians, who used 0 simply as a numerical placeholder, not a distinct number. The first people known to have used 0 as a number were the mathematicians of India, and even they did not fully embrace the concept of 0 as a number. For more information on the history of the number 0, see *ZERO: The Biography of a Dangerous Idea* by Charles Seife (2000).

The set of *whole numbers* is formed by simply adding the number 0 to the set of natural numbers. The definition of the whole numbers is given below.

Definition 4.2.2: The set of *whole numbers* is denoted by $\mathbb{W}$ and

$$\mathbb{W} = \{\omega : \omega \in \mathbb{N} \text{ or } \omega = 0\}$$

Note that $\mathbb{W} = \{0, 1, 2, 3, 4, \ldots\}$, and $\mathbb{W}$ contains all the natural numbers along with the additive identity 0. Thus, the whole numbers will share the same properties as the natural numbers along with the additional property of containing the additive identity. Several facts concerning the whole numbers are listed below.

Facts about $\mathbb{W}$:

a. $\mathbb{W}$ is closed under addition and multiplication (Axioms A8 and A9).
b. $\mathbb{W}$ is not closed under subtraction since $1 - 4 = -3$ and $-3 \notin \mathbb{W}$.
c. $\mathbb{W}$ is not closed under division since $1 \div 4 = 0.25$ and $0.25 \notin \mathbb{W}$.
d. Since 0 is the additive identity (Axiom A3) and $0 \in \mathbb{W}$, the whole numbers do contain the additive identity.
e. Since 1 is the multiplicative identity (Axiom A7) and $1 \in \mathbb{W}$, the whole numbers do contain the multiplicative identity.
f. Since $-a$ is the additive inverse of a (Axiom A4) and $-a \in \mathbb{W}$ only for the element 0, none of the other elements in the whole numbers have additive inverses.
g. Since $1/a$ is the multiplicative inverse of a and $1/a \notin \mathbb{W}$ except for $a = 1$, only $a = 1$ has a multiplicative inverse.
h. $\mathbb{W}$ is made up of the even whole numbers and the odd whole numbers.
i. $\mathbb{N}$ is contained within $\mathbb{W}$ ($\mathbb{N} \subset \mathbb{W}$).

While the set of whole numbers contains both the additive and multiplicative identities, it is not closed under subtraction or division, nor does it contain the additive and multiplicative inverses. Thus, even the early mathematicians needed to consider larger classes of numbers than just the natural numbers and whole numbers.

4.2.3 The Integers

In the Western civilizations, one reason for the slow development of negative numbers was that much of Western mathematics was developed by the Greeks, whose concept of numbers was geometrically based and generally tied to lengths, areas, and volumes. Since, negative lengths, areas, or volumes were ridiculous ideas in the eyes of the Greeks, there was no need for the negative numbers. The strong influence of the Greek development and philosophy of mathematics was longlasting with European mathematicians, lasting even into the sixteenth, seventeenth, and eighteenth centuries. In fact, many sixteenth/seventeenth-century mathematicians called negative numbers "fictitious," "absurd," or "false numbers." On the other hand, in the Eastern civilizations of India and China, which were far removed from the influence of the mathematics of the Greeks, negative numbers were embraced as early as the seventh century.

Now, the set of *integers* is formed by adding the negative whole numbers to the set of whole numbers. The definition of the integers is given below.

Definition 4.2.3: The set of *integers* is denoted by $\mathbb{Z}$ and

$$\mathbb{Z} = \{z : z \in \mathbb{N},\ z = 0, \text{ or } -z \in \mathbb{N}\}$$

Note that $\mathbb{N} = \{1, 2, 3, 4, \ldots\}$ is contained in $\mathbb{W} = \{0, 1, 2, 3, \ldots\}$, which is contained in $\mathbb{Z} = \{0, \pm 1, \pm 2, \pm 3, \ldots\}$. Furthermore, all the elements in $\mathbb{Z}$ have additive inverses in $\mathbb{Z}$. Several important facts concerning the properties of the integers under the basic arithmetic operations are given below.

Facts about $\mathbb{Z}$:

a. $\mathbb{Z}$ is closed under addition and multiplication (Axioms A8 and A9).
b. $\mathbb{Z}$ is closed under subtraction since $a - b = a + -b$ and $\mathbb{Z}$ is closed under addition.
c. $\mathbb{Z}$ is not closed under division since $1 \div 4 = 0.25$ and $0.25 \notin \mathbb{Z}$.
d. Since 0 is the additive identity (Axiom A3) and $0 \in \mathbb{Z}$, the integers do contain the additive identity.
e. Since 1 is the multiplicative identity (Axiom A7) and $1 \in \mathbb{Z}$, the integers do contain the multiplicative identity.
f. Since $-a$ is the additive inverse of a (Axiom A4) and $-a \in \mathbb{Z}$ for every integer a, all the integers have additive inverses.
g. Since $1/a$ is the multiplicative inverse of a and $1/a \notin \mathbb{Z}$ except for $a = \pm 1$, only the integers $a = 1$ and $a = -1$ have multiplicative inverses.
h. $\mathbb{Z}$ is made up of both the even and odd integers.
i. The natural numbers are contained in the whole numbers that are contained in the integers ($\mathbb{N} \subset \mathbb{W} \subset \mathbb{Z}$).

Theorem 4.2.2: Let $\mathbb{Z}_E$ and $\mathbb{Z}_O$ be the collection of even and odd integers, respectively. Then

(i) $\mathbb{Z}_E$ is closed under addition.

(ii) $\mathbb{Z}_E$ is closed under multiplication.

(iii) $\mathbb{Z}_O$ is closed under multiplication.

Proof:

Proof of part (i): Let $z_1, z_2 \in \mathbb{Z}_E$ be ABF. Then, $\exists\ k, j \in \mathbb{Z}$ such that $z_1 = 2k$ and $z_2 = 2j$. Consider $z_1 + z_2$

$$z_1 + z_2 = 2k + 2j = 2(k + j) = 2l$$

where $l = k + j \in \mathbb{Z}$ since $\mathbb{Z}$ is closed under addition. Thus, $z_1 + z_2$ is even and therefore, $z_1 + z_2 \in \mathbb{Z}_E$ whenever $z_1, z_2 \in \mathbb{Z}_E$.

Hence, $\mathbb{Z}_E$ is closed under addition.

Proof of part (ii): The proof of part (ii) is left as an exercise.

Proof of part (iii): The proof of part (iii) is left as an exercise.

■

Thus, the set of integers has many important structural properties, including (1) closure under addition, multiplication, and subtraction; (2) additive and multiplicative identities; and (3) all additive inverses. However, the integers do not contain the multiplicative inverses; hence, a system of numbers that contains all the integers and their multiplicative inverses will need to be considered next.

4.2.4 The Rational Numbers

Since $\mathbb{Z}$ is not closed under division, the natural extension of the integers is to create a new set of numbers by adding the ratios of the integers to create a new set that contains $\mathbb{Z}$. In part, this was done by the ancient Greek mathematicians. In fact, to the followers of Pythagoras, the Pythagoreans, the only numbers that could make any sense at all were numbers of the form a/b, where a and b were counting numbers. Zero was not allowed, because it did not exist as a number in the Pythagorean train of thought, and ratios of the form $1/1$, $2/1$, $3/1, \ldots$ allowed for the counting numbers. The set of numbers consisting of the ratios of integers is called the *rational numbers* and is defined below.

Definition 4.2.4: The set of *rational numbers* is denoted by $\mathbb{Q}$. $\mathbb{Q}$ is formed by taking all possible ratios of the integers (denominator not equal to zero, of course) and is often represented by

$$\mathbb{Q} = \left\{ r : \ r = \frac{p}{q} \text{ where } p \text{ and } q \neq 0 \text{ are integers} \right\}$$

Note that in the creation of a rational number division by zero is not allowed. In fact, it is the definition of division that excludes division by 0, not a constraint of the rational numbers. Thus, there will be no set of numbers discussed in this course for which division by 0 is allowed. Several facts concerning the set of rational numbers are given below.

Facts about $\mathbb{Q}$: The list of facts on the rational numbers given below is considerably shorter than the previous lists for $\mathbb{N}, \mathbb{W}$, and $\mathbb{Z}$, since many of the properties concerning the rational numbers will be proved in the following theorems of this section.

a. Since 0 is the additive identity (Axiom A3) and $0 \in \mathbb{Q}$, the rational numbers do contain the additive identity.
b. Since 1 is the multiplicative identity (Axiom A7) and $1 \in \mathbb{Q}$, the rational numbers do contain the multiplicative identity.
c. The natural numbers are contained in the whole numbers that are contained in the integers that are contained in the rational numbers; that is
$$\mathbb{N} \subset \mathbb{W} \subset \mathbb{Z} \subset \mathbb{Q}$$

Thus, the rational numbers contain both the additive and multiplicative identities, and the following theorem shows that $\mathbb{Q}$ is closed under each of the four basic arithmetic binary operators.

Theorem 4.2.3: $\mathbb{Q}$ is closed under

(i) Addition
(ii) Multiplication
(iii) Subtraction
(iv) Nonzero division

Proof (Closure Proof): Let $a, b \in \mathbb{Q}$. Then, there exist integers p, q, r, s such that $a = \frac{p}{q}$ and $b = \frac{r}{s}$ with $q \neq 0$ and $s \neq 0$.

Proof of part (i): Consider $a + b$.

$$a + b = \frac{p}{q} + \frac{r}{s} = \frac{ps + rq}{qs}$$

Now, ps, rq and qs are integers since $\mathbb{Z}$ is closed under multiplication, $ps + rq$ is an integer since $\mathbb{Z}$ is closed under addition, and $qs \neq 0$ since $q \neq 0$ and $s \neq 0$.

Thus, $a+b$ is the ratio of two integers, with the denominator nonzero, and hence is a rational number.

Therefore, $\mathbb{Q}$ is closed under addition.

Proof of part (ii): Consider $a \times b$:

$$a \times b = \frac{p}{q} \times \frac{r}{s} = \frac{pr}{qs}$$

Now, pr and qs are integers since $\mathbb{Z}$ is closed under multiplication, and $qs \neq 0$ since $q \neq 0$ and $s \neq 0$.

Thus, $a \times b$ is the ratio of two integers, with the denominator non-esro, and hence is a rational number.

Therefore, $\mathbb{Q}$ is closed under multiplication.

Proof of part (iii): The proof of part (iii) is left as an exercise.

Proof of part (iv): The proof of part (iv) is left as an exercise.

■

The following theorem shows that $\mathbb{Q}$ contains all the additive inverses of its elements.

Theorem 4.2.4: If $a \in \mathbb{Q}$, then the additive inverse of a is in $\mathbb{Q}$.

Proof (Existence Proof): Let $a \in \mathbb{Q}$ be ABF. Then, $-1 \in \mathbb{Q}$ since $-1 = \frac{-1}{1}$ and therefore, $-1(a) = -a \in \mathbb{Q}$ since $\mathbb{Q}$ is closed under multiplication.

Now, $-a+a = 0$ and thus, $-a$ is the additive inverse of a. Therefore, if $a \in \mathbb{Q}$, then its additive inverse $-a$ is also in $\mathbb{Q}$.

■

Finally, the theorem stated below shows that $\mathbb{Q}$ contains the multiplicative inverse of each of its elements with the exception of 0.

Theorem 4.2.5: If $a \in \mathbb{Q}$ and $a \neq 0$, then the multiplicative inverse of a is in $\mathbb{Q}$.

Proof (Existence Proof): Let $a \in \mathbb{Q}$ with $a \neq 0$ be ABF. Then, $1 \in \mathbb{Q}$ since $1 = \frac{1}{1}$ and therefore, $\frac{1}{a} \in \mathbb{Q}$ since $\mathbb{Q}$ is closed under division.

Now, $a \cdot \frac{1}{a} = 1$ and thus, $\frac{1}{a}$ is the multiplicative inverse of a. Therefore the multiplicative inverse of a is in $\mathbb{Q}$ for every nonzero $a \in \mathbb{Q}$.

■

Another interesting property of the rational numbers is that the same rational number can be written in several equivalent ways. Thus, every rational number has multiple representations. For example, consider the rational number 1/2:

$$\frac{1}{2} = \frac{2}{4} = \frac{13}{26} = \frac{191}{382} = 0.5$$

A rational number may be expressed as the ratio of two integers in many ways and also in a decimal representation. Furthermore, a rational number will have either a terminating decimal representation or an infinite repeating decimal representation. For example, $1/4 = 0.25$ has a terminating decimal expansion and $2/3 = 0.666\ldots$ has an infinite repeating decimal expansion. The mathematical shorthand for an infinitely repeating decimal expansion is to place a bar over the repeating pattern in the decimal expansion. For example, $0.666\ldots$ can be written as $0.\overline{6}$ and $3.4712712712\ldots$ can be written as $3.4\overline{712}$. The following example shows that $0.\overline{9} = 1$, contrary to popular belief.

Example 4.2.1: Prove that $0.\overline{9} = 1$.

Proof: Let $x = 0.\overline{9}$. Then

$$\begin{aligned} x &= 0.\overline{9} \\ 10x &= 9.\overline{9} \end{aligned}$$

Subtracting x from $10x$ yields

$$\begin{aligned} 10x &= 9.\overline{9} \\ -\ \ x &= 0.\overline{9} \\ 9x &= 9. \end{aligned}$$

Thus, $9x = 9$ and hence $x = 1$.

Therefore, $0.\overline{9} = 1$.

■

Now, given a rational number in decimal form a rational representation of the decimal form can always be found. The key to finding the rational form of a number is to find multiples of the decimal form that when subtracted result in a whole number. The following example illustrates the process of finding a rational representation for a rational number from its decimal representation.

Example 4.2.2: Let $x = 12.30\overline{34}$. Find a rational form for x:

$$x = 12.30\overline{34}$$

$$100x = 1230.\overline{34}$$

$$10000x = 123,034.\overline{34}$$

Now

$$\begin{aligned} 10000x &= 123,034.\overline{34} \\ -\quad 100x &= 1230.\overline{34} \\ 9900x &= 121,804 \end{aligned}$$

Thus, $x = \dfrac{121,804}{9900}$.

The following theorem shows that between every two distinct rational numbers there is another rational number.

Theorem 4.2.6: Between every two rational numbers there is at least one other rational number.

Proof (Existence Proof): Let $a, b \in \mathbb{Q}$ with $a < b$. Now

$$a = \frac{a+a}{2} < \frac{a+b}{2} < \frac{b+b}{2} = b$$

Furthermore, $\dfrac{a+b}{2}$ is a rational number since the rational numbers are closed under addition and division.

Therefore, between every two rational numbers there is at least one rational number.

■

Another way of stating the result of the previous theorem is "The set of rational numbers is everywhere dense." However, the previous theorem does not state that there are only rational numbers between any two rational numbers, and Theorem 4.2.7 shows that there are infinitely many rational numbers between every two rational numbers.

Theorem 4.2.7: Between every two rational numbers there are infinitely many other rational numbers.

Proof (Existence Proof): Let $a, b \in \mathbb{Q}$ with $a < b$. Now, for $k \in \mathbb{N}$

$$a = \frac{(k+1)a}{k+1} = \frac{ka+a}{k+1} < \frac{ka+b}{k+1} < \frac{kb+b}{k+1} = b$$

Furthermore, $\dfrac{ka+b}{k+1}$ is a rational number since the rational numbers are closed under addition and division, $\forall\ k \in \mathbb{N}$.

Therefore, between every two rational numbers there are infinitely many rational numbers.

■

Now, $\mathbb{Q}$ is closed under addition, multiplication, subtraction, and division and contains the additive and multiplicative identities, a natural question to consider next is "Are the rational numbers the final set of numbers that a mathematician needs to consider?" To see that an even larger collection of numbers is needed in mathematics consider the following problem: "Let $a \circ b$ be the binary operator that provides the solutions to $ax^2 - b = 0$. Is $\mathbb{Q}$ closed under $\circ$?" Consider $1 \circ 2$, this provides the solutions to the equation $x^2 - 2 = 0$, which are $\pm\sqrt{2}$. Now, is $\sqrt{2}$ a rational number? The following theorem shows that $\sqrt{2}$ is not a rational number, and therefore $\mathbb{Q}$ is not closed under the binary operator $\circ$.

Theorem 4.2.8: $\sqrt{2}$ is not a rational number.

Proof (by Contradiction): Assume that $\sqrt{2}$ is rational. Then, $\exists\ p, q \in \mathbb{Z}$ such that $\sqrt{2} = \dfrac{p}{q}$.

Without loss of generality (WLOG), assume that p and q have no common factors.

Now consider $(\sqrt{2})^2$. First, $(\sqrt{2})^2 = 2$ and

$$2 = (\sqrt{2})^2 = \left(\frac{p}{q}\right)^2 = \frac{p^2}{q^2}$$

Cross-multiplying yields $2q^2 = p^2$. Thus, by Example 3.3.4 it follows that p^2 is even and therefore, so is p. Thus, $\exists\, k \in \mathbb{Z}$ such that $p = 2k$.

Now, substituting $p = 2k$ into $2q^2 = p^2$ yields $2q^2 = 4k^2$, and from $2q^2 = 4k^2$ it follows that $q^2 = 2k^2$. Hence, q^2 and q are even. Furthermore, this means that both p and q are even. Thus, p and q have common factor 2 which contradicts the assumption that p and q have no common factors.

Therefore, $\sqrt{2}$ cannot be a rational number.

■

The fact that $\sqrt{2}$ is not rational is often attributed to Hippasus of the Pythagoreans and was one of the first mathematical theorems to be deductively proved. In reward for this discovery, the Pythagoreans supposedly took Hippasus out to sea and threw him overboard, leading to his death. The idea that some numbers could not be expressed in the form integer over integer seems to have been unacceptable to the Pythagoreans, regardless of the proof. For more information on Hippasus, the Pythagoreans, and $\sqrt{2}$, see *ZERO: The Biography of a Dangerous Idea* (Seife 2000).

Examples of some other numbers that are not rational include $\sqrt{3}$, $\sqrt{5}$, $\sqrt{61}$, e, π, and $e + \pi$. Thus, the set of rational numbers is not an adequate set of numbers for solving simple second-degree equations, and thus, a larger more complete set of numbers will be required for solving equations.

4.2.5 The Real Numbers

Now, the rational numbers form a system of numbers that is closed under the four basic arithmetic operations containing both the additive identity and the multiplicative identity and contains both the additive and multiplicative inverses. However, the result in Theorem 4.2.8 shows that there exist numbers that are not rational numbers. Thus, the set of numbers that form the basis for solving equations is made up of the rational numbers and another set of numbers that are not rational, such as $\sqrt{2}$, which are called the *irrational numbers*. The set of numbers consisting of both the rational numbers and the irrational numbers is called the set of *real numbers*. The set of real numbers consists of the collection of all of the rational numbers and the collection of the limits of all convergent sequences of rational numbers. The definitions of the real numbers and the irrational numbers are given below.

Definition 4.2.5: The set of *real numbers* consists of all numbers between $-\infty$ and ∞ (i.e., $-\infty < x < \infty$) and is denoted by $\mathbb{R}$ or by the interval $(-\infty, \infty)$.

Definition 4.2.6: Any real number that is not a rational number is called an *irrational number*. The set of irrational numbers will be denoted by $\mathbb{I}$.

Note that any real number that is not a rational number is by definition a real number; therefore, the real numbers consist of only the rational numbers and the irrational numbers. Also, recall that a rational number has either a finite decimal representation or an infinite repeating decimal representation. On the other hand, every irrational number has a nonrepeating infinite decimal representation, and therefore, the exact decimal value of any irrational number can never be known. The following theorem shows that a basic arithmetic mixture of a rational and an irrational number will produce an irrational number.

Theorem 4.2.9: Let $a \in \mathbb{Q}$ and $b \in \mathbb{I}$. Then

(i) $a + b \in \mathbb{I}$.

(ii) $a \cdot b \in \mathbb{I}$, provided that $a \neq 0$.

(iii) $a - b \in \mathbb{I}$.

(iv) $b - a \in \mathbb{I}$.

(v) $\frac{a}{b} \in \mathbb{I}$, provided that $a \neq 0$.

(vi) $\frac{b}{a} \in \mathbb{I}$, provided that $a \neq 0$.

Proof: Let $a \in \mathbb{Q}$ and $b \in \mathbb{I}$ be ABF.

Proof of part (i) (by Contradiction): Suppose that $a + b$ is a rational number.

Since a and $a + b$ are rational numbers, and $\mathbb{Q}$ is closed under subtraction, it follows that $b = a + b - a \in \mathbb{Q}$. Thus, b is a rational number, which contradicts the hypothesis that b is irrational.

Therefore, $a + b$ must be an irrational number.

Proof of parts (ii)–(vi): The proofs of parts (ii)–(vi) are left as exercises.

■

The following list contains several facts concerning the properties of the real numbers under the four basic arithmetic operations: addition, multiplication, subtraction, and division.

Facts about $\mathbb{R}$:

a. $\mathbb{R}$ is closed under addition and multiplication (Axioms A8 and A9).

b. $\mathbb{R}$ is closed under subtraction.

c. $\mathbb{R}$ is closed under nonzero division.

d. Since 0 is the additive identity (Axiom A3) and $0 \in \mathbb{R}$, the real numbers do contain the additive identity.

e. Since 1 is the multiplicative identity (Axiom A7) and $1 \in \mathbb{R}$, the real numbers do contain the multiplicative identity.

f. Since $-a$ is the additive inverse of a (Axiom A4) and $-a \in \mathbb{R}$ for every real number a, all the real numbers have additive inverses.

g. Since $1/a$ is the multiplicative inverse of a and $1/a \in \mathbb{R}$ except for $a = 0$, all nonzero real numbers have multiplicative inverses.

h. $\mathbb{N}$ is contained in $\mathbb{W}$, which is contained in $\mathbb{Z}$, which is contained in $\mathbb{Q}$, which is contained in $\mathbb{R}$, and $\mathbb{R}$ also contains $\mathbb{I}$; that is

$$\mathbb{N} \subset \mathbb{W} \subset \mathbb{Z} \subset \mathbb{Q} \subset \mathbb{R} \text{ and } \mathbb{I} \subset \mathbb{R}$$

The following theorem provides two essential tools that are often used in solving a system of mathematical equations. In particular, the cancellation properties for the arithmetic operators addition and multiplication are proved in Theorem 4.2.10.

Theorem 4.2.10: Let $a, b, c \in \mathbb{R}$.

(i) If $a + c = b + c$, then $a = b$.

(ii) If $ac = bc$ and $c \neq 0$, then $a = b$.

Proof: Let $a, b, c \in \mathbb{R}$ be ABF.

Proof of part (i): Suppose that $a + c = b + c$. Then, $-c \in \mathbb{R}$ since $\mathbb{R}$ contains the additive inverses of its elements. Now, adding $-c$ to $a + c$ yields

$$(a + c) + -c \underbrace{=}_{\text{By A2}} a + (c + -c) = a$$

Similarly, adding $-c$ to $b + c$ yields

$$(b + c) + -c \underbrace{=}_{\text{By A2}} b + (c + -c) = b$$

Thus,

$$a = (a + c) + -c = \underbrace{(b + c) + -c}_{\text{Since } a + c = b + c} = b$$

Therefore, $a = b$ whenever $a + c = b + c$.

Proof of part (ii): The proof of part (ii) is left as an exercise.

■

In Chapter 5, the distance between real numbers is of primary importance in the study of sequences of real numbers, limits of real-valued functions, continuity of real-valued functions, and the derivatives of real-valued functions. The standard measure of the distance between two real numbers is the *absolute value* of the difference of the two numbers. The absolute value of a real number is defined below.

Definition 4.2.7: The *absolute value* of a real number a, denoted by $|a|$, is defined as follows:

$$|a| = \begin{cases} a & \text{when } a \geq 0 \\ -a & \text{when } a < 0 \end{cases}$$

For example, $|-3.12| = 3.12$, $|6.76| = 6.76$. Note that the absolute value of a real number is always nonnegative and hence $a \leq |a|$, $\forall\, a \in \mathbb{R}$. Furthermore, $|a| \leq b$ if and only if $-b < a < b$ and $|a| > b$ if and only if $a > b$ or $a < -b$; similar results hold for the strict inequalities $<$ and $>$. For example, $|x| \leq 4$ if and only if $-4 \leq x \leq 4$ and $|x| > 4$ if and only if $x > 4$ or $x < -4$. Moreover, the distance between two real numbers x and y is $|x - y|$, and thus, the distance between $x = -3.12$ and $y = 6.76$ is $|-3.12 - 6.76| = |-9.88| = 9.88$.

The following theorems provide important results concerning the absolute value of two real numbers. The theorem stated below shows that the absolute value of a product can also be computed as the product of the individual absolute values.

Theorem 4.2.11: If $x, y \in \mathbb{R}$, then $|xy| = |x| \cdot |y|$.

Proof (by cases): Let $x, y \in \mathbb{R}$ be ABF. Then, either (1) $x \geq 0$ and $y \geq 0$, (2) $x < 0$ and $y < 0$, (3) $x \geq 0$ and $y < 0$, or (4) $x < 0$ and $y \geq 0$.

Case 1: Suppose that $x \geq 0$ and $y \geq 0$. Then, $xy \geq 0$, $|xy| = xy$, $|x| = x$, and $|y| = y$. Hence

$$|xy| = xy = |x| \cdot |y|$$

Case 2: Suppose that $x < 0$ and $y < 0$. Then, it follows that $xy > 0$, $|xy| = xy, |x| = -x$, and $|y| = -y$. Hence

$$|xy| = xy = (-x)(-y) = |x| \cdot |y|$$

Cases 3 and 4: By symmetry, the same proofs can be used for the cases $x \geq 0$ and $y < 0$ and $x < 0$ and $y \geq 0$. Thus, WLOG suppose that $x \geq 0$ and $y < 0$. Then, $xy \leq 0$, $|xy| = -xy, |x| = x$, and $|y| = -y$. Hence

$$|xy| = -xy = x(-y) = |x| \cdot |y|$$

Thus, in each of the four cases it follows that $|xy| = |x| \cdot |y|$ and hence, $|xy| = |x| \cdot |y|, \forall\, x, y \in \mathbb{R}$.

■

The following corollary of theorem 4.2.11 shows that $|-x| = |x|, \forall\, x \in \mathbb{R}$.

Corollary to Theorem 4.2.11: If $x \in \mathbb{R}$, then $|-x| = |x|$.

Proof: The corollary to Theorem 4.2.11 follows directly from that theorem with $x = x$ and $y = -1$.

■

The distance between two real numbers will be play an important role in the Chapter 5 discussion of sequences or real numbers, limits of real-valued functions, continuity, and differentiation. The following theorem and its corollary provide two important results for absolute values that will often be utilized in Chapter 5. In particular, of the following two results, it is the corollary to Theorem 4.2.12 that will be the more frequently used result.

Theorem 4.2.12 (The Triangle Inequality): If $x, y \in \mathbb{R}$, then

$$|x + y| \leq |x| + |y|$$

Proof (by cases): Let $x, y \in \mathbb{R}$. Then, either (1) both x and y are greater than or equal to 0, (2) both x and y are less than 0, and

(3) one of the values, x and y, is greater than or equal to 0 and the other is less than or equal to 0.

Case 1: Let $x, y \geq 0$. Then, $|x| = x$, $|y| = y$, and $|x+y| = x+y$.

Thus, $|x+y| = x+y = |x|+|y|$ and hence, $|x+y| \leq |x|+|y|$ for $x, y \geq 0$.

Case 2: Let $x, y < 0$. Then, $|x| = -x$, $|y| = -y$, and

$$|x+y| = -(x+y) = -x-y$$

Thus, $|x+y| = -x-y = -x+-y = |x|+|y|$ and it follows that $|x+y| \leq |x|+|y|$ for $x, y < 0$.

Case 3: By symmetry of argument, WLOG let $x < 0$ and $y \geq 0$. In this case, there are two subcases to consider, namely, $x+y \geq 0$ and $x+y < 0$.

Subcase 1: Let $x < 0$ and $y \geq 0$ and assume that $x+y \geq 0$. In this case, $|x| = -x$, $|y| = y$, and $|x+y| = x+y$. Thus

$$|x+y| = x+y \underbrace{\leq}_{\text{since } x < 0 < -x} -x+y = |x|+|y|$$

and hence, for $x < 0$, $y \geq 0$, and $x+y \geq 0$ it follows that $|x+y| \leq |x|+|y|$.

Subcase 2: Let $x < 0$ and $y \geq 0$ and assume that $x+y < 0$. In this case, $|x| = -x$, $|y| = y$, and $|x+y| = -(x+y) = -x-y$. Thus

$$|x+y| = -x-y \underbrace{\leq}_{\text{since } -y < 0 < y} -x+y = |x|+|y|$$

and hence, for $x < 0$, $y \geq 0$, and $x+y < 0$ it follows that $|x+y| \leq |x|+|y|$.

Hence, in each case it follows that $|x+y| \leq |x|+|y|$ and therefore, $|x+y| \leq |x|+|y|$, $\forall\, x, y \in \mathbb{R}$.

■

Corollary to Theorem 4.2.12 (The Triangle Inequality): If $x, y, z \in \mathbb{R}$, then

$$|x-y| \leq |x-z|+|z-y|$$

Proof: Let $x, y, z \in \mathbb{R}$ be ABF. Then

$$|x - y| = |x + \underbrace{(-z + z)}_{0} - y| = |x - z + z - y|$$

$$\underbrace{\leq |x - z| + |z - y|}_{\text{By the triangle inequality}}$$

■

The following three results also provide useful bounds on the distance between two real numbers.

Theorem 4.2.13: If $x, y \in \mathbb{R}$, then $|x - y| \geq |x| - |y|$.

Proof: Let $x, y \in \mathbb{R}$. Consider $|x|$:

$$|x| = |x + 0| = |x + (-y + y)| = |x - y + y|$$

Applying Theorem 4.2.11, it follows that $|x - y + y| \leq |x - y| + |y|$.

Now, $|x| \leq |x - y| + |y|$ and subtracting $|y|$ from both sides of this equation yields

$$|x| - |y| \leq |x - y|$$

and therefore, $|x - y| \geq |x| - |y|, \forall\, x, y \in \mathbb{R}$.

■

Corollary to Theorem 4.2.13: If $x, y \in \mathbb{R}$, then $|x| - |y| \leq |x - y| \leq |x| + |y|$.

Proof: This result follows directly from Theorems 4.2.12 and 4.2.13.

■

Theorem 4.2.14: If $x, y \in \mathbb{R}$, then $\big||x| - |y|\big| \leq |x - y|$.

Proof: The proof of Theorem 4.2.14 is left as an exercise.

■

Finally, the real numbers are closed under the arithmetic operations of addition, multiplication, subtraction, and division; $\mathbb{R}$ contains both the additive and multiplicative identities, and $\mathbb{R}$ contains the additive and multiplicative inverses. However, the real numbers are not the final set of numbers that a mathematician requires for solving algebraic equations. Recall that the set of real numbers is not closed under the binary operator defined by $a \circ b$ that provides the solutions to the equation $ax^2 - b = 0$. For example, $1 \circ -2$ has no real solutions since the solutions to $x^2 + 2 = 0$ are $\pm\sqrt{-2}$; the square root of a negative number is a complex number, not a real number. However, complex numbers and their properties will not be considered any further in this text.

4.3 Elementary Number Theory

The study of numbers and their properties is an area of mathematics called *number theory*, which is one of the oldest areas of mathematical research dating as far back as circa 300 B.C. with Books VII–IX of Euclid's *Elements* In fact, many of the most famous mathematicians of all time have worked in the area of number theory. One of the most famous mathematicians, Carl Friedrich Gauss (1777–1855), said "mathematics is the queen of the sciences, but number theory is the queen of mathematics." Moreover, there are many classic problems in number theory, including Fermat's Last Theorem, which was more recently proved by Andrew Wiles and Richard Taylor. Some of the topics that are studied in number theory are odd and even numbers, divisibility, prime numbers, factorization of numbers, polynomial congruences, number theoretic functions, modulo arithmetic, and continued fractions. Furthermore, many of the topics in number theory deal with only the natural numbers and the integers since these are the oldest and most easily understood numbers. The real beauty of number theory is that the definitions and mathematical results are often very simple, utilizing only the basic arithmetic operations of addition, subtraction, multiplication, and subtraction. However, many of the results in number theory have proofs that are extremely long and complex, such as Wiles and Taylor's proof of Fermat's Last Theorem.

4.3.1 Odd and Even Numbers

The first type of specialized numbers that will be studied are the "odd" and "even" numbers. As early as 300 B.C., the Greek mathematician Euclid noted that natural numbers were composed of both the even and odd numbers. Moreover, the concept of odd and even numbers is easily understood and in fact, is one of the first number theoretic topics that is taught to grade-school students. Thus, a good reason to begin the discussion of number theory with odd and even numbers is that they are not a new to students studying mathematics. The definitions of even and odd integers are given below.

Definition 4.3.1: An integer n is said to be an *even* integer if and only if n can be written as $n = 2k$ for some integer k.

Definition 4.3.2: An integer n is said to be an *odd* integer if and only if n can be written as $n = 2k + 1$ for some integer k.

Note that an alternative definition of an odd number is sometimes stated with "$n = 2k-1$ for some integer k" in place of "$n = 2k+1$ for some integer k." These definitions are equivalent since $2k-1 = 2k+1-2 = 2(k-1)+1 = 2k'+1$ and k' is an integer. Furthermore, proving that an integer z is even requires only showing that $z = 2k$ for some $k \in \mathbb{Z}$ (i.e., an existence proof). Similarly, proving that an integer z is odd requires showing that $z = 2k + 1$ for $k \in \mathbb{Z}$.

Example 4.3.1: The number 3 is odd because $3=2(1)+1$, and the number -6 is even because $-6 = 2(-3)$. However, the real number 3.45 is neither even nor odd since it is not an integer.

Note that integers are made up of even integers and odd integers. Let the set of even integers be denoted by $\mathbb{Z}_E = \{0, \pm 2, \pm 4, \ldots\}$ and the set of odd integers by $\mathbb{Z}_O = \{\pm 1, \pm 3, \pm 5, \ldots\}$. Several elementary theorems concerning odd and even numbers are given below.

Theorem 4.3.1: Let $\mathbb{Z}_E$ be the set of even integers. Then

(i) $\mathbb{Z}_E$ is closed under addition.

(ii) $\mathbb{Z}_E$ is closed under multiplication.

Proof (Closure Proof):

Proof of part (i): Let $a, b \in \mathbb{Z}_E$ be ABF. Then, there exist integers k and j such that $a = 2k$ and $b = 2j$.

Consider $a + b$

$$a + b = 2k + 2j = 2(k + j) = 2l$$

where $l = k + j$ and $l \in \mathbb{Z}$ since $\mathbb{Z}$ is closed under addition. Hence, $a + b$ is even, and therefore $\mathbb{Z}_E$ is closed under addition.

Proof of part (ii): The proof of (ii) is left as an exercise.

■

Note that $\mathbb{Z}_O$ is not closed under addition since $3 + 5 = 8$, which is even; however, the following theorem shows that $\mathbb{Z}_O$ is closed under multiplication.

Theorem 4.3.2: If m and n are odd integers, then nm is odd.

> ***Proof:*** Let m and n be ABF odd integers. Then, $n = 2k + 1$ for some integer k and $m = 2j + 1$ for some integer j.
>
> Consider nm
>
> $$nm = (2k + 1)(2j + 1) = 4kj + 2k + 2j + 1$$
>
> $$= 2(2kj + k + j) + 1 = 2l + 1$$
>
> where $l = 2kj + k + j$, which is an integer since $\mathbb{Z}$ is closed under multiplication and addition.
>
> Thus, $nm = 2l + 1$ and hence, n is an odd number. Therefore nm is odd whenever m and n are odd.
>
> ∎

Theorem 4.3.3: If n is an odd integer, then so is n^2.

> ***Proof:*** Let n be an odd integer. Then, by Definition 4.2.2, $\exists\ k \in \mathbb{Z}$ such that $n = 2k + 1$. Now consider n^2.
>
> $$n^2 = (2k + 1)^2 = 4k^2 + 4k + 1 = 2(2k^2 + 2k) + 1$$
>
> $$= 2k' + 1, \quad \text{where } k' = 2k^2 + 2k$$
>
> Furthermore, $2k^2 + 2k \in \mathbb{Z}$ since $\mathbb{Z}$ is closed under multiplication and addition, and thus n^2 is odd.
>
> Therefore, n^2 is odd whenever n is odd.
>
> ∎

Note that Theorem 4.3.3 could have been stated as a corollary of Theorem 4.3.2 since $n^2 = n \cdot n$. Specifically, since n^2 is the product of two odd numbers (i.e., n and n), it follows from Theorem 4.3.3 that n^2 is also odd. Moreover, the following theorem shows that whenever n^2 is an odd number, then so is n.

Theorem 4.3.4: Let n be an integer. If n^2 is odd, then so is n.

Proof (by Contrapositive): Note that the contrapositive of this theorem is "If n is even, then so is n^2," and since the contrapositive of this theorem was proved in Example 3.3.2, this theorem follows by the method of contrapositive.

■

Note that Theorems 4.3.4 and 4.3.5 could now be stated as the following biconditional theorem:

Theorem: Let n be an integer. Then n is odd if and only if n^2 is odd.

An analogous biconditional theorem for even integers is also stated below:

Theorem: Let n be an integer. Then n is even if and only if n^2 is even.

The next three theorems state results concerning sums of integers. In particular, the first theorem states that the sum of two consecutive integers is odd, the second concerns the sum of the first n odd natural numbers, and the last theorem concerns the sum of the first n even natural numbers.

Theorem 4.3.5: The sum of two consecutive integers is an odd integer.

Proof: Let n and $n+1$ be ABF consecutive integers. Without loss of generality, assume that n is even.

Then, $\exists\ k \in \mathbb{Z}$ such that $n = 2k$ and $n + 1 = 2k + 1$. Now

$$n + (n + 1) = 2k + (2k + 1) = 4k + 1 = 2k' + 1$$

where $k' = 4k$. Furthermore, $k' \in \mathbb{Z}$ since $\mathbb{Z}$ is closed under multiplication, and hence $n + (n + 1)$ is odd.

Therefore, the sum of two consecutive integers is an odd integer.

■

Theorem 4.3.6: $\sum_{j=1}^{n}(2j-1) = n^2,\ \forall\ n \in \mathbb{N}$.

Proof: (by Induction): Let $\mathcal{P}_n := \sum_{j=1}^{n}(2j-1) = n^2$.

For $n = 1$, $\sum_{j=1}^{1}(2j-1) = 2(1) - 1 = 1 = 1^2$. Thus, $\mathcal{P}_1$ is true.

Now, assume that $\mathcal{P}_k$ is true for some arbitrary but fixed (ABF) natural number k. This means that $\sum_{j=1}^{k}(2j-1) = k^2$. Now, if $\mathcal{P}_{k+1}$ is true, then

$$\sum_{j=1}^{k+1}(2j-1) = (k+1)^2$$

Consider $\sum_{j=1}^{k+1}(2j-1)$.

$$\sum_{j=1}^{k+1}(2j-1) = \sum_{j=1}^{k}(2j-1) + 2(k+1) - 1 = \underbrace{k^2}_{\text{By } \mathcal{P}_k} + 2k + 1$$

$$= (k+1)^2$$

Thus, $\mathcal{P}_{k+1}$ is true when $\mathcal{P}_k$ is true, and therefore, $\mathcal{P}_n$ is true $\forall\, n \in \mathbb{N}$. Hence, $\sum_{j=1}^{n}(2j-1) = n^2$, $\forall\, n \in \mathbb{N}$.

■

For example, the sum of the first 10 odd natural numbers is

$$\sum_{i=1}^{10}(2i-1) = (1 + 3 + 5 + \cdots + 19 = 10^2 = 100$$

and the sum of the first 1231 odd numbers is $1231^2 = 1,515,361$.

Theorem 4.3.7: $\sum_{j=1}^{n} 2j = n(n+1)$, $\forall\, n \in \mathbb{N}$.

Proof: Let $n \in \mathbb{N}$ and consider $\sum_{j=1}^{n} 2j$. Using the result of Example 3.4.1, the sum of the first $2n$ natural numbers is

$$\sum_{j=1}^{2n} j = \frac{2n(2n+1)}{2} = 2n^2 + n$$

Now, since Theorem 4.3.6 states that the sum of the first n odd natural numbers is n^2, it follows that

$$2n^2 + n = \sum_{j=1}^{2n} j = \sum_{j=1}^{n} (2j - 1) + \sum_{j=1}^{n} 2j$$

$$= n^2 + \sum_{j=1}^{n} 2j$$

Thus

$$\sum_{j=1}^{n} 2j = 2n^2 + n - n^2 = n^2 + n = n(n + 1)$$

■

Thus, from Theorem 4.3.7 it follows that the sum of the first 25 even numbers

$$2 + 4 + 6 + \cdots + 50 = 50(51) = 2550$$

and the sum of the first 100 even numbers is $100(101) = 10,100$.

Finally, note that every even number is a multiple of 2. In the next section, multiples of numbers other than 2 will be considered.

4.3.2 Divisibility

Since the even numbers are simply integers that are multiples of 2, it is also true that the even integers consist of all those integers that are evenly divisible by 2. A natural extension from studying the multiples of 2 is to study multiples of other integers. For example, one might consider the multiples of 3 or 7 and their properties. Furthermore, the ideas of multiplicity and divisibility play a key role in the study of composite and prime numbers, which will be discussed in Section 4.3.3. The definition of the *divisibility* of an integer b by an integer a is given below.

Definition 4.3.3: An integer a is said to *divide* an integer b, denoted by $a|b$, if and only if $b = ak$ for some integer k. When a divides b, a is called a *divisor* of b and b is said to be a *multiple* of a.

For example, $5|315$ since $315 = 5(63)$ and $3 \not| 542$ since there is no integer k such that $542 = 3k$. Thus, 5 is a divisor of 315 and 315 is a multiple of

5, however, 3 is not a divisor of 542 and 542 is not a multiple of 3. Also, a divides b can also be stated as b is *divisible* by a.

Example 4.3.2: Show that each of the following numbers is divisible by 7:

a. 441

b. 1057

c. −784

Solutions:

a. 441 is divisible by 7 since $441 = 7 \cdot 63$

b. 1057 is divisible by 7 since $1057 = 7 \cdot 151$

c. −784 is divisible by 7 since $-784 = 7 \cdot (-112)$

Now, when $a|b$, this simply means that b is a multiple of a and thus, to prove that $a|b$ simply requires showing that b is a multiple of a (i.e., $b = ak$ for some $k \in \mathbb{Z}$). On the other hand, when $a \nmid b$ it follows that a is not a divisor of b and that b is not a multiple of a. A more precise meaning for $a \nmid b$ will follow from Theorem 4.3.8, the *Division Algorithm*, which is given without proof; a proof of the *Division Algorithm* can be found in *Modern Algebra* by Jimmie and Linda Gilbert (1996).

Theorem 4.3.8 (The Division Algorithm): If $a, b \in \mathbb{Z}$ and $a \neq 0$, then there exist unique integers q and r such that $b = qa + r$ where $0 \leq r < a$.

Thus, if a does not divide b (i.e., $a \nmid b$), then it follows from the Division Algorithm that $b = qa + r$ for some integers q and r where $0 < r < a$. For example, if $3 \nmid b$, then b is of either form $3k + 1$ or $3k + 2$. Similarly, if $7 \nmid b$, then b has one of the following forms: $7k + 1, 7k + 2, 7k + 3, 7k + 4, 7k + 5$, or $7k + 6$.

Now, several theorems concerning divisibility results will be stated and proved. Note that Theorem 4.3.9 is analogous to the theorem proved in Chapter 3 showing that if n is even, then so is n^2.

Theorem 4.3.9: Let $a, b \in \mathbb{Z}$. If $a|b$, then $a|b^2$.

Proof: Let $a, b \in \mathbb{Z}$ and suppose that $a|b$.

Then, $b = ak$ for some integer k. Now, consider b^2

$$b^2 = (ak)^2 = a^2k^2 = a(ak^2) = ak'$$

where $k' = ak^2 \in \mathbb{Z}$ since $\mathbb{Z}$ is closed under multiplication. Hence, b^2 is a multiple of a, and therefore, $a|b^2$ whenever $a|b$.

■

The following theorem shows that when a divides both b and c, then it is also true that $a|(b+c)$.

Theorem 4.3.10: Let $a, b, c \in \mathbb{Z}$. If $a|b$ and $a|c$, then $a|(b+c)$.

Proof: Let $a, b, c \in \mathbb{Z}$ and suppose $a|b$ and $a|c$.

Then, $b = ak$ for some integer k and $c = aj$ for some integer j. Consider $b + c$:

$$b + c = ak + aj = a(k + j) = ak', \quad \text{where } k' = k + j.$$

Furthermore, $k + j \in \mathbb{Z}$ since $\mathbb{Z}$ is closed under addition, and hence $b + c$ is a multiple of a.

Therefore, if $a|b$ and $a|c$, then $a|(b+c)$.

■

A further generalization of Theorem 4.3.10 is given below. In particular, Theorem 4.3.11 extends Theorem 4.3.10 to more general linear combinations.

Theorem 4.3.11: Let $a, b, c \in \mathbb{Z}$. If $a|b$ and $a|c$, then $a|(bx + cy)$, $\forall\ x, y \in \mathbb{Z}$.

Proof: The proof of theorem 4.3.11 is left as an exercise.

■

An important corollary to Theorem 4.3.11 is the special case where $x = 1$ and $y = -1$; that is, if $a|b$ and $a|c$, then $a|(b - c)$. This result is stated in the following corollary.

Corollary to Theorem 4.3.11: Let $a, b, c \in \mathbb{Z}$. If $a|b$ and $a|c$, then $a|(b-c)$.

Proof: The corollary to Theorem 4.3.11 follows directly from Theorem 4.3.11 with $x = 1$ and $y = -1$.

■

Note that had Theorem 4.3.11 preceded Theorem 4.3.10, the latter could have been stated as a corollary to Theorem 4.3.11 since it is the special case of Theorem 4.3.11 with $x = 1$ and $y = 1$. The following example shows that when $a|(b+c)$, it is not necessarily the case that $a|b$ and $a|c$.

Example 4.3.3: Consider the following conjecture:

Conjecture: Let a, b, and c be integers. If $a|(b+c)$, then $a|b$ and $a|c$.

Solution: A counterexample to this conjecture is $6|(13+5)$ since $18 = 6 \times 3$, but $6 \nmid 13$ and $6 \nmid 5$. Thus, this conjecture is false.

Example 4.3.4: Prove or disprove the following conjecture:

Conjecture: Let a, b, and c be integers. If $a|bc$, then $a|b$ or $a|c$.

Solution: A counterexample to this conjecture is $6|(9 \times 4)$ since $9 \times 4 = 36 = 6 \times 6$, but $6 \nmid 9$ and $6 \nmid 4$. Thus, this conjecture is false.

Thus, the previous two examples show that knowledge of the fact that a divides $b + c$ does not necessarily imply that $a|b$ and $a|c$. However, the following theorem shows that when a divides both b and $b + c$, then it follows that a must divide c, also. For example, clearly $3|39$ and $3|27$ and therefore, since $39 = 27 + 12$, it follows that $3|12$.

Theorem 4.3.12: Let $a, b, c \in \mathbb{Z}$. If $a|b$ and $a|(b+c)$, then $a|c$.

Proof: Let $a, b, c \in \mathbb{Z}$ and suppose $a|b$ and $a|(b+c)$.

Then, there exist integers k, j such that $b = ak$ and $b + c = aj$.

Now, since $b + c = aj$, it follows that $c = aj - b$. Thus

$$c = aj - b = aj - ak = a(j - k) = al$$

where $l = j - k \in \mathbb{Z}$ since $\mathbb{Z}$ is closed under subtraction. Thus, c is a multiple of a and therefore $a|c$ whenever $a|b$ and $a|(b+c)$.

■

Theorem 4.3.13: Let $a, b \in \mathbb{Z}$. If $a|b$, then $a|b^n$, $\forall\, n \in \mathbb{N}$.

Proof: The proof of Theorem 4.3.13 is left as an exercise.

■

Recall that Theorem 4.3.9 states that "If $a|b$, then $a|b^2$." This theorem is a direct result of Theorem 4.3.13 and thus could be stated as a corollary to Theorem 4.3.13, also.

Corollary to Theorem 4.3.13: Let $a, b \in \mathbb{Z}$. If $a|b$, then $a|b^2$, $\forall\ k \in \mathbb{N}$.

Proof: This result follows directly from Theorem 4.3.13 with $k = 2$.

■

Now, suppose that $a|b$; then it is also true that $a|b^2$. However, it is not necessarily true that if $a|b^2$, then a will also divide b. A counterexample to $a|b^2$ implying $a|b$ is given in the following example.

Example 4.3.5: Disprove the following conjecture:

Conjecture: Let $a, b \in \mathbb{Z}$. If $a|b^2$, then $a|b$.

Solution: A counterexample to this conjecture is $4|6^2$ since $6^2 = 36 = 9 \times 4$, but $4 \not| 6$. Thus, this conjecture is false.

The previous example shows that it is not always true that when a divides b^2 it also follows that a will divide b. In other words, there is at least one a such that when a divides b^2, it is not true that a will divide b as the counterexample illustrates. However, it was shown in Chapter 3 that when $2|b^2$, it does follow that $2|b$. The following theorem shows that when $3|b^2$, it follows that $3|b$, also.

Theorem 4.3.14: Let $b \in \mathbb{Z}$. If $3|b^2$, then $3|b$.

Proof (by Contrapositive): Note that the contrapositive of this theorem is "If $3 \not| b$, then $3 \not| b^2$."

Let $b \in \mathbb{Z}$ and suppose that $3 \not| b$. Then, according to the Division Algorithm

$$b = \begin{cases} 3k + 1 \\ \text{or} \\ 3k + 2 \end{cases}$$

for some $k \in \mathbb{Z}$.

Case 1: Suppose that $b = 3k + 1$. Then

$$b^2 = (3k + 1)^2 = 9k^2 + 6k + 1 = 3(3k^2 + 2k) + 1 = 3k' + 1$$

where $k' = 3k^2 + 2k$. Furthermore, $3k^2 + 2k \in \mathbb{Z}$ since $\mathbb{Z}$ is closed under multiplication and addition. Thus, b^2 is of the form $3k+1$ and hence, by the Division Algorithm it follows that b^2 is not divisible by 3.

Case 2: Suppose that $b = 3k + 2$. Then

$$b^2 = (3k+2)^2 = 9k^2 + 12k + 4 = 3(3k^2 + 4k + 1) + 1 = 3k' + 1$$

where $k' = 3k^2 + 4k + 1$. Furthermore, $3k^2 + 2k + 1 \in \mathbb{Z}$ since $\mathbb{Z}$ is closed under multiplication and addition. Thus, b^2 is of the form $3k + 1$ and hence, by the Division Algorithm it follows that b^2 is not divisible by 3.

Thus, in both cases it follows that when $3 \not| b$, then $3 \not| b^2$, either. Therefore, when $3|b^2$, it follows that $3|b$.

■

Now, there is a quick way to test whether an integer is divisible by 3, based on the sum of the digits making up the integer. In particular, when the sum of digits making up an integer is divisible by 3, then so is the integer. Note that a positive integer a (i.e., a natural number) may be expressed in base-10 form as

$$a = \sum_{k=0}^{n} c_k 10^k$$

For example, the integer $a = 1217$ can be written as

$$1257 = 1 \cdot 10^3 + 2 \cdot 10^2 + 5 \cdot 10^1 + 7 \cdot 10^0$$

Similarly, a negative integer a can be written as

$$-(-a) = -\sum_{k=0}^{n} c_k 10^k,$$

where $-a = \sum_{k=0}^{n} c_k 10^k$ is a positive integer. The base-10 representation of an integer is especially useful in proving theorems such as Theorem 4.3.15 on divisibility tests for the numbers 3 and 9. In particular, Theorem 4.3.15 states that a natural number a is divisible by 3(9) if and only if the sum of the digits making up the base-10 representation of a is divisible by 3(9). For example, 3 divides 177 since $3|(1+7+7)$ and 9 divides 12,609 since $9|(1+2+6+0+0)$.

Theorem 4.3.15: Let $a \in \mathbb{N}$ have base-10 representation $\sum_{k=0}^{n} c_k 10^k$, and let $S = \sum_{k=0}^{n} c_k$. Then

(i) $3|a$ if and only if $3|S$.

(ii) $9|a$ if and only if $9|S$.

Before proceeding with the proof of Theorem 4.3.15, the following two lemmas must be proved since the proof of Theorem 4.3.15 is based on the results of these lemmas.

Lemma 4.3.1: $3|10^{n-1} - 1$, for $n \in \mathbb{N}$.

Proof (by Induction): Let $\mathcal{P}_n := 3|10^{n-1} - 1$.

For $n = 1$, it follows that $10^0 - 1 = 0$, which is divisible by 3. Thus $\mathcal{P}_1$ is true. Also, for $n = 2$, $10^1 - 1 = 9$, which is divisible by 3, and thus $\mathcal{P}_2$ is true.

Assume that $\mathcal{P}_k$ is true for some ABF $k \in \mathbb{N}$; that is, $3|10^{k-1} - 1$ or $10^{k-1} - 1 = 3j$ for some $j \in \mathbb{Z}$. Now, if $\mathcal{P}_{k+1}$ is true, then 3 will divide $10^k - 1$.

Consider $10^k - 1$

$$
\begin{aligned}
10^k - 1 &= 10^k - 1 + \underbrace{10^{k-1} - 10^{k-1}}_{0} \\
&= (10^k - 10^{k-1}) + (10^{k-1} - 1) \\
&= 10^{k-1}(10 - 1) + (10^{k-1} - 1) \\
&= 9 \cdot 10^{k-1} + (10^{k-1} - 1) \\
&= 3(3 \cdot 10^{k-1}) + \underbrace{3j}_{\text{By } \mathcal{P}_k} = 3m
\end{aligned}
$$

where $m = 3 \cdot 10^{k-1} + j$, which is in $\mathbb{Z}$ since $\mathbb{Z}$ is closed under multiplication and addition. Thus, $10^k - 1$ is a multiple of 3, and hence $3|10^k - 1$.

Thus, $\mathcal{P}_{k+1}$ is true whenever $\mathcal{P}_k$ is true, and therefore $3|10^n - 1$ $\forall\, n \in \mathbb{N}$.

■

Lemma 4.3.2: $9|10^{n-1} - 1$, for $n \in \mathbb{N}$.

Proof: The proof of Lemma 4.2.2 is similar to the proof of Lemma 4.2.1 with 9 substituted in place of 3.

■

The proof of theorem 4.3.15 is now presented.

Proof of Theorem 4.3.15 (Biconditional Proof): Let $a \in \mathbb{N}$ have base-10 representation $\sum_{k=0}^{n} c_k 10^k$, and let $S = \sum_{k=0}^{n} c_k$.

Proof of part (i): To prove part (i) of this theorem, it must be shown that $3|a \rightarrow 3|S$ and $3|S \rightarrow 3|a$.

First, suppose that $3|a$ and consider a.

$$a = \sum_{k=0}^{n} c_k 10^k = \sum_{k=0}^{n} c_k(10^k + \underbrace{(-1+1)}_{0})$$

$$= \sum_{k=0}^{n} c_k(10^k - 1) + \sum_{k=0}^{n} c_k = \sum_{k=0}^{n} c_k(10^k - 1) + S$$

Now, by Lemma 4.3.1 it follows that $3|c_k(10^k - 1)$ for $k = 0, 1, 2, \ldots, n$ and thus $3|\sum_{k=0}^{n} c_k(10^k - 1)$. Furthermore, since $3|\sum_{k=0}^{n} c_k(10^k - 1)$ and $3|a$, it follows from Theorem 4.3.12 that $3|S$.

Conversely, suppose that $3|S$. Consider a:

$$a = \sum_{k=0}^{n} c_k 10^k = \sum_{k=0}^{n} c_k(10^k - 1) + S$$

By Lemma 4.3.1 it follows that $3|c_k(10^k - 1)$ for $k = 0, 1, 2, \ldots, n$ and thus, $3|\sum_{k=0}^{n} c_k(10^k - 1)$. Now, $3|S$ and $3|\sum_{k=0}^{n} c_k(10^k - 1)$. Thus, since $a = \sum_{k=0}^{n} c_k(10^k - 1) + S$, it follows from Theorem 4.3.10 that $3|a$. Therefore, $3|a$ if and only if $3|S$.

Proof of part (ii): The proof of part (ii) is left as an exercise.

■

Example 4.3.6: Determine whether 3 and 9 divide each of the following numbers:

a. 2137
b. 99,089
c. 314,299,806
d. 20,314,299,807

Solutions:

a. Neither 3 nor 9 divides 2137 since neither 3 nor 9 divides $S = 13$.
b. Neither 3 nor 9 divides 99,089 since neither 3 nor 9 divides $S = 35$.
c. 3 divides 314,299,806 since 3 divides $S = 42$. However, 9 does not divide 314,299,806 since $9 \nmid 42$.
d. Both 3 and 9 divide 20,314,299,807 since both 3 and 9 divide $S = 2+0+3+1+4+2+9+9+8+0+7 = 45 = 3\cdot 15 = 9\cdot 5$.

4.3.3 Prime Numbers

As the study of numbers and the divisibility properties of numbers progressed, it soon became apparent that some natural numbers had the special property of being indivisible by any of their predecessors other than 1. For example, 11 is not divisible by any of its predecessors $2, 3, 4, \ldots, 10$. These special numbers are called *prime numbers* and have fascinated mathematicians since at least the publication of Euclid's *Elements*, which was published around 300 B.C. The definition of a prime number is given below.

Definition 4.3.4: A natural number $p \geq 2$ is said to be a *prime number* if and only if the only divisors of p are 1 and p. A natural number $c \geq 2$ that is not a prime number is called a *composite number*.

Let $\mathcal{P}$ be the collection of all prime numbers. Then $\mathcal{P}$ is a subset of the natural numbers, and in fact, the natural numbers consist of three types of numbers: the number 1, the prime numbers, and the composite numbers. The first 20 primes are:

$$2, 3, 5, 7, 11, 13, 17, 19, 23, 29, 31, 37, 41, 43, 47, 53, 59, 61, 67, 71$$

A table containing all of the prime numbers less than 10,000 is given in *Elementary Introduction to Number Theory* (Long 1972), and as of March 2006, the largest known prime number was $2^{30,402,457}-1$. The base-10 representation of this prime number requires 9,152,052 digits to be written out completely.

Now, if a natural number is not a prime number, then it is either the number 1 or it is a composite number. Note that since a composite number is not a prime number, it must be divisible by at least one of its predecessors. Furthermore, it also follows that a composite number must be divisible by at least one of its prime predecessors. For example, 15 is a composite number and is divisible by the prime numbers 3 and 5.

Definition 4.3.5: A prime number p is said to be a *prime factor* of a composite number c if and only if $p|c$.

For example, 3 is a prime factor of 12,345,081 (3 divides the sum of the digits of this number). Note that a prime number only has 2 factors, 1 and p, while a composite number may have many prime factors. For example, the composite number $420 = 2^2 \times 3 \times 5 \times 7$, so 420 has the 4 prime factors 2, 3, 5, and 7. Greek mathematicians were fascinated by the prime numbers, and the Greek mathematician Euclid stated, with proofs, several important theorems concerning the prime numbers in his book *Elements*. In particular, the next two theorems were first stated in Euclid's *Elements*.

Theorem 4.3.16 (Euclid's First Theorem): Let $a, b \in \mathbb{N}$. If p is a prime number and $p|ab$, then $p|a$ or $p|b$.

The following theorem, the *Fundamental Theorem of Arithmetic*, which is also a corollary of Euclid's First Theorem, shows that there is one and only one prime factorization of a composite number.

Theorem 4.3.17 (The Fundamental Theorem of Arithmetic): Every natural number $n \geq 2$ is either a prime number or the product of prime numbers, and the product is unique up to the order in which the prime factors appear.

The unique representation of a composite number c as the product of primes is known as the *prime factorization* of c. For example, the unique prime factorization of 180 is $2^2 \cdot 3^2 \cdot 5$. However, the Fundamental Theorem of Arithmetic does not provide any direct help in determining whether or not a number is a prime or a composite number. For example, the Fundamental Theorem of Arithmetic does not answer the question "Is $n = 1,578,301$ a prime number or a composite number?" In general, it is difficult to determine whether a very large number is a prime number or a composite number. However, the following theorem does provide some help in determining whether a number is a prime number by restricting the set of possible prime factors of the number. In particular, the following theorem shows that if a natural number n is composite, then it must have a prime factor that is less than or equal to $\sqrt{n}$.

Theorem 4.3.18: If c is a composite number, then c has at least one prime factor that is less than or equal to $\sqrt{c}$.

> ***Proof (by Contradiction):*** Let c be a composite number, and suppose that c has no prime factors less than or equal to $\sqrt{c}$.
>
> Now, since c is a composite number it follows that c is the product of at least two prime numbers. Let p be the smallest prime number greater than $\sqrt{c}$. Now, the smallest product of two prime numbers greater than $\sqrt{c}$ is p^2, but $p^2 > c$. Thus, no prime greater than or equal to p can be a prime factor of c. Thus, c has no prime factors less than or equal to $\sqrt{c}$, nor does c have any prime factors greater than $\sqrt{c}$. Hence, c is a prime number, contradicting the fact that c is a composite number.
>
> Therefore, c must have at least one prime factor less than or equal to $\sqrt{c}$.
>
> ■

Example 4.3.7: Use Theorem 4.3.18 to determine whether each of the following numbers is a prime or a composite number:

a. 1217

b. 9983

Solutions:

a. First, $\sqrt{1217} = 34.885$. Now, if 1217 is not divisible by any prime less than 34, then 1217 is a prime number. Thus, 1217 will be a prime number if and only if it is not divisible by the prime numbers

$$2, 3, 5, 7, 11, 13, 17, 19, 23, 29, 31$$

$1217 = 2 \times 608 + 1$	$1217 = 3 \times 405 + 2$	$1217 = 5 \times 243 + 2$
$1217 = 7 \times 173 + 6$	$1217 = 11 \times 110 + 7$	$1217 = 13 \times 93 + 8$
$1217 = 17 \times 71 + 10$	$1217 = 19 \times 64 + 1$	$1217 = 23 \times 52 + 21$
$1217 = 29 \times 41 + 28$	$1217 = 31 \times 39 + 8$	

Thus, since 1217 is not divisible by any of the prime numbers less than $\sqrt{1217}$, it follows that 1217 is a prime number.

b. The solution to part (b) is left as an exercise.

An algorithm, based on the result of Theorem 4.3.18, was proposed by Eratosthenes (276–194 B.C.) for finding all the prime numbers less than a particular natural number n. This algorithm is called the "Sieve of Eratosthenes" after its founder. Eratosthenes' algorithm for finding prime numbers is given below.

Sieve of Eratosthenes: Begin with a sequential list of integers from 2 to the largest number n to be studied.

1. Cross out all multiples of 2 (every second number in the list).
2. Determine the smallest remaining number in the list, which is 3. Now, cross out all multiples of 3 (every third number in the list).
3. Determine the smallest remaining number in the list, which is 5. Now, cross out all multiples of 5 (every fifth number in the list).
4. Repeat the process outlined in steps 2 and 3 until all multiples of the smallest integer less than or equal to $\sqrt{n}$ are crossed out.
5. The numbers remaining in the list from 2 to n are the prime numbers less than or equal to n.

Example 4.3.8: Use the Sieve of Eratosthenes to determine the prime numbers less than 116.

Solution: First list the numbers less than 116:

2	3	4	5	6	7	8	9	10	11	12	13	14	15	16	17	18	19	20
21	22	23	24	25	26	27	28	29	39	31	32	33	34	35	36	37	38	39
40	41	42	43	44	45	46	47	48	49	50	51	52	53	54	55	56	57	58
59	60	61	62	63	64	65	66	67	68	69	70	71	72	73	74	75	76	77
78	79	80	81	82	83	84	85	86	87	88	89	90	91	92	93	94	95	96
97	98	99	100	101	102	103	104	105	106	107	108	109	110	111	112	113	114	115

Now, after crossing out the multiples of all the prime numbers less than $\sqrt{116}$ (i.e., 2, 3, 5, and 7), the following prime numbers remain:

2 3 5 7 11 13 17 19

23 29 31 37

41 43 47 53

59 61 67 71 73

79 83 89

97 101 103 107 109 113

In general, the problem of checking for the primality of a large number such as 1,578,301 is a computationally difficult problem, since it involves testing for divisibility by the prime numbers less than $\sqrt{1,578,301}$. In this example, divisibility by all the 214 prime numbers less than $\sqrt{1,578,301}$ must be considered. If any one of these 214 prime numbers divides 1,578,301, then it is a composite number. On the other hand, if none of these 214 prime numbers divides 1,578,301, then it is a prime number. It turns out that 1,578,301 is not a prime number because $1,578,301 = 653 \cdot 2417$

Other interesting questions concerning the prime numbers are "What is the largest prime number?" and "How many prime numbers are there?" Euclid's *Second Theorem* in Book VII of *Elements* answers both of these questions. In particular, Euclid shows that there are an infinite number of primes using a simple yet elegant mathematical proof. Thus, since there are infinitely many prime numbers, there is no largest prime number.

Theorem 4.3.19 (Euclid's Second Theorem:) There are infinitely many prime numbers.

> ***Proof (by Contradiction):*** Let $\mathcal{P}$ be the collection of all prime numbers, and suppose that $\mathcal{P}$ does not contain an infinite number of elements. Then, $\mathcal{P}$ contains a finite number of elements. WLOG let the elements of $\mathcal{P}$ be $p_1, p_2, p_3, \ldots, p_n$.
>
> Now, let $N = p_1 \times p_2 \times p_3 \cdots \times p_n + 1$. Then, there are two possibilities for N, namely, N is either a prime number or a composite number.
>
> **Case 1:** Suppose that N is a prime number. Then, since N is prime and N greater than any of the primes in $\mathcal{P}$, $\mathcal{P}$ does not contain all the prime numbers. However, this contradicts the assumption that $\mathcal{P}$ does contain all the prime numbers. Thus, in this case $\mathcal{P}$ must be infinite.
>
> **Case 2:** Suppose that N is a composite number. Since N is composite, it follows that N must have at least one prime factor in $\mathcal{P}$. Now, clearly
>
> $$N = p_1 \times p_2 \times p_3 \cdots \times p_n + 1$$

is not divisible by any of the primes in $\mathcal{P}$, since $N = qp_i + 1$ for every $p_i \in \mathcal{P}$.

Therefore, there must be some other prime number, say, $p_{n+1} \notin \mathcal{P}$ that is a factor of N. However, $\mathcal{P}$ contains all the prime numbers, but $p_{n+1} \notin \mathcal{P}$. This is a contradiction. Hence, in this case, $\mathcal{P}$ must be infinite, also.

Therefore in either case, $\mathcal{P}$ is infinite and hence, there are infinitely many prime numbers.

■

Many mathematicians, including Fermat and Marin Mersenne (1588–1648), have proposed algorithms for the creation of prime numbers. Fermat conjectured that $2^{2^n} + 1$ is prime for every whole number n; primes of the form $p = 2^{2^n} + 1$ are called *Fermat primes*. In particular

$$
\begin{aligned}
2^{2^1} + 1 &= 4 + 1 = 5 \quad \text{which is prime} \\
2^{2^2} + 1 &= 16 + 1 = 17 \quad \text{which is prime} \\
2^{2^3} + 1 &= 256 + 1 = 257 \quad \text{which is prime} \\
2^{2^4} + 1 &= 65,536 + 1 = 65,537 \quad \text{which is prime}
\end{aligned}
$$

Unfortunately for Fermat, in 1731 Leonhard Euler showed that for $n = 5$, $2^{2^5} + 1 = 4,294,967,297$ is divisible by 641, and hence $2^{2^5} + 1$ is not a prime number. Mersenne conjectured that there are infinitely many primes of the form $2^p - 1$, where p is a prime number; prime numbers of this form, such as $2^2 - 1 = 3$ and $2^3 - 1 = 7$, are called *Mersenne primes*. Mersenne's conjecture has never been proved or disproved; however, whenever a new prime number is found, it invariably turns out to be a Mersenne prime.

Another classic problem concerning prime numbers is one that deals with the frequency of the prime numbers. Let $\pi(n)$ be the function that counts the number of prime numbers less than or equal to a natural number n. Then

$$\pi(n) \; = \; \text{the number of primes less than or equal to } n$$

For example, $\pi(200) = 46$, $\pi(300) = 62$, and $\pi(400) = 78$. Table 4.3.1 illustrates the frequency of the prime numbers for $n = 10^2$, 10^3, 10^4, 10^5, 10^6 and 10^{12}.

Table 4.3.1 Frequencies of the Prime Numbers

N	No. of Primes	Percentage of Primes
10^2	25	25%
10^3	168	16.8%
10^4	1,229	12.3%
10^5	9,592	9.6%
10^6	78,498	7.8%
10^{12}	37,607,912,018	3.8%

Examining Table 4.3.1, it appears that the frequency of the prime numbers is decreasing as n gets larger. Gauss was one of the first mathematicians to note that $\pi(n)/n$ appeared to approach the value $1/\ln n$ as n grew larger; however, he did not provide a proof of this result. Jacques Hadamard (1865–1963) and Charles de la Vallée Poussin (1866–1962) independently discovered proofs of the following theorem in 1896. This theorem is known as the *Prime Number Theorem* and is provided without proof. For more information on the Prime Number Theorem, see *The Mathematical Universe* by William Dunham (1997) or *Prime Obsession: Bernhard Riemann and the Greatest Unsolved Problem in Mathematics* by John Derbyshire (2003).

Theorem 4.3.20 (The Prime Number Theorem): If $\pi(n)$ is the number of prime numbers less than or equal to n, then

$$\lim_{n\to\infty} \frac{\pi(n)}{n} \bigg/ \frac{1}{\ln n} = 1.$$

Finally, a classic and still unproven conjecture concerns pairs of prime numbers of the form p and $p+2$. Pairs of primes of this nature are called *twin primes*. For example, the first five pairs of twin primes are (3,5), (5,7), (11,13), (17,19), and (29,31). It has been conjectured that there are infinitely many pairs of twin primes; however, this conjecture, which many mathematicians believe is true, has never been proved. The final theorem of this section provides an interesting result concerning the natural number lying between twin prime numbers.

Theorem 4.3.21: If $p \geq 5$ and $(p, p+2)$ are twin primes, then $6|(p+1)$.

Proof: The proof of Theorem 4.3.21 is left as an exercise.

■

4.3.4 Recursively Defined Numbers

This last section of the chapter deals with a special way of generating a sequence of numbers. In particular, this section deals with sequences of recursively defined numbers. Many interesting sequences of numbers arise from a recursive formulation. In fact, real-world applications where a recursively defined sequence of numbers are used include the iterative solution of an nonlinear equation, birth–death models, and mortgage payments. The definitions of a *sequence* and a *recursive sequence* are given below.

Definition 4.3.6: A *sequence* of real numbers is a function whose domain is $\mathbb{N}$.

Definition 4.3.7: A sequence of numbers is said to be a *recursive sequence* or *recursively defined sequence* when each element of the sequence is based on the previous elements in the sequence.

For example, the sequence defined by $a_1 = 2$ and $a_n = a_{n-1}^2 - 1$, for $n \in \mathbb{N}$, is a recursive sequence since the value of a_n is based on the preceding terms of the sequence. On the other hand, $a_n = \dfrac{n}{n+1}$ for $n \in \mathbb{N}$ is not a recursive sequence since no knowledge of preceding terms of the sequence are used in the definition of a_n.

Example 4.3.9: Let a_n be defined by $a_1 = 2$ and $a_{n+1} = \sqrt{2 + \sqrt{a_n}}$, for $n \in \mathbb{N}$. The first five terms in this sequence are $a_1 = 2$, $a_2 = 1.8478$, $a_3 = 1.8328$, $a_4 = 1.8313$, and $a_5 = 1.8312$.

Two very famous recursively defined sequences of numbers are the Fibonacci and Lucas numbers. A surprisingly common set of recursively defined numbers was introduced in the thirteenth century by Leonardo of Pisa. Leonardo of Pisa (1170–1250), also known as Fibonacci, was a one of the most important mathematicians of the Middle Ages, and one of Fibonacci's main accomplishments was the introduction of the Indo-Arabic numeration system and Indo-Arabic computational methods to the European community. Fibonacci is most famous for the following problems that he posed and solved around 1202 A.D. in his book *Liber Abaci*.

Suppose that there are two newborn rabbits, a male and a female. Determine the number of rabbits produced in a year given that

1. Each rabbit takes one month to become mature.
2. Each pair produces a mixed pair of rabbits (i.e., a male and a female) every month, from the second month on.
3. No rabbits die during the course of the year.

Now, in month 1, pair 1 is born so there is one pair of rabbits. In month 2, pair 1 becomes mature so there is still only one pair of rabbits; in month 3, pair 1 produces a second pair of rabbits (pair 2), which is still immature, so at the end month 3 there are 2 pairs of rabbits. In month 4, pair 1 produces a pair of rabbits (pair 3) and pair 2 matures, so at the end of month 4 there are three pairs of rabbits. The process continues in this way. Let F_n be the number of pairs of rabbits after n months. Then, the solution to this problem is based on the recursive relationship $F_n = F_{n-1} + F_{n-2}$; thus, the number of rabbits after n months is the sum of the number of rabbits after $n-1$ and $n-2$ months. This sequence of numbers is called the *Fibonacci sequence.* The first 21 values of F_n are tabulated in Table 4.3.2.

Table 4.3.2 The First 21 Fibonacci Numbers

n	F_n		n	F_n		n	F_n
1	1		8	21		15	610
2	1		9	34		16	987
3	2		10	55		17	1,597
4	3		11	89		18	2,584
5	5		12	144		19	4,181
6	8		13	233		20	6,565
7	13		14	377		21	10,946

The Fibonacci pattern is often seen in sunflower and daisy florets; the scale patterns of pinecones pineapples, and artichokes; and even with mating habits of bees and rabbits. Thomas Koshy (2001) gives many applications of the Fibonacci pattern in his book *Fibonacci and Lucas Numbers with Applications.* Moreover, since their introduction, Fibonacci numbers and their related properties have been studied extensively by number theorists, and hence there are a tremendous number of mathematical results concerning the Fibonacci numbers. The definition of the Fibonacci sequence is given below.

Definition 4.3.8: The *Fibonacci sequence* of numbers is generated by the recursive formula $F_1 = 1, F_2 = 1$, and $F_{n+2} = F_{n+1} + F_n$ for $n \in \mathbb{N}$.

Note that every Fibonacci number is simply the sum of the two preceding Fibonacci numbers. Also, occasionally there will be a need for the 0th Fibonacci number, F_0, and when needed, $F_0 = 0$ can be used. The following theorems provide some of the most basic results on Fibonacci numbers. For example, in Table 4.3.2 the first 21 Fibonacci numbers were listed. Note that $F_{21} = 10,946$ and thus, the Fibonacci numbers grow large very rapidly. The following theorem puts an upper bound on the value of F_n, namely, $F_n < 2^n$.

Theorem 4.3.22: If F_n is the nth Fibonacci number, then $F_n < 2^n$, $\forall\, n \in \mathbb{N}$.

Proof: This theorem can be proved with strong induction. Let $\mathcal{P}_n := F_n < 2^n$.

For $n = 1$, $F_1 = 1$ and clearly $1 < 2^1 = 2$. Therefore, $\mathcal{P}_1$ is true.

Now, assume that $\mathcal{P}_1, \ldots, \mathcal{P}_k$ are all true for some ABF $k \in \mathbb{N}$. This means that $F_i < 2^i$ whenever $i = 1, 2, \ldots, k$. Now, if $\mathcal{P}_{k+1}$ is true, then it will be the case that F_{k+1} is less than 2^{k+1}.

Consider F_{k+1}:

$$F_{k+1} = F_k + F_{k-1} \underbrace{< 2^k + 2^{k-1}}_{\text{By } \mathcal{P}_k \text{ and } \mathcal{P}_{k-1}} < 2^k + 2^k$$

$$= 2 \cdot 2^k = 2^{k+1}$$

Thus, $\mathcal{P}_k \rightarrow \mathcal{P}_{k+1}$ and therefore, $F_n < 2^n$, $\forall\, n \in \mathbb{N}$.

∎

The next theorem provides several interesting results concerning the sums of Fibonacci numbers. In particular, Theorem 4.3.23 provides results for the sums of the first n Fibonacci numbers, the first n Fibonacci numbers with odd indices, the first n Fibonacci numbers with even indices, and the first n squared Fibonacci numbers.

Theorem 4.3.23: Let F_n be the nth Fibonacci number. Then, for $n \in \mathbb{N}$

(i) $\sum_{i=1}^{n} F_i = F_{n+2} - 1.$

(ii) $\sum_{i=1}^{n} F_{2i-1} = F_{2n}.$

(iii) $\sum_{i=1}^{n} F_{2i} = F_{2n+1} - 1.$

(iv) $\sum_{i=1}^{n} F_i^2 = F_n F_{n+1}.$

Proof: Each part of this theorem will be proven using mathematical induction.

Proof of part (i): Let $\mathcal{P}_n := \sum_{i=1}^{n} F_i = F_{n+2} - 1.$

For $n = 1$, $\sum_{i=1}^{1} F_i = F_1 = 1$ and $F_{1+2} - 1 = F_3 - 1 = 2 - 1 = 1.$
Therefore, $\mathcal{P}_1$ is true.

Now, suppose that $\mathcal{P}_k$ is true for some ABF $k \in \mathbb{N}$. This means that $\sum_{i=1}^{k} F_i = F_{k+2} - 1$. Now, if $\mathcal{P}_{k+1}$ is true, then it will be true that

$$\sum_{i=1}^{k+1} F_i = F_{k+1+2} - 1 = F_{k+3} - 1$$

Consider $\sum_{i=1}^{k+1} F_i$:

$$\sum_{i=1}^{k+1} F_i = \sum_{i=1}^{k} F_i + F_{k+1} = \underbrace{F_{k+2} - 1}_{\text{By } \mathcal{P}_k} + F_{k+1}$$

$$= F_{k+1} + F_{k+2} - 1 = F_{k+3} - 1$$

Thus, $\mathcal{P}_{k+1}$ is true whenever $\mathcal{P}_k$ is true, and therefore

$$\sum_{i=1}^{n} F_i = F_{n+2} - 1, \forall\, n \in \mathbb{N}$$

Proof of part (ii): The proof of part (ii) is left as an exercise.

Proof of part (iii): The proof of part (iii) is left as an exercise.

Proof of part (iv): Let $\mathcal{P}_n := \sum_{i=1}^{n} F_i^2 = F_n \cdot F_{n+1}$.

For $n = 1$, $\sum_{i=1}^{1} F_i^2 = F_1^2 = 1^2 = 1$ and $F_1 \cdot F_2 = 1 \cdot 1 = 1$. Therefore, $\mathcal{P}_1$ is true.

Now, suppose that $\mathcal{P}_k$ is true for some ABF $k \in \mathbb{N}$. This means that $\sum_{i=1}^{k} F_i^2 = F_k \cdot F_{k+1}$. Now, if $\mathcal{P}_{k+1}$ is true, then $\sum_{i=1}^{k+1} F_i^2 = F_{k+1} \cdot F_{k+2}$.

Consider $\sum_{i=1}^{k+1} F_i^2$:

$$\sum_{i=1}^{k+1} F_i^2 = \sum_{i=1}^{k} F_i^2 + F_{k+1}^2 = \underbrace{F_k \cdot F_{k+1}}_{\text{By } \mathcal{P}_k} + F_{k+1}^2$$

$$= F_{k+1}\left[F_k + F_{k+1}\right] = F_{k+1} \cdot F_{k+2}$$

Thus, $\mathcal{P}_{k+1}$ is true whenever $\mathcal{P}_k$ is true. Therefore, for $n \in \mathbb{N}$

$$\sum_{i=1}^{n} F_i^2 = F_n \cdot F_{n+1}$$

■

Theorem 4.3.23 was first proved by Edouard Lucas (1842–1891). Lucas is also credited with naming the Fibonacci sequence and for proving many other results concerning the Fibonacci numbers. Lucas also introduced a recursive sequence of numbers that is closely related to the Fibonacci numbers, namely, the *Lucas sequence* of numbers. The definition of the Lucas sequence is given below.

Definition 4.3.9: The *Lucas sequence* of numbers is generated by the recursive formula $L_1 = 1$, $L_2 = 3$, and $L_{n+2} = L_{n+1} + L_n$, $\forall\ n \in \mathbb{N}$.

The first 21 Lucas numbers are listed in Table 4.3.3.

Table 4.3.3 The First 21 Lucas Numbers

n	L_n		n	L_n		n	L_n
1	1		8	47		15	1,364
2	3		9	76		16	2,207
3	4		10	123		17	3,571
4	7		11	199		18	5,778
5	11		12	322		19	9,349
6	18		13	521		20	15,127
7	29		14	843		21	24,476

Like the Fibonacci numbers, the Lucas numbers grow large very quickly. In fact, examining Tables 4.3.2 and 4.3.3, it becomes clear that for $n > 1$, $L_n \geq 2F_n$. Theorem 4.3.24 shows that $L_n \geq 2F_n$, for $n > 1$ and also, that the Lucas and Fibonacci numbers are closely related.

Theorem 4.3.24: Let F_n be the nth Fibonacci number and L_n the nth Lucas number. Then, for $n \in \mathbb{N}$

(i) $L_{n+1} = F_{n+2} + F_n$.

(ii) $L_{n+2} + L_n = 5F_{n+1}$.

(iii) $L_{n+1} \geq 2 \cdot F_{n+1}$.

Proof: Each part of this theorem will be proved with mathematical induction.

Proof of part (i): Let $\mathcal{P}_n := F_{n+2} + F_n = L_{n+1}$.

For $n = 1$, $L_{1+1} = L_2 = 3, F_{1+2} = F_3 = 2$ and $F_1 = 1$. Therefore, $L_2 = 3 = 2 + 1 = F_3 + F_1$, and hence $\mathcal{P}_1$ is true.

Now, suppose that $\mathcal{P}_1, \ldots, \mathcal{P}_k$ are all true for some ABF $k \in \mathbb{N}$. This means that $L_{j+1} = F_{j+2} + F_j$, for $j = 1, 2, \ldots, k$. Now, if $\mathcal{P}_{k+1}$ is true, then $L_{k+2} = F_{k+3} + F_{k+1}$.

Consider $L_{k+1+1} = L_{k+2}$:

$$L_{k+2} = L_{k+1} + L_k = \underbrace{(F_{k+2} + F_k)}_{\text{By } \mathcal{P}_k} + \underbrace{(F_{k+1} + F_{k-1})}_{\text{By } \mathcal{P}_{k-1}}$$

$$= (F_{k+2} + F_{k+1}) + (F_k + F_{k-1})$$

$$= F_{k+3} + F_{k+1}$$

Thus, $\mathcal{P}_{k+1}$ is true whenever $\mathcal{P}_1, \ldots, \mathcal{P}_k$ are true and therefore, $L_{n+1} = F_{n+2} + F_n$, $\forall\, n \in \mathbb{N}$.

Proof of part (ii): The proof of part (ii) is left as an exercise.

Proof of part (iii): Let $\mathcal{P}_n := L_{n+1} \geq 2F_{n+1}$.

For $n = 1$, $L_2 = 3$ and $F_2 = 1$ and since $L_2 = 3 \geq 2 \cdot F_1 = 2$, $\mathcal{P}_1$ is true.

Now, suppose that $\mathcal{P}_2, \ldots, \mathcal{P}_k$ are all true for some ABF $k \in \mathbb{N}$. This means that $L_{j+1} > 2F_{j+1}$, for $j = 1, 2, \ldots, k$. Now, if $\mathcal{P}_{k+1}$ is true, then $L_{k+2} \geq 2 \cdot F_{k+2}$.

Consider L_{k+2}:

$$L_{k+2} = L_{k+1} + L_k \geq \underbrace{2F_{k+1} + 2F_k}_{\text{By } \mathcal{P}_k \text{ and } \mathcal{P}_{k-1}}$$

$$= 2\left[F_{k+1} + F_k\right] = 2 \cdot F_{k+2}$$

Thus, $\mathcal{P}_{k+1}$ is true whenever $\mathcal{P}_1, \ldots, \mathcal{P}_k$ are true, and therefore $L_{n+1} \geq 2 \cdot F_{n+1}$ whenever $n \in \mathbb{N}$.

■

The following theorem provides results concerning the sums of Lucas numbers, analogous to the sums of the Fibonacci numbers given in Theorem 4.3.23.

Theorem 4.3.25: Let L_n be the nth Lucas number. Then, for $n \in \mathbb{N}$

(i) $\displaystyle\sum_{i=1}^{n} L_i = L_{n+2} - 3.$

(ii) $\sum_{i=1}^{n} L_{2i-1} = L_{2n} - 2.$

(iii) $\sum_{i=1}^{n} L_{2i} = L_{2n+1} - 1.$

(iv) $\sum_{i=1}^{n} L_i^2 = L_n L_{n+1} - 2.$

Proof: The proof of each part of theorem 4.3.25 is left as an exercise.

■

Finally, a more general sequence of recursive numbers that is related to both the Fibonacci and Lucas numbers is the *generalized Fibonacci sequence* of numbers. The generalized version of the Fibonacci numbers is denoted by G_n and is generated by simply altering the starting values of G_1 and G_2, while still utilizing the recursive relationship $G_{n+2} = G_n + G_{n+1}$ for $n \in \mathbb{N}$. The formal definition of a generalized Fibonacci sequence is given below.

Definition 4.3.10: Let $a, b \in \mathbb{R}$. The *generalized Fibonacci sequence* of numbers is generated by the recursive relationship $G_1 = a$, $G_2 = b$, and $G_{n+2} = G_{n+1} + G_n$ for $n \in \mathbb{N}$.

Note that the Fibonacci sequence can be generated as a generalized Fibonacci sequence by letting $a = b = 1$, and the Lucas sequence is the generalized Fibonacci sequence generated by taking $a = 1$ and $b = 3$.

Example 4.3.10: Generate the first 10 terms for a generalized Fibonacci sequences with the following starting values:

a. $a = -1$ and $b = 4$

b. $a = 1$ and $b = 2$

Solutions: The first 10 terms of the generalized Fibonacci sequence with the starting values given above are as follows:

a. $a = -1$ and $b = 4$ are $-1, 4, 3, 7, 10, 17, 27, 44, 71, 115$.

b. $a = 1$ and $b = 2$ are $1, 2, 3, 5, 8, 13, 21, 34, 55, 89$.

Note that in Example 4.3.10, for starting values $a = 1$ and $b = 2$, it turns out that $G_i = F_{i+1}$, for $i \in \mathbb{N}$. The following example reveals an interesting pattern in the generation of the values of a generalized Fibonacci sequence.

Example 4.3.11: Determine the nth term (G_n) in a generalized Fibonacci sequence with $G_1 = a$ and $G_2 = b$.

Solution: Table 4.3.4 lists the first 15 terms in a generalized Fibonacci sequence.

Table 4.3.4 The First 15 Generalized Fibonacci Numbers

n	G_n		n	G_n		n	G_n
1	a		6	$3a+5b$		11	$34a+55b$
2	b		7	$5a+8b$		12	$55a+89b$
3	$a+b$		8	$8a+13b$		13	$89a+144b$
4	$a+2b$		9	$13a+21b$		14	$144a+233b$
5	$2a+3b$		10	$21a+34b$		15	$233a+377b$

Note that the coefficients of the a and b terms in G_n in Table 4.3.4 are Fibonacci numbers. In fact, on close inspection of Table 4.3.4, it appears that the relationship between G_{n+2} and the Fibonacci sequence of numbers might be $G_{n+2} = aF_n + bF_{n+1}$; this conjecture is proved in the following theorem.

Theorem 4.3.26: Let F_n be the nth Fibonacci number and G_n the nth generalized Fibonacci number with starting values $G_1 = a$ and $G_2 = b$. Then, $G_{n+2} = aF_n + bF_{n+1}$, $\forall\, n \in \mathbb{N}$.

Proof (by Induction): Let $\mathcal{P}_n := aF_n + bF_{n+1} = G_{n+2}$.

For $n = 1$, $G_3 = G_1 + G_2 = a + b = aF_1 + bF_2$ since $F_1 = F_2 = 1$; therefore $\mathcal{P}_1$ is true.

Now, suppose that $\mathcal{P}_1, \ldots, \mathcal{P}_k$ are all true for some ABF $k \in \mathbb{N}$. This means that $G_{j+2} = aF_j + bF_{j+1}$ for $j = 1, 2, \ldots, k$. Now, if $\mathcal{P}_{k+1}$ is true, then $G_{k+3} = aF_{k+1} + bF_{k+2}$.

Consider G_{k+3}:

$$G_{k+3} = G_{k+2} + G_{k+1} = \underbrace{(aF_k + bF_{k+1})}_{\text{By } \mathcal{P}_k} + \underbrace{(aF_{k-1} + bF_k)}_{\text{By } \mathcal{P}_{k-1}}$$

$$= a\,(F_k + F_{k-1}) + b\,(F_{k+1} + F_k) = aF_{k+1} + bF_{k+2}$$

Thus, $\mathcal{P}_{k+1}$ is true whenever $\mathcal{P}_1, \ldots, \mathcal{P}_k$ are all true, and therefore $G_{n+2} = aF_n + bF_{n+1}$, $\forall\, n \in \mathbb{N}$.

■

Theorem 4.3.27: Let F_n be the nth Fibonacci number and G_n the nth generalized Fibonacci number. If $G_1 = 1$ and $G_2 = 2$, then $G_n = F_{n+1}$, $\forall\, n \in \mathbb{N}$.

Proof: The proof of Theorem 4.3.27 is left as an exercise.

■

As with the Fibonacci and Lucas sequences, there are many results that have been proven concerning generalized Fibonacci sequences, see Koshy (2001). The following theorem provides three results concerning the sums of the terms in a generalized Fibonacci sequence that are analogous to those given in Theorems 4.3.23 and 4.3.25 on Fibonacci and Lucas numbers.

Theorem 4.3.28: If G_n is the nth Generalized Fibonacci number when $G_1 = a$ and $G_2 = b$, then

(i) $\sum_{i=1}^{n} G_i = G_{n+2} - b, \forall\, n \in \mathbb{N}$

(ii) $\sum_{i=1}^{n} G_{2i-1} = G_{2n} + a - b, \forall\, n \in \mathbb{N}$

(iii) $\sum_{i=1}^{n} G_{2i} = G_{2n+1} - a, \forall\, n \in \mathbb{N}$.

Proof: The proof of Theorem 4.3.28 is left as an exercise.

■

Finally, note that one could have started with the definition of the generalized Fibonacci sequence; then the results for the Fibonacci and Lucas sequences given in Theorems 4.3.23 and 4.3.25 would simply be corollaries of Theorem 4.3.28.

EXERCISES

4.1 Let $\circ$ be defined on $[0, \infty)$ by $a \circ b = |a - b|$.

a. Show that $\circ$ is an Abelian operator on $[0, \infty)$.

b. Show that $[0, \infty)$ is closed under $\circ$.

c. Show that there exists an identity element in $[0, \infty)$ under $\circ$.

4.2 Let $\circ$ be defined on $\mathbb{R}^+$ by $a \circ b = a^b b^a$.

a. Prove that $\circ$ is an Abelian operator on $\mathbb{R}^+$.

b. Prove that $\mathbb{R}^+$ is closed under $\circ$.

c. Prove that there exists an identity element in $\mathbb{R}^+$ under $\circ$.

d. Find 2^{-1}.

e. Prove that $\mathbb{R}^+$ contains all the inverse elements under $\circ$.

4.3 Let $\Omega = \mathbb{R}$ and for $a, b \in \mathbb{R}$ define $a \circ b = ab - a - b + 2$.

a. Show that $\mathbb{R}$ is closed under $\circ$.

b. Determine whether there is an identity element in $\mathbb{R}$ under $\circ$.

c. Determine whether there is an inverse element in $\mathbb{R}$ under $\circ$ for each $a \in \mathbb{R}$.

d. Solve the equation $3 \circ x = 13$ for x.

4.4 Let $\Omega = \{\omega : \omega = 3k + 1 \text{ for some } k \in \mathbb{Z}\}$.

a. Prove that Ω is closed under multiplication.

b. Show that Ω is not closed under addition, subtraction, or division.

4.5 Let $\mathbb{Q}$ be the set of rational numbers. Prove that $\mathbb{Q}$ is closed under subtraction.

4.6 Let $\mathbb{Q}_3 := \left\{r \in \mathbb{Q} : r = \frac{p}{3} \text{ for some } p \in \mathbb{Z}\right\}$. Show that $\mathbb{Q}_3$ is closed under addition, subtraction, multiplication, and nonzero division.

4.7 Let $\mathbb{Z}_E$ be the collection of even integers, and let $\circ$ be the binary operator defined on $\mathbb{Z}_E$ by $a \circ b = ab + 2$.

a. Prove that $\mathbb{Z}_E$ is closed under $\circ$.

b. Is $\circ$ an Abelian operator on $\mathbb{Z}_E$?

c. Is $\circ$ an associative operator on $\mathbb{Z}_E$?

4.8 Let $\mathbb{Z}_O$ be the collection of odd integers, and let $\circ$ be the binary operator defined on $\mathbb{Z}_O$ by $a \circ b = ab + 2$.

a. Show that $\mathbb{Z}_O$ is closed under $\circ$.

b. Is $\circ$ an Abelian operator on $\mathbb{Z}_O$?

c. Is $\circ$ an associative operator on $\mathbb{Z}_O$?

4.9 Prove that $\mathbb{R}$ is closed under subtraction.

4.10 Prove each of the following theorems:

a. **Theorem:** If $z \in \mathbb{Z}$, then $z^n \in \mathbb{Z}$ for every natural number n.

b. **Theorem:** If $z \in \mathbb{Z}$, then $rz + s \in \mathbb{Z}$ for all $r, s \in \mathbb{Z}$.

4.11 For each of the following sets, determine whether the set is closed under the prescribed binary operator. If the set is closed, provide a formal closure proof. If the set is not closed under the binary operator, provide a counterexample to show that it is not closed.

a. Let $\Omega = \{\omega : \omega = 4k + 1 \text{ for some } k \in \mathbb{Z}\}$, and let $\circ$ be defined by $a \circ b = ab$.

b. Let $\Omega = \{\omega : \omega = 4k + 1 \text{ for some } k \in \mathbb{Z}\}$, and let $\circ$ be defined by $a \circ b = a + b$.

c. Let $5\mathbb{Z} = \{z : z = 5k \text{ for some } k \in \mathbb{Z}\}$, and let $\circ$ be defined by $a \circ b = a - b$.

d. Let $5\mathbb{Z} = \{z : z = 5k \text{ for some } k \in \mathbb{Z}\}$, and let $\circ$ be defined by $a \circ b = 2a + 3b$.

e. Let $\Omega = \{z : z = 2k \text{ or } z = 3k \text{ for some } k \in \mathbb{Z}\}$, and let $\circ$ be defined by $a \circ b = a + b$.

f. Let $\Omega = \{z : z = 2k \text{ and } z = 3k \text{ for some } k \in \mathbb{Z}\}$, and let $\circ$ be defined by $a \circ b = ab$

4.12 Let Ω be a set and $\circ$ a binary operator defined on Ω. Prove each of the following theorems:

a. **Theorem:** If Ω is closed under $\circ$, $\circ$ is an associative binary operator $\circ$, and $a^{-1} \in \Omega$ whenever $a \in \Omega$, then a^{-1} is unique.

b. **Theorem:** If Ω is closed under $\circ$, $\circ$ is an associative binary operator $\circ$, and $a^{-1} \in \Omega$ whenever $a \in \Omega$, then $(a \circ b)^{-1} = b^{-1} \circ a^{-1}$ for all $a, b \in \Omega$.

c. **Theorem:** If Ω is closed under $\circ$ and $a^{-1} \in \Omega$ whenever $a \in \Omega$, then $\left(a^{-1}\right)^{-1} = a$.

4.13 Determine the rational forms of

a. $12.67\overline{83}$

b. $0.98\overline{76}$

c. $10.3\overline{15}$

d. $213.81\overline{651}$

4.14 Let $\mathbb{Q}$ be the set of rational numbers:

a. Prove that $\mathbb{Q}$ is closed under subtraction.

b. Prove that $\mathbb{Q}$ is closed under division.

4.15 Prove that

a. $\sqrt{3}$ is an irrational number.

b. $\sqrt{5}$ is an irrational number.

c. $\sqrt{p}$ is an irrational number for every prime number p.

4.16 Prove that $\sum_{k=1}^{n+1} \frac{1}{k}$ is not equal to an integer for any $n \in \mathbb{N}$.

4.17 Let $a_n = \left(1 + \frac{1}{n}\right)^n, \ \forall\ n \in \mathbb{N}$.

a. Prove that $a_n \in \mathbb{Q}, \forall\ n \in \mathbb{N}$.

b. Let $a = \lim_{n \to \infty} a_n$. Is $a \in \mathbb{Q}$?

4.18 Find pairs of irrational numbers showing that the irrationals are not closed under addition, subtraction, multiplication, and division.

4.19 Let $a \neq 0$ be a real number, and let the reciprocal of a be the value x such that $a \times x = 1$. Prove that if $a \in \mathbb{R}$ and $a \neq 0$, then there exists a unique real number x such that $ax = 1$.

4.20 For real numbers x, y, z, prove that

a. $\big|\, |x| - |y| \,\big| \leq |x - y|$

b. $|x + y + z| \leq |x| + |y| + |z|$

c. $|x| = \max(-x, x)$.

d. $|x + y| = |x| + |y|$ if and only if $ab \geq 0$.

e. $\left|\frac{x}{y}\right| = \frac{|x|}{|y|}$.

f. $\sqrt{x^2} = |x|$.

4.21 Prove that if $a \in \mathbb{Q}$, $b \in \mathbb{I}$, and $a \neq 0$, then $ab - a \notin \mathbb{Q}$.

4.22 Prove that

a. If $a \in \mathbb{Q}$ and $b \in \mathbb{I}$, then $a - b \notin \mathbb{Q}$.
b. If $a \neq 0 \in \mathbb{Q}$ and $b \in \mathbb{I}$, then $ab \notin \mathbb{Q}$.
c. If $a \neq 0 \in \mathbb{Q}$ and $b \in \mathbb{I}$, then $\frac{a}{b} \notin \mathbb{Q}$.
d. If $a \neq 0 \in \mathbb{Q}$ and $b \in \mathbb{I}$, then $\frac{b}{a} \notin \mathbb{Q}$.

4.23 Prove that if $a, b, c \in \mathbb{R}$, $c \neq 0$, and $ac = bc$, then $a = b$.

4.24 Prove each of the following theorems:

a. **Theorem:** Let $n \in \mathbb{N}$. If p is a prime number and $p|n^2$, then $p|n$.
b. **Theorem:** If p is a prime number, then there exists $k \in \mathbb{Z}$ such that $p = 4k + 1$ or $p = 4k + 3$.
c. **Theorem:** If $n \in \mathbb{N}$, then $3|(n + 1)$ or $3|(n + 3)$ or $3|(n + 5)$.
d. **Theorem:** If $2^n - 1$ is prime, then n is a prime number.
e. **Theorem:** Let $m, n \in \mathbb{N}$, and let p be a prime number. If $p|mn$, then $p|m$ or $p|n$.
f. **Theorem:** If p_1, p_2, p_3 and p_4 are odd prime numbers, then $p_1p_2 + p_3p_4$ is not a prime number.

4.25 Let $a, b, c \in \mathbb{Z}$. Prove that

a. If $a|b$, then $a|b^n$, $\forall\ n \in \mathbb{N}$.
b. If $a|b$ and $a|c$, then $a|(b^2 + bc + c^2)$.
c. If $a|b$ and $b|a$, then $|a| = |b|$.
d. If $a^2|b$, then $a|b$.
e. If $a|b$ and $a|c$, then $\forall\ x, y \in \mathbb{Z}$, $a|(bx + cy)$.

4.26 Prove that $9|(10^{n-1} - 1)$ for $n \in \mathbb{N}$.

4.27 Suppose $a \in \mathbb{N}$ and $a = \sum_{k=0}^{n} c_k \cdot 10^k$. Prove that if $S = \sum_{k=0}^{n} c_k$, then $9|a$ if and only if $9|S$.

4.28 Let $a \in \mathbb{N}$ and suppose that $b_i \in \mathbb{N}$, $\forall\ i \in \mathbb{N}$. Prove that if $a|b_i\ \forall\ i \in \mathbb{N}$, then a divides $\sum_{i=1}^{n+1} b_i$ for every $n \in \mathbb{N}$.

4.29 Let $a \in \mathbb{N}$ and suppose that $a = \sum_{i=0}^{n} a_i 10^i$. Prove

a. $4|a$ if and only if $4|(10a_1 + a_0)$.
b. $8|a$ if and only if $8|(100a_2 + 10a_1 + a_0)$.
c. $11|a$ if and only if $11\left|\left(\sum_{i=0}^{n} (-1)^i a_i\right)\right.$.

4.30 Let F_n be the nth Fibonacci number. Prove that

a. $\sum_{i=1}^{n} F_{2i+1} = F_{2n},\ \forall\ n \in \mathbb{N}$.
b. $\sum_{i=1}^{n} F_{2i} = F_{2n+1} - 1,\ \forall\ n \in \mathbb{N}$.
c. $1 \leq \frac{F_{n+1}}{F_n} \leq 2,\ \forall\ n \in \mathbb{N}$.
d. $5|F_{5n},\ \forall\ n \in \mathbb{N}$.
e. $F_n \leq \left(\frac{7}{4}\right)^{n-1},\ \forall\ n \in \mathbb{N}$.
f. $\sum_{i=1}^{n+1} (-1)^{i-1} F_{i+1} = (-1)^n F_{n+1},\ \forall\ n \in \mathbb{N}$.

4.31 Let L_n be the nth Lucas number. Prove that

a. $L_{n+2} + L_n = 5F_{n+1},\ \forall\ n \in \mathbb{N}$.
b. $\sum_{i=1}^{n} L_i = L_{n+2} - 3,\ \forall\ n \in \mathbb{N}$.
c. $\sum_{i=1}^{n} L_{2i-1} = L_{2n} - 2,\ \forall\ n \in \mathbb{N}$.
d. $\sum_{i=1}^{n} L_{2i} = L_{2n+1} - 1,\ \forall\ n \in \mathbb{N}$.
e. $\sum_{i=1}^{n} L_i^2 = L_n L_{n+1} - 2,\ \forall\ n \in \mathbb{N}$.

4.32 Let F_n and G_n be the nth terms in the Fibonacci Sequence and the Generalized Fibonacci Sequence generated by $G_1 = 1$ and $G_2 = 2$, respectively. Prove that $G_n = F_{n+1}$, $\forall\, n \in \mathbb{N}$.

4.33 Let G_n be the nth term in a generalized Fibonacci sequence generated by $G_1 = a$ and $G_2 = b$. Prove that

a. $\sum_{i=1}^{n} G_i = G_{n+2} - b$, $\forall\, n \in \mathbb{N}$.

b. $\sum_{i=1}^{n} G_{2i-1} = G_{2n} = a - b$, $\forall\, n \in \mathbb{N}$.

c. $\sum_{i=1}^{n} G_{2i} = G_{2n+1} - a$, $\forall\, n \in \mathbb{N}$.

Chapter 5
The Foundations of Calculus

The area of mathematics known as *calculus of a single variable* is one of the best known areas of modern mathematics. The credit for developing *calculus* is generally given to Sir Isaac Newton (1643–1727) and Wilhelm Gottfried Leibniz (1646–1716), although many other mathematicians actually played an important role in the development of calculus. For an outstanding history of the development of calculus, see *The Calculus Gallery: Masterpieces from Newton to Lebesgue* by William Dunham (2005).

Now, calculus deals with the study of the real-valued functions and their properties. In particular, the two main properties of a function that are investigated in the typical first course on calculus are (1) how the function behaves near the point $x = x_0$ (i.e., limits) and (2) what the graph of the function looks like (i.e., continuity and derivatives). In both cases, the properties of interest are based on the limiting behavior of the function. Since the foundation of calculus is built on the idea of limits, throughout this chapter the key idea being studied is the limiting behavior of a real-valued function on $\mathbb{R}$. In particular, limits of real-valued sequences, limits of real-valued functions, continuity, and the derivative of a real-valued function will be discussed in Chapter 5.

5.1 Functions

Since the focus of calculus is the study of real-valued functions and their behavior, it is important to understand what a function actually is. In lay terms a real-valued function f on $\mathbb{R}$ is a well-defined rule that maps each point $x \in \mathbb{R}$ to a point $y \in \mathbb{R}$ according to the rule f. The definition of a real-valued function is given below.

Definition 5.1.1: A function f defined on a subset $\mathcal{D}$ of $\mathbb{R}$ is called a *real-valued function* if and only if f is a rule that assigns to each x in $\mathcal{D}$ one and only one real number y. The set $\mathcal{D}$ is called the *domain* of the function f.

For example, the function $f(x) = x^2$ takes a value in $\mathbb{R}$ and maps it to the square of x. Note that in this example, there is no ambiguity concerning what happens to each value of x once it is inserted into the function f; when there is ambiguity concerning what happens to a value of x when inserted into f, then f is most likely not a function.

Example 5.1.1: Let $f(x) = x^2$, $g(x) = \sin(x)$, and $h(x) = |x|$. Then f, g and h are all functions. However

$$k(x) = \begin{cases} x^2 & \text{if } x \geq 0 \\ |x| & \text{if } x < 1 \end{cases}$$

is not a function since $k(x)$ assigns two values to each value of $x \in [0, 1)$.

Definition 5.1.2: Two functions f and g are said to be *equal* if and only if they have the same domain $\mathcal{D}$ and $f(x) = g(x)$, $\forall\, x \in \mathcal{D}$.

Note that when two functions are equal they must have the same domain. Also, note that there are many functions that appear to be equal, yet according to Definition 5.1.2 are not equal. For example, let $f(x) = (x-1)/(x^2-1)$ and $g(x) = 1/(x+1)$. Clearly, when $x \neq 1$, then

$$\frac{x-1}{x^2-1} = \frac{x-1}{(x-1)(x+1)} = \frac{1}{x+1}$$

However, the domain of f is $\mathcal{D}_f = \{x \in \mathbb{R} : x \neq \pm 1\}$, while the domain of g is $\mathcal{D}_g = \{x \in \mathbb{R} : x \neq -1\}$. Thus, $\mathcal{D}_f \neq \mathcal{D}_g$ and therefore the functions f and g are not equal.

Now, there are many different types of functions, some of which are extremely complicated in nature; however, the elementary functions are the polynomial, rational, power, exponential, logarithmic, and trigonometric functions. The definitions of the polynomial, rational, and exponential functions are given below.

Definition 5.1.3: A real-valued function f, defined on $\mathbb{R}$, is called

a. A *polynomial* if and only if it is of the form

$$f(x) = a_n x^n + a_{n-1}x^{n-1} + \ldots + a_1 x + a_0 = \sum_{i=0}^{n} a_i x^i,$$

where $n \in \mathbb{N}$ and $a_0, a_1, \ldots, a_n \in \mathbb{R}$.

b. A *rational* function if and only if it is of the form $f(x) = \dfrac{p(x)}{q(x)}$, where $p(x)$ and $q(x)$ are polynomials.

c. An *exponential* function if and only if it is of the form $f(x) = a^x$, where $a \in \mathbb{R}^+$.

Note that the domain of a polynomial function or an exponential function is $\mathbb{R}$; however, the domain of a rational function may not be all of $\mathbb{R}$. In particular, for a rational function of the form $r(x) = \frac{p(x)}{q(x)}$, all points in $\mathbb{R}$ for which $q(x) = 0$ must be excluded from the domain of $r(x)$; thus, the domain of a rational function $r(x)$ is $\mathcal{D} = \{x \in \mathbb{R} : q(x) \neq 0\}$.

Example 5.1.2: Examples of each of type of function defined in Definition 5.1.3 are given below:

a. $p(x) = 3x^2 - x + 11$ and $q(x) = x^7 - 3x^5 + 4x^2 - 13x + 1$ are polynomials.

b. $r(x) = \dfrac{3x^2 - x + 11}{x^7 - 3x^5 + 4x^2 - 13x + 1}$ and $s(x) = \dfrac{x - 1}{x^2 + 1}$ are rational functions.

c. $t(x) = 2^x$ and $u(x) = e^x$ are exponential functions.

Functions, like numbers, are often combined using the standard arithmetic operators (i.e., $+, -, \times, \div$) to create new functions. For instance, if f and g are functions, then so are $f + g, f - g, f \cdot g$, and f/g on the appropriate domains. Another important way of creating a new function from two functions f and g is to form the composition of the two functions. The definition of the composition of two functions is given below.

Definition 5.1.4: Let f and g be real-valued functions on $\mathbb{R}$. The *composition* of the functions f and g is defined to be the function $f \circ g(x) = f(g(x))$.

Thus, given two real-valued functions f and g defined on $\mathbb{R}$, the composition of f and g is the function $c(x)$ formed by applying the function f to the value of $g(x)$. It is important to note that $f \circ g$ and $g \circ f$ are nearly always different functions. The domain of the function $f \circ g$, namely, $\mathcal{D}_{f \circ g}$, is the set of all values in $\mathcal{D}_g$ that produce values of $g(x) \in \mathcal{D}_f$, while the domain of the function $g \circ f$, specifically, $\mathcal{D}_{g \circ f}$, is the set of all values in $\mathcal{D}_g$ that produce values of $f(x) \in \mathcal{D}_g$. For example, if $f(x) = \sqrt{x}$ and $g(x) = x^2$, then the domain of $f \circ g$ is $\mathcal{D}_{f \circ g} = \mathbb{R}$, however, the domain of $g \circ f$ is $[0, \infty)$. Furthermore, $f \circ g(x) = |x|$, while $g \circ x(x) = x$, which are clearly not the same functions.

Example 5.1.3: Let $f(x) = \sqrt{x}$ and $g(x) = \dfrac{x - 1}{x + 1}$. Determine

a. $\mathcal{D}_f$ and $\mathcal{D}_g$.

b. $f \circ g$ and its domain.

c. $g \circ f$ and its domain.

Solutions: Let $f(x) = \sqrt{x}$ and $g(x) = \dfrac{x - 1}{x + 1}$.

a. $\mathcal{D}_f = [0, \infty)$ and $\mathcal{D}_g = \{x \in \mathbb{R} : x \neq -1\}$.

b. $f \circ g(x) = \sqrt{\dfrac{x-1}{x+1}}$, and $\mathcal{D}_{f \circ g}$ is equal to all those points in $\mathcal{D}_g$ that produce values of $g(x) \geq 0$. Thus, $\mathcal{D}_{f \circ g} = \{x \in \mathbb{R} : x < -1 \text{ or } x \geq 1\}$.

b. $f \circ g(x) = \dfrac{\sqrt{x}-1}{\sqrt{x}+1}$, and $\mathcal{D}_{g \circ f}$ is equal to all those points in $[0, \infty)$ that produce values of $f(x) \neq -1$. Thus, $\mathcal{D}_{f \circ g} = \{x \in \mathbb{R} : x \geq 0\} = [0, \infty)$.

5.2 Sequences of Real Numbers

In this section a special type of function, functions known as *real-valued sequences*, will be studied. The definition of a real-valued sequence is given below.

Definition 5.2.1: A *real-valued sequence* is a real-valued function a whose domain is a set of the form $\{n \in \mathbb{Z} : n \geq m\}$. For $n \in \mathcal{D}_a$, $a(n)$ will be denoted by a_n and is called an *element* of the sequence a.

Thus, a real-valued sequence is simply a list of real numbers generated by a function whose domain is a subset of the integers. In Chapter 4, two sequences of natural numbers were introduced and studied, namely, the Fibonacci and Lucas sequences. The Fibonacci and Lucas sequences represent commonly occurring sequences of real numbers. Now, a sequence a actually generates a sequence of real numbers of the form $a_m, a_{m+1}, a_{m+2}, \ldots$, which is often written as $\{a_n\}_{n=m}^{\infty}$ or simply a_n. Most sequences start with $m = 0$ or $m = 1$; however for simplicity, all of the sequences in this text will be assumed to have the common domain $\mathbb{N}$ (i.e., start at $m = 1$). Thus, a sequence a_n consists of the elements $a_1, a_2, a_3, a_4, \ldots$

Example 5.2.1: Let the sequence a_n be defined by $a_n = 2^n$. Then, the first 5 terms of this sequence are

$$a_1 = 2^1 = 2, \ a_2 = 2^2 = 4, \ a_3 = 2^3 = 8, \ a_4 = 2^4 = 16, \ a_5 = 2^5 = 32$$

Example 5.2.2: Let the sequence a_n be defined by $a_n = \dfrac{(-1)^n}{n}$. Determine the first 10 terms in this sequence.

Solution: The first 10 terms of this sequence are

$$a_1 = -1, \ a_2 = \frac{1}{2}, \ a_3 = \frac{-1}{3}, \ a_4 = \frac{1}{4}, \ a_5 = \frac{-1}{5},$$

$$a_6 = \frac{1}{6}, \ a_7 = \frac{-1}{7}, \ a_8 = \frac{1}{8}, \ a_9 = \frac{-1}{9}, \ a_{10} = \frac{1}{10}$$

5.2.1 Convergent Sequences and Limit Theorems

The most interesting mathematical question concerning a sequence a_n is "How does the sequence a_n behave as the index n grows large?" In mathematical terminology this question becomes "Does the sequence a_n converge to a limit, or does it diverge?" The ϵ-N definition for the convergence of a sequence of real numbers is given below.

Definition 5.2.2: A sequence of real numbers a_n is said to *converge* to a limit a if and only if for every $\epsilon > 0$, there exists $N \in \mathbb{N}$ such that $|a_n - a| < \epsilon$, whenever $n \geq N$. When the sequence a_n converges to a limit a, this will be denoted by either $\lim_{n\to\infty} a_n = a$ or $a_n \to a$; if a sequence does not converge, it is said to *diverge*.

Note that the convergence of a sequence a_n is based entirely on the tail behavior (i.e., large values of n) of the sequence. In fact, first m terms in a sequence play absolutely no role in whether the sequence converges, no matter how large m. Also, it is important to note that the value of N required to satisfy the condition "$|a_n - a| < \epsilon$ whenever $n \geq N$" depends on the particular value of ϵ. In fact, the smaller the value of ϵ, the larger the value of N will be. For example, the value of N required for $\epsilon = 0.01$ will be less than or equal to the value of N required for $\epsilon = 0.001$.

Now, a sequence a_n can diverge only when it either increases without bound, decreases without bound, or oscillates in a nonconvergent fashion. For example, $a_n = 2^n$ increases without bound and therefore cannot converge. An oscillating sequence a_n that does not converge is given below:

$$a_n = \begin{cases} 1 & \text{when } n \text{ is a multiple of 3} \\ \frac{1}{n} & \text{otherwise} \end{cases}$$

The first 10 terms of a_n are $1, \frac{1}{2}, 1, \frac{1}{4}, \frac{1}{5}, 1, \frac{1}{7}, \frac{1}{8}, 1, \frac{1}{10}$. This sequence cannot converge since no matter how large n is, every third term in the sequence continues to jump up to the value 1, while the remaining terms are decreasing to 0. Thus, this sequence diverges since there is no way to force $|a_n - 0|$ or $|a_n - 1|$ to be arbitrarily small from some point n on. A plot illustrating the behavior of the first 100 terms in this sequence is given in Figure 5.2.1.

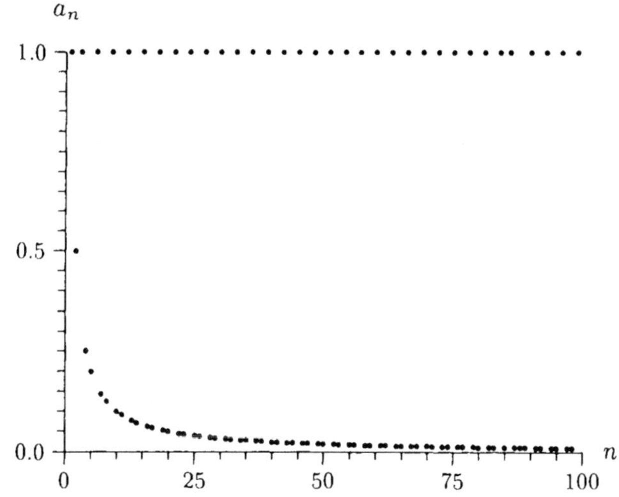

Figure 5.2.1 The values of a_n for n =1,2,. . ., 100.

Example 5.2.3: Consider the following two sequences:

$$a_n = \frac{1}{n+3} \qquad b_n = (-1)^n \cdot \frac{n}{n+3}$$

Plots of the first 100 terms in each of these two sequences are given in Figures 5.2.2 and 5.2.3.

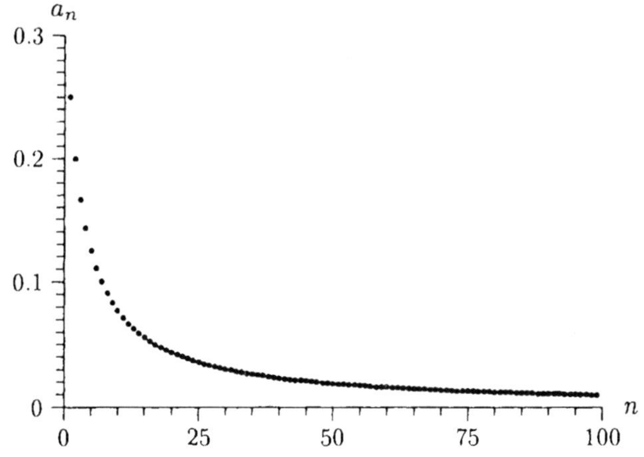

Figure 5.2.2 A plot of the first 100 terms of the sequence a_n.

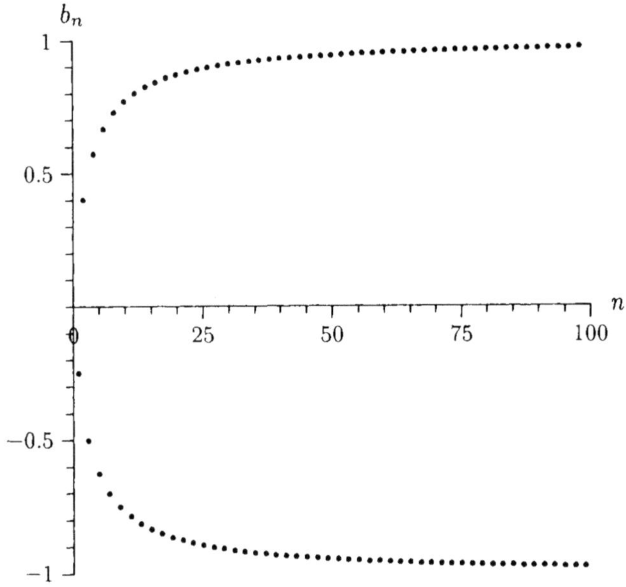

Fig. 5.2.3 A plot of the first 100 terms of the sequence b_n.

From the plots in Figures 5.2.2 and 5.2.3

a. Which of these two sequences appear to converge?
b. What are the apparent limits for the convergent sequences?

Solutions:

a. Clearly, a_n appears to be converging, but b_n does not appear to be converging to a limit.
b. a_n appears to be converging to a limit of $a = 0$.

Note that the key steps in proving that a sequence a_n converges to a limit a are (1) let $\epsilon > 0$ be arbitrary but fixed and (2) determine the value of $N \in \mathbb{N}$ such that $|a_n - a| < \epsilon$, $\forall\ n \geq N$. Note that the most difficult step in proving convergence is finding or constructing the value of N that works for a fixed value of ϵ. Most convergence proofs begin with the consideration of $|a_n - a|$ and then require algebraic manipulation so that $|a_n - a|$ can be related to the hypotheses of the theorem. A convergence proof of this type is called an ϵ-N proof, and an algorithm for proving $a_n \to a$ with an ϵ-N proof follows.

The ϵ-N Algorithm for Proving a Sequence Converges: Let a_n be a sequence of real numbers. Then, to prove that $a_n \to a$, perform the following steps:

1. Let $\epsilon > 0$ be ABF.
2. Consider $|a_n - a|$ and relate $|a_n - a|$ to the hypotheses of the theorem.
3. Determine how $|a_n - a|$ can be made arbitrarily small. Specifically, begin work on determination of the value of N so that $|a_n - a| < \epsilon$ whenever $n \geq N$.
4. From step 3, determine the value of $N \in \mathbb{N}$ so that $|a_n - a| < \epsilon$ whenever $n \geq N$.
5. Conclude $|a_n - a| < \epsilon$ whenever $n \geq N$ and therefore $a_n \to a$.

The following example illustrates a typical proof of the convergence for a sequence of real numbers a_n to a limit a. Again, the key to this proof, and all convergence proofs, is to determine a value of N so that $|a_n - a| < \epsilon$ whenever $n \geq N$.

Example 5.2.4: Let $a_n = \dfrac{n-1}{n+1}$. Prove that $\lim_{n\to\infty} a_n = 1$.

Scratchwork: Let $\epsilon > 0$ be ABF and consider $|a_n - a|$:

$$|a_n - a| = \left|\frac{n-1}{n+1} - 1\right| = \left|\frac{n-1}{n+1} - \frac{n+1}{n+1}\right| = \left|\frac{-2}{n+1}\right|$$

Now, the key is to find a value of $N \in \mathbb{N}$ such that $\left|\dfrac{-2}{n+1}\right| < \epsilon$ whenever $n \geq N$. To find the value of N, solve the inequality $\left|\dfrac{-2}{n+1}\right| < \epsilon$ for n.

$$\left|\frac{-2}{n+1}\right| < \epsilon \quad \text{iff} \quad \frac{2}{n+1} < \epsilon \quad \text{iff} \quad \frac{2}{\epsilon} - 1 < n$$

Thus, whenever $n > \dfrac{2}{\epsilon} - 1$, then $|a_n - a| < \epsilon$. Therefore, take N to be the smallest natural number greater than $n > \dfrac{2}{\epsilon} - 1$.

Proof: Let $\epsilon > 0$ be ABF. Consider $|a_n - 1|$:

$$|a_n - 1| = \left|\frac{n-1}{n+1} - 1\right| = \left|\frac{-2}{n+1}\right|$$

Now, the value of $N \in \mathbb{N}$ such that $|a_n - 1| < \epsilon$, $\forall\, n \geq N$ is found by solving the inequality $\left|\dfrac{-2}{n+1}\right| < \epsilon$ for n. Note that $\dfrac{2}{n+1} < \epsilon$ if and only if $n > \dfrac{2}{\epsilon} - 1$. Hence, let N be the smallest natural number greater than $\frac{2}{\epsilon} - 1$. Then, $\left|\frac{n-1}{n+1} - 1\right| < \epsilon$ whenever $n \geq N$ and therefore $\frac{n-1}{n+1} \to 1$.

■

Example 5.2.5: The elementary sequences listed in Table 5.2.1 are well known and commonly used convergent sequences.

Table 5.2.1 Elementary Sequences

Sequence	Domain	Limit	Restrictions
$\left(1+\frac{\gamma}{n}\right)^n$	$\mathbb{N}$	e^γ	$\gamma \in \mathbb{R}$
$\frac{1}{n^p}$	$\mathbb{N}$	0	$p>0$
β^n	$\mathbb{N}$	0	$\|\beta\| < 1$
$n^{\frac{1}{n}}$	$\mathbb{N}$	1	None
$\delta^{\frac{1}{n}}$	$\mathbb{N}$	1	$\delta > 0$

Example 5.2.6: Using Table 5.2.1, determine the limits of the following sequences:

a. $a_n = \left(1 - \frac{3}{n}\right)^n$.

b. $b_n = \left(1 + \frac{1}{2n}\right)^n$.

c. $c_n = 2^{\frac{1}{n}}$.

d. $d_n = 0.99^n$.

Solutions: From table 5.2.1 it follows that

a. For $a_n = \left(1 - \frac{3}{n}\right)^n$, $\lim\limits_{n\to\infty} a_n = e^{-3}$.

b. For $b_n = \left(1 + \frac{1}{2n}\right)^n$, $\lim\limits_{n\to\infty} b_n = e^{\frac{1}{2}}$.

c. For $c_n = 2^{\frac{1}{n}}$, $\lim_{n \to \infty} c_n = 1$.

d. For $d_n = 0.99^n$, $\lim_{n \to \infty} d_n = 0$.

One of the most frequently used mathematical tools in an ϵ-N proof of convergence is the *triangle inequality* (Theorem 4.2.12). The particular version of the triangle inequality that will be used in most convergence proofs is given in the corollary to Theorem 4.2.12, which is stated below.

Corollary to Theorem 4.2.12 (The Triangle Inequality): If $x, y, z \in \mathbb{R}$, then $|x - y| \leq |x - z| + |z - y|$.

The key to using the triangle inequality is to determine the appropriate form of $0 = z - z$ which is then added to $x - y$:

$$|x - y| = |x - y + 0| = |x - y + z - z| = |x - z + z - y| \underbrace{\leq |x - z| + |z - y|}_{\text{By Theorem 4.2.12}}$$

The proof of the following theorem illustrates a simple application of the triangle inequality. The following theorem states that the limit of a sequence is unique.

Theorem 5.2.1: Let a_n be a sequence of real numbers. If $\lim_{n \to \infty} a_n = a$, then the limit a is unique.

Proof:* *(Uniqueness Proof): Let a_n be a sequence of real numbers with $\lim_{n \to \infty} a_n = a$, and suppose that the limit a is not unique. Furthermore, suppose that $\lim_{n \to \infty} a_n = b$ and $b \neq a$. Thus, $a_n \to a$, $a_n \to b$, and $a \neq b$.

Let $\epsilon > 0$ be ABF. Since $a_n \to a$, there exists $N_1 \in \mathbb{N}$ such that $|a_n - a| < \frac{\epsilon}{2}$, whenever $n \geq N_1$. Similarly, since $a_n \to b$, there exists $N_2 \in \mathbb{N}$ such that $|a_n - b| < \frac{\epsilon}{2}$ whenever $n \geq N_2$.

Consider $|a - b|$:

$$|a - b| = |a - b + 0| = |a \underbrace{- a_n + a_n}_{0} - b|$$

$$\underbrace{\leq |a - a_n| + |a_n - b|}_{\text{By the triangle inequality}}$$

Let $n \geq N = \max(N_1, N_2)$. Then, it follows that both $|a_n - a| < \frac{\epsilon}{2}$ and $|a_n - b| < \frac{\epsilon}{2}$, $\forall\, n \geq N$. Hence, for $n \geq N$, it also follows that

$$|a - b| = |a - a_n + a_n - b| < |a - a_n| + |a_n - b| < \frac{\epsilon}{2} + \frac{\epsilon}{2} = \epsilon$$

Now, since $|a - b| < \epsilon$ for every $\epsilon > 0$, it follows that $a = b$, which contradicts $a \neq b$. Therefore, if $a_n \to a$, then the limit a is unique.

■

It is important to note that most ϵ-N convergence proofs are somewhat similar in nature and that the key steps for showing that $a_n \to a$ are (1) let $\epsilon > 0$ be ABF and (2) determine the value of $N \in \mathbb{N}$ such that $|a_n - a| < \epsilon$ for all $n \geq N$.

Example 5.2.7: Prove the following result. If $\lim_{n\to\infty} a_n = a$, then $\lim_{n\to\infty} ca_n + l = ca + l$ for all $c, l \in \mathbb{R}$.

Solution: The scratchwork for solving this problem is given below.

Step 1: Let $\epsilon > 0$ be ABF and let c and l be real numbers. Suppose $a_n \to a$.

Step 2: The hypothesis is $a_n \to a$. This means that $\exists\, n \in \mathbb{N}$ such that $|a_n - a| < \epsilon$ whenever $n \geq N$.

Step 3: Consider $|ca_n + l - (ca + l)|$. First, if $c = 0$, then $ca_n = 0$ for all $n \in \mathbb{N}$ and thus, $|ca_n + l - (ca + l)| = |l - l| = 0$. In this case, $ca_n + l = l = ca + l$ for each n and therefore, $ca_n + l \to ca + l$. Now, if $c \neq 0$, then

$$|ca_n + l - (ca + l)| = |ca_n - ca| = |c| \cdot |a_n - a|$$

Step 4: Using steps 2 and 3, the goal is to make $|ca_n + l - (ca + l)|$ arbitrarily small. Note that this can be done by making $|a_n - a|$ small, since c is a constant. Thus, for $n \geq N$

$$|ca_n + l - (ca + l)| = |ca_n - ca| = |c| \cdot |a_n - a| < |c| \cdot \epsilon$$

Now, when $|a_n - a| < \frac{\epsilon}{|c|}$, it follows that

$$|ca_n + l - (ca + l)| = |ca_n - ca| = |c| \cdot |a_n - a| < |c| \cdot \frac{\epsilon}{|c|} = \epsilon$$

Thus, the value of N that will make $|a_n - a| < \dfrac{\epsilon}{|c|}$ whenever $n \geq N$ is also the value of N such that $|ca_n + l - (ca + l)| < \epsilon$ whenever $n \geq N$.

Therefore, let $N \in \mathbb{N}$ be the value such that $|a_n - a| < \dfrac{\epsilon}{|c|}$ whenever $n \geq N$.

Now, this scratchwork is rewritten below in the form of a formal proof.

Proof: Let c and l be real numbers and suppose that $a_n \to a$. Let $\epsilon > 0$ be ABF. Now, $c = 0$ or $c \neq 0$.

Case 1: Suppose that $c = 0$; then $ca + l = l$, $ca_n = 0$ and $ca_n + l = l$ $\forall\, n \in \mathbb{N}$. Therefore, when $c = 0$, it follows that $|ca_n + l - (ca + l)| = 0$, $\forall\, n \in \mathbb{N}$. Thus, for $N = 1$, it follows that $|ca_n + l - (ca + l)| = 0 < \epsilon$ whenever $n \geq 1$. Therefore, $ca_n + l \to ca + l$ when $c = 0$.

Case 2: Suppose that $c \neq 0$. Then, since $a_n \to a$, $\exists\, n \in \mathbb{N}$ such that $|a_n - a| < \epsilon^{\star} = \dfrac{\epsilon}{|c|}$ whenever $n \geq N$. Consider $|ca_n + l - (ca + l)|$:

$$|ca_n + l - (ca + l)| = |ca_n - ca| = |c| \cdot |a_n - a|$$

Now, since c is a constant and $a_n \to a$, it follows that $|ca_n - ca|$ can be made arbitrarily small by making $|a_n - a|$ small. Furthermore, since $|a_n - a| < \epsilon^{\star} = \dfrac{\epsilon}{|c|}$ whenever $n \geq N$, it follows that for $n \geq N$

$$|ca_n + l - (ca + l)| = |ca_n - ca| = |c| \cdot |a_n - a|$$

$$< |c| \cdot \epsilon^{\star} = |c| \cdot \frac{\epsilon}{|c|} = \epsilon$$

Thus, $ca_n + l \to ca + l$ when $c \neq 0$.

Therefore, in either case $\lim\limits_{n\to\infty} ca_n + l = ca + l$ for all $c, l \in \mathbb{R}$ whenever $\lim\limits_{n\to\infty} a_n = a$.

■

The following two theorems provide important results concerning the convergence of a sequence created from the sum of two convergent sequences. In particular, these two theorems provide rules for determining the limits of sequences of the form $a_n + b_n$ and $ra_n + sb_n$ (i.e., linear combinations of a_n and b_n) for convergent sequences a_n and b_n.

Theorem 5.2.2: Let a_n and b_n be sequences of real numbers. If $\lim_{n\to\infty} a_n = a$ and $\lim_{n\to\infty} b_n = b$, then

(i) $\lim_{n\to\infty} (ca_n + l) = ca + l$ for all $c, l \in \mathbb{R}$.

(ii) $\lim_{n\to\infty} (a_n + b_n) = a + b$.

(iii) $\lim_{n\to\infty} (ra_n + sb_n) = ra + sb$ for all $r, s \in \mathbb{R}$.

Proof: Let a_n and b_n be sequences of real numbers with $\lim_{n\to\infty} a_n = a$ and $\lim_{n\to\infty} b_n = b$. Let $\epsilon > 0$ be arbitrary but fixed.

Proof of part (i): Part (i) was proved in Example 5.2.7.

Proof of part (ii): Since $a_n \to a$, there exists $N_1 \in \mathbb{N}$ such that $|a_n - a| < \frac{\epsilon}{2}$ whenever $n \geq N_1$. Similarly, since $b_n \to b$ there exists $N_2 \in \mathbb{N}$ such that $|b_n - b| < \frac{\epsilon}{2}$ whenever $n \geq N_2$.

Consider $|a_n + b_n - (a + b)|$:

$$|a_n + b_n - (a + b)| = |a_n - a + b_n - b| \quad \underbrace{\leq |a_n - a| + |b_n - b|}_{\text{By the triangle inequality}}$$

Let $N = \max(N_1, N_2)$. For $n \geq N$ it follows that

$$|a_n + b_n - (a + b)| \leq |a_n - a| + |b_n - b| < \frac{\epsilon}{2} + \frac{\epsilon}{2} = \epsilon$$

Therefore, $a_n + b_n \to a + b$.

Proof of part (iii): The proof of part(iii) is left as an exercise.

■

Corollary to Theorem 5.2.2: Let a_n and b_n be sequences of real numbers. If $\lim_{n\to\infty} a_n = a$ and $\lim_{n\to\infty} b_n = b$, then, $\lim_{n\to\infty} a_n - b_n = a - b$.

Proof: The corollary to Theorem 5.2.2 follows directly from Theorem 5.2.2 (iii) with $r = 1$ and $s = -1$.

■

Example 5.2.8: Let the sequences a_n and b_n be defined by $a_n = \dfrac{n^2}{n^2+2n+2}$ and $b_n = 1 + \dfrac{(-1)^n}{n}$. Then, $\lim_{n\to\infty} a_n = 1$ and $\lim_{n\to\infty} b_n = 1$. Use these limits to determine

a. $\lim_{n\to\infty} [3a_n + 2]$.

b. $\lim_{n\to\infty} [a_n + b_n]$.

c. $\lim_{n\to\infty} [3a_n - 4b_n]$.

Solutions: Since $\lim_{n\to\infty} a_n = 1$ and $\lim_{n\to\infty} b_n = 1$, it follows that

a. By theorem 5.2.2 (i), $3a_n + 2 \to 3(1) + 2 = 5$.

b. By theorem 5.2.2 (ii), $a_n + b_n \to 1 + 1 = 2$.

c. By theorem 5.2.2 (iii), $3a_n - 4b_n \to 3 - 4 = -1$.

The following two lemmas provide important results concerning bounds on a convergent sequence, and these two lemmas will be used in proving the theorems concerning the product and ratio of two convergent sequences. In particular, Lemma 5.2.1 shows that every convergent sequence is bounded, and Lemma 5.2.2 shows that a convergent sequence whose limit is nonzero is also bounded from below from some point on.

Lemma 5.2.1: Let a_n be a sequence of real numbers. If $\lim_{n\to\infty} a_n = a$, then there exists a real number M such that $|a_n| \leq M$, $\forall\, n \in \mathbb{N}$.

Proof: Let a_n be a sequence of real numbers with $a_n \to a$. Since $a_n \to a$, there exists N_1 such that for $\epsilon = 1$, $|a_n - a| < 1$ whenever $n \geq N_1$. Now, consider $|a_n|$.

$$|a_n| = |a_n - a + a| \quad \underbrace{\leq |a_n - a| + |a|}_{\text{By the triangle inequality}} \quad < 1 + |a|$$

whenever $n \geq N_1$. Thus, for $n \geq N_1$, $|a_n|$ is bounded by $1 + |a|$.

When $n < N_1$, $|a_n|$ is bounded by $L = \max\{|a_1|, |a_2|, \ldots, |a_{N_1-1}|\}$. Thus, let $M = \max\{L, 1 + |a|\}$. Then, for all $n \in \mathbb{N}$, it follows that M is an upper bound for a_n.

Therefore, a_n is bounded.

∎

Lemma 5.2.2: Let b_n be a sequence of real numbers. If $\lim_{n\to\infty} b_n = b$ and $b \neq 0$, then there exists $N \in \mathbb{N}$ such that $|b_n| \geq \frac{|b|}{2}$, $\forall\, n \geq N$.

Proof: Let b_n be a sequence of real numbers with $b_n \to b$ and $b \neq 0$. Now, since $b \neq 0$, there exists $N \in \mathbb{N}$ such that $|b_n - b| < \frac{|b|}{2}$, whenever $n \geq N$. Consider $|b_n|$:

$$|b_n| = |b_n + b - b| = |b - (b - b_n)| \quad \underbrace{\geq |b| - |b_n - b|}_{\text{By Theorem 4.2.13}}$$

Now, for $n \geq N$ it follows that

$$|b_n| \geq |b| - |b_n - b| > |b| - \frac{|b|}{2} = \frac{|b|}{2} > 0$$

and therefore, $|b_n| \geq \dfrac{|b|}{2}$ whenever $n \geq N$.

■

Now, Lemmas 5.2.1 and 5.2.2 will be used in proving the following theorem on the product and ratio of convergent sequences. In particular, this theorem states that if $a_n \to a$ and $b_n \to b$, then it follows that $a_n b_n \to ab$ and $\dfrac{a_n}{b_n} \to \dfrac{a}{b}$, also.

Theorem 5.2.3: Let a_n and b_n be sequences of real numbers. If $\lim_{n\to\infty} a_n = a$ and $\lim_{n\to\infty} b_n = b$, then

(i) $\lim_{n\to\infty} (a_n \cdot b_n) = a \cdot b$.

(ii) $\lim_{n\to\infty} \dfrac{a_n}{b_n} = \dfrac{a}{b}$, provided that $b \neq 0$.

Proof: Let a_n and b_n be sequences of real numbers with $\lim_{n\to\infty} a_n = a$ and $\lim_{n\to\infty} b_n = b$. Let $\epsilon > 0$ be ABF.

Proof of part (i): Now since $a_n \to a$, there exists

(1) A bound M such that $|a_n| \leq M$, $\forall\, n \in \mathbb{N}$.

(2) $N_1 \in \mathbb{N}$ such that $|a_n - a| < \dfrac{\epsilon}{2(1 + |b|)}$ whenever $n \geq N_1$.

Also, since $b_n \to b$, there exists $N_2 \in \mathbb{N}$ such that $|b_n - b| < \dfrac{\epsilon}{2M}$ whenever $n \geq N_2$.

Consider $|a_n \cdot b_n - a \cdot b|$:

$$|a_n \cdot b_n - a \cdot b| = |a_n \cdot b_n \underbrace{- a_n \cdot b + a_n \cdot b}_{0} - a \cdot b|$$

$$= |a_n(b_n - b) + b(a_n - a)| \underbrace{\leq |a_n(b_n - b)| + |b(a_n - a)|}_{\text{By the triangle inequality}}$$

$$\leq M \cdot |b_n - b| + |b| \cdot |a_n - a|$$

Let $N = \max(N_1, N_2)$. Then, for $n \geq N$, it follows that

$$|a_n \cdot b_n - a \cdot b| \leq M \cdot |b_n - b| + |b| \cdot |a_n - a|$$

$$< M \cdot \frac{\epsilon}{2M} + |b| \cdot \frac{\epsilon}{2(1 + |b|)}$$

$$= \frac{\epsilon}{2} + \frac{|b|\epsilon}{2(1 + |b|)} < \frac{\epsilon}{2} + \frac{\epsilon}{2} = \epsilon$$

Therefore, $a_n \cdot b_n \to a \cdot b$ whenever $a_n \to a$ and $b_n \to b$.

Proof of part (ii): Let $\epsilon > 0$ be arbitrary but fixed and assume that $b \neq 0$.

Now, since $b_n \to b \neq 0$, there exists

(1) $N_1 \in \mathbb{N}$ such that $|b_n| > \dfrac{|b|}{2}$ whenever $n \geq N_1$.

(2) $N_2 \in \mathbb{N}$ such that $|b_n - b| < \epsilon \cdot \dfrac{|b|^2}{4(|a| + 1)}$ whenever $n \geq N_2$.

Also, since $a_n \to a$, there exists $N_3 \in \mathbb{N}$ such that $|a_n - a| < \epsilon \cdot \dfrac{|b|}{4}$ whenever $n \geq N_3$.

Now consider $\left|\frac{a_n}{b_n} - \frac{a}{b}\right|$.

$$\left|\frac{a_n}{b_n} - \frac{a}{b}\right| = \left|\frac{a_n b - a b_n}{b b_n}\right| = \left|\frac{a_n b - ab + ab - a b_n}{b b_n}\right|$$

$$\leq \frac{|b| \cdot |a_n - a| + |a| \cdot |b_n - b|}{|b b_n|}$$

$$= \frac{|a_n - a|}{|b_n|} + \frac{|a| \cdot |b_n - b|}{|b b_n|}$$

Let $N = \max(N_1, N_2, N_3)$. Then for $n > N$ it follows that

$$\left|\frac{a_n}{b_n} - \frac{a}{b}\right| < \underbrace{2 \cdot \frac{|a_n - a|}{|b|} + 2 \cdot \frac{|a| \cdot |b_n - b|}{|b|^2}}_{\text{Since } |b_n| > \frac{|b|}{2}}$$

$$< 2 \cdot \frac{|b| \cdot \epsilon}{4|b|} + 2 \cdot \frac{|a| \cdot |b|^2 \epsilon}{4|b|^2 \cdot (|a| + 1)} < \frac{\epsilon}{2} + \frac{\epsilon}{2} = \epsilon$$

Therefore, $\frac{a_n}{b_n} \rightarrow \frac{a}{b}$ whenever $a_n \rightarrow a$ and $b_n \rightarrow b \neq 0$.

■

Example 5.2.9: Let $a_n = \frac{n+4}{2n+1}$ and $b_n = 2 - \frac{1}{n}$. Determine

a. $\lim_{n \to \infty} a_n$

b. $\lim_{n \to \infty} b_n$.

c. $\lim_{n \to \infty} a_n b_n$.

d. $\lim_{n \to \infty} \frac{a_n}{b_n}$.

Solutions: Since $\lim_{n \to \infty} a_n = \frac{1}{2}$ and $\lim_{n \to \infty} b_n = 2$, it follows that

a. $\lim_{n \to \infty} a_n = \frac{1}{2}$.

b. $\lim_{n \to \infty} b_n = 2.$

c. $\lim_{n \to \infty} a_n b_n = \frac{1}{2} \cdot 2 = 1.$

d. $\lim_{n \to \infty} \frac{a_n}{b_n} = \frac{\frac{1}{2}}{2} = \frac{1}{4}.$

The following corollary considers two special cases of Theorem 5.2.3. In particular, this corollary shows that when $a_n \to a$, it follows that a_n^2 converges to a^2 and $\frac{1}{a_n}$ converges to $\frac{1}{a}$, provided that $a \neq 0$.

Corollary to Theorem 5.2.3: Let a_n be a sequence of real numbers with $\lim_{n \to \infty} a_n = a$. Then

(i) $\lim_{n \to \infty} a_n^2 = a^2$.

(ii) $\lim_{n \to \infty} \frac{1}{a_n} = \frac{1}{a}$, provided that $a \neq 0$.

Proof: Let a_n be a sequence of real numbers with $\lim_{n \to \infty} a_n = a$.

Proof of (i): Part (i) follows directly from Theorem 5.2.3 part (i) since $a_n^2 = a_n \cdot a_n$.

Proof of (ii): Part (ii) follows directly from Theorem 5.2.3 part (ii) with $a_n = 1$ and $b_n = a_n$.

■

The following theorem provides a more general result than part (i) of the corollary to Theorem 5.2.3. In particular, the following theorem shows that if $a_n \to a$, then $a_n^m \to a^m$, $\forall\, m \in \mathbb{N}$.

Theorem 5.2.4: Let a_n be a sequence of real numbers. If $a_n \to a$, then $a_n^m \to a^m$, $\forall\, m \in \mathbb{N}$.

Proof: The proof of Theorem 5.2.4 is left as an exercise.

■

While the previous theorem states that $a_n^m \to a^m$ for all natural numbers m, this result can be further generalized to hold for all positive real numbers

m (i.e., $m \in \mathbb{R}^+$) for a nonnegative sequence a_n. The proof for $m \in \mathbb{R}^+$ will not be considered here; however, the following theorem shows that for $m = \frac{1}{2}$ that $\sqrt{a_n} \to \sqrt{a}$ whenever $a_n \geq 0$ and $a_n \to a$.

Theorem 5.2.5: Let a_n be a sequence of nonnegative real numbers with $\lim_{n\to\infty} a_n = a$. Then, $\sqrt{a_n} \to \sqrt{a}$.

Proof: Let a_n be a sequence of nonnegative real numbers with $\lim_{n\to\infty} a_n = a$, and let $\epsilon > 0$ be ABF. Now, either $a = 0$ or $a > 0$.

Case 1: Suppose that $a = 0$. Since $a_n \to a$, there exists $N \in \mathbb{N}$ such that $|a_n - 0| < \epsilon^2$. Consider $|\sqrt{a_n} - 0|$:

$$|\sqrt{a_n} - 0| = |\sqrt{a_n}| = \sqrt{a_n}$$

Now, for $n \geq N$

$$|\sqrt{a_n} - 0| = \sqrt{a_n} \quad \underbrace{< \epsilon}_{\text{Since } |a_n| < \epsilon^2}$$

Therefore, $\sqrt{a_n} \to 0$.

Case 2: Suppose that $a > 0$. Then

(1) $\exists\, N_1 \in \mathbb{N}$ such that $|a_n - a| < \epsilon\left(\sqrt{\frac{a}{2}} + \sqrt{a}\right)$ whenever $n \geq N$.

(2) $\exists\, N_2 \in \mathbb{N}$ such that $|a_n| \geq \frac{a}{2}$ whenever $n \geq N$.

Now consider $|\sqrt{a_n} - \sqrt{a}|$.

$$|\sqrt{a_n} - \sqrt{a}| = |\sqrt{a_n} - \sqrt{a}| \cdot \frac{|\sqrt{a_n} + a\sqrt{a}|}{|\sqrt{a_n} + \sqrt{a}|} = \frac{|a_n - a|}{|\sqrt{a_n} + \sqrt{a}|}$$

Let $n \geq N = \max(N_1, N_2)$. Then for $n \geq N$

$$|\sqrt{a_n} - \sqrt{a}| = \frac{|a_n - a|}{|\sqrt{a_n} + \sqrt{a}|} \underbrace{< \frac{|a_n - a|}{|\sqrt{\frac{a}{2}} + \sqrt{a}|}}_{\text{By condition 2}}$$

$$\underbrace{< \frac{\left(\sqrt{\frac{a}{2}} + \sqrt{a}\right)\epsilon}{\left[\sqrt{\frac{a}{2}} + \sqrt{a}\right]}}_{\text{By condition 1}} = \epsilon$$

Hence, $\sqrt{a_n} \to \sqrt{a}$ when $a > 0$.

Therefore, in either case $\sqrt{a_n} \to \sqrt{a}$ whenever a_n is a nonnegative sequence and $a_n \to a$.

■

Corollary to Theorem 5.2.5: Let a_n be a sequence of nonnegative real numbers. If $\lim_{n\to\infty} a_n = a$, then $a_n^{m/2} \to a^{m/2}$, $\forall\, m \in \mathbb{N}$.

Proof: The proof of the corollary to Theorem 5.2.5 follows directly from Theorems 5.2.4 and 5.2.5 since $a_n^{m/2} = \sqrt{a_n^m}$.

■

Example 5.2.10: Let $a_n = \dfrac{n+1}{4n-3}$. Determine

a. $\displaystyle\lim_{n\to\infty} \frac{n+1}{4n-3}$.

b. $\displaystyle\lim_{n\to\infty} \left(\frac{n+1}{4n-3}\right)^3$.

c. $\displaystyle\lim_{n\to\infty} \sqrt{\frac{n+1}{4n-3}}$.

Solutions:

a. Consider $\dfrac{n+1}{4n-3}$.

$$\frac{n+1}{4n-3} = \frac{n(1+1/n)}{n(4-3/n)} = \frac{(1+\frac{1}{n})}{(4-\frac{3}{n})}$$

Now, $\lim_{n\to\infty} (1+\dfrac{1}{n}) = 1$ and $\lim_{n\to\infty} (4-\dfrac{3}{n}) = 4$. Hence

$$\lim_{n\to\infty} \frac{n+1}{4n-3} = \frac{1}{4}$$

by Theorem 5.2.3.

b. By Theorem 5.2.4, $\displaystyle\lim_{n\to\infty} \left(\frac{n+1}{4n-3}\right)^3 = \left(\frac{1}{4}\right)^3 = \frac{1}{64}$.

c. By theorem 5.2.5, $\lim_{n \to \infty} \sqrt{\frac{n+1}{4n-3}} = \sqrt{\frac{1}{4}} = \frac{1}{2}$.

The next theorem is often useful in determining the convergence or divergence of a sequence. However, using this theorem often requires some mathematical ingenuity. In particular, the following theorem provides a useful result for determining the limit of a sequence by comparing it to two other sequences that have the same limit. The following theorem is known as the *squeeze theorem.*

Theorem 5.2.6 (The Squeeze Theorem): Let a_n, b_n and c_n be sequences with $a_n \leq b_n \leq c_n$, $\forall\, n \in \mathbb{N}$. If $\lim_{n \to \infty} a_n = \alpha$ and $\lim_{n \to \infty} c_n = \alpha$, then $\lim_{n \to \infty} b_n = \alpha$.

Proof: Let a_n, b_n and c_n be sequences with $a_n \leq b_n \leq c_n$, $\forall\, n \in \mathbb{N}$. Suppose that $a_n \to \alpha$ and $c_n \to \alpha$, and let $\epsilon > 0$ be ABF.

Since $a_n \to \alpha$, then $\exists\, N_1 \in \mathbb{N}$ such that $|a_n - \alpha| < \epsilon$ whenever $n \geq N_1$. This means that whenever $n \geq N_1$

$$\alpha - \epsilon < a_n < \alpha + \epsilon$$

Similarly, since $c_n \to \alpha$, $\exists\, N_2 \in \mathbb{N}$ such that $|c_n - \alpha| < \epsilon$ whenever $n \geq N_2$, which means that whenever $n \geq N_2$

$$\alpha - \epsilon < c_n < \alpha + \epsilon$$

Now, let $N = \max(N_1, N_2)$. Then, for $n \geq N$ it follows that

$$\alpha - \epsilon < a_n \leq b_n \leq c_n < \alpha + \epsilon$$

Thus, $|b_n - \alpha| < \epsilon$ whenever $n \geq N$ and therefore $b_n \to \alpha$.

■

Example 5.2.11: Show that $\lim_{n \to \infty} \frac{\sin(n)}{n} = 0$.

Solution: Since $-1 \leq \sin(n) \leq 1$, it follows that $-\frac{1}{n} \leq \frac{\sin(n)}{n} \leq \frac{1}{n}$, $\forall\, n \in \mathbb{N}$.

Thus, let $a_n = -\frac{1}{n}$, $b_n = \frac{\sin(n)}{n}$, and $c_n = \frac{1}{n}$.

Now, $\lim_{n\to\infty} a_n = \lim_{n\to\infty} c_n = 0$. Thus, by the squeeze theorem it follows that $\lim_{n\to\infty} \frac{\sin(n)}{n} = 0$.

Since the convergence of a sequence does not depend on its first N terms, it is not necessary that $a_n \le b_n \le c_n$, $\forall\, n \in \mathbb{N}$, in order for the squeeze theorem to work. A less restrictive hypothesis that can be used in the squeeze theorem in place of $a_n \le b_n \le c_n$, $\forall\, n \in \mathbb{N}$ is "there exists $N \in \mathbb{N}$ such that $a_n \le b_n \le c_n$ whenever $n \ge N$." The more general version of the squeeze theorem is stated below.

Theorem 5.2.7 (The Generalized Squeeze Theorem): Let a_n, b_n, and c_n be sequences and suppose that there exists $N \in \mathbb{N}$ such that $a_n \le b_n \le c_n$ whenever $n \ge N$. If $\lim_{n\to\infty} a_n = \alpha$ and $\lim_{n\to\infty} c_n = \alpha$, then $\lim_{n\to\infty} b_n = \alpha$.

Proof: Let a_n, b_n and c_n be sequences. Suppose there exists $N \in \mathbb{N}$ such that $a_n \le b_n \le c_n$ whenever $n \ge N$. Assume $\lim_{n\to\infty} a_n = \alpha$ and $\lim_{n\to\infty} c_n = \alpha$. Let $\epsilon > 0$ be ABF. Now, $\exists\ N_1 \in \mathbb{N}$ such that $a_n \le b_n \le c_n$ whenever $n \ge N_1$ and since $a_n \to \alpha$, $\exists\ N_2 \in \mathbb{N}$ such that $|a_n - \alpha| < \epsilon$ whenever $n \ge N_2$. Then, for $n \ge N_1$ it follows that

$$\alpha - \epsilon < a_n < \alpha + \epsilon$$

Also, since $c_n \to \alpha$, $\exists\ N_3 \in \mathbb{N}$ such that $|c_n - \alpha| < \epsilon$ whenever $n \ge N_3$. Then, for $n \ge N_2$ it follows that

$$\alpha - \epsilon < c_n < \alpha + \epsilon$$

Now, let $N = \max(N_1, N_2, N_3)$. Then, for $n \ge N$ it follows that

$$\alpha - \epsilon < a_n \le b_n \le c_n < \alpha + \epsilon$$

Thus, $|b_n - \alpha| < \epsilon$ whenever $n \ge N$, and therefore $b_n \to \alpha$.

■

Example 5.2.12: Prove the following result. If $a > 0$, then $\lim_{n\to\infty} a^{1/n} = 1$.

Solution: The second version of the squeeze theorem, Theorem 5.2.7, will be used to prove this result.

Proof: Let $a > 0$ be ABF. Then, either $a \ge 1$ or $a < 1$.

Case 1: Suppose that $a \geq 1$. Then, $\exists\ N \in \mathbb{N}$ such that $a \leq N$ and thus $1 \leq a^{1/n} \leq n^{1/n}$ for $n \geq N$. Also, $\lim\limits_{n\to\infty} 1 = \lim\limits_{n\to\infty} n^{1/n} = 1$.

Therefore, by Theorem 5.2.7, $\lim\limits_{n\to\infty} a^{1/n} = 1$ whenever $a \geq 1$.

Case 2: Suppose that $0 < a < 1$. Then, $\dfrac{1}{a} > 1$ and thus, by case 1 it follows that $\lim\limits_{n\to\infty} \left(\dfrac{1}{a}\right)^{1/n} = 1$. Now, by the corollary to Theorem 5.2.3 part (ii)

$$\lim_{n\to\infty} a^{1/n} = \lim_{n\to\infty} \frac{1}{\left(\frac{1}{a}\right)^{1/n}} = \frac{1}{\lim\limits_{n\to\infty} \left(\frac{1}{a}\right)^{1/n}} = \frac{1}{1} = 1$$

Thus, $\lim\limits_{n\to\infty} a^{1/n} = 1$ whenever $0 < a < 1$.

Therefore, $\lim\limits_{n\to\infty} a^{1/n} = 1,\ \forall\ a > 0$.

■

5.2.2 Monotone Sequences

Often the most difficult question concerning a sequence of real numbers is the question of whether the sequence converges or diverges. Once it is determined that a sequence does converge, finding the value of its limit is often fairly easy. In fact, the limit of a convergent sequence a_n can be approximated by simply looking at the value of a_n for a "very large" value of n; note that the actual size of n will depend on the rate of convergence of the sequence.

In this section, the two special types of sequences that will be studied are the nondecreasing and nonincreasing sequences. A sequence that is either nondecreasing and nonincreasing is called a *monotone sequence.* In Theorem 5.2.16 it will be shown that a monotone sequence converges if and only if it is bounded.

Definition 5.2.3: A sequence of real numbers a_n is called

(i) A *nondecreasing* sequence when $a_{n+1} \geq a_n,\ \forall n \in \mathbb{N}$, and a *strictly increasing* sequence when $a_{n+1} > a_n,\ \forall\ n \in \mathbb{N}$.

(ii) A *nonincreasing* sequence when $a_{n+1} \leq a_n,\ \forall n \in \mathbb{N}$, and a *strictly decreasing* sequence when $a_{n+1} < a_n,\ \forall\ n \in \mathbb{N}$.

(iii) A *monotone* sequence or a *monotonic* sequence if it is either nondecreasing, strictly increasing, nonincreasing, or strictly decreasing.

When a_n is a monotonic sequence, then $a_n \uparrow$ is used to denote that a_n is a nondecreasing sequence and $a_n \downarrow$ is used to denote that a_n is a nonincreasing sequence. Note that for a nondecreasing sequence a_n, it follows that

$$a_1 \leq a_2 \leq a_3 \leq \cdots$$

and analogously, for a nonincreasing sequence a_n

$$a_1 \geq a_2 \geq a_3 \geq \cdots$$

For example, the sequence $a_n = \frac{1}{n}$ is a strictly decreasing sequence and

$$a_1 = 1 > a_2 = \frac{1}{2} > a_3 = \frac{1}{3} > \ldots$$

Examples of three different monotone sequences are shown in Example 5.2.13.

Example 5.2.13: For each of the following sequences write out the first five terms and determine whether the sequence is nondecreasing, strictly increasing, nonincreasing, or strictly decreasing:

a. $a_n = \dfrac{n}{n+1}$

b. $b_n = 1 + \dfrac{1}{n}$

c. $c_n = \left(\dfrac{1}{2}\right)^n$

Solutions:

a. For $a_n = \frac{n}{n+1}$, $a_1 = \frac{1}{2} < a_2 = \frac{2}{3} < a_3 = \frac{3}{4} < a_4 = \frac{4}{5} < a_5 = \frac{5}{6}$ and a_n is a nondecreasing sequence (i.e., $a_n \uparrow$). In fact, a_n is strictly increasing.

b. For $b_n = 1 + \frac{1}{n}$, $b_1 = 2 > b_2 = \frac{3}{2} > b_3 = \frac{4}{3} > b_4 = \frac{5}{4} > b_5 = \frac{6}{5}$; therefore, $b_n \downarrow$, and in fact b_n is strictly decreasing.

c. For $c_n = \left(\frac{1}{2}\right)^n$, $c_1 = \frac{1}{2} > b_2 = \frac{1}{4} > b_3 = \frac{1}{8} > b_4 = \frac{1}{16} > b_5 = \frac{1}{32}$. Thus, c_n is strictly decreasing.

Now, two very important questions concerning a sequence a_n are (1) "Is a_n a monotone sequence?" and (2) "If a_n is a monotone sequence, is $a_n \uparrow$ or $a_n \downarrow$?" The answer to both of these questions can often be found by comparing the values of a_n and a_{n+1} for an arbitrary value of $n \in \mathbb{N}$. The following three theorems provide useful tests for determining whether a sequence of real numbers is monotone. The monotonicity test in Theorem 5.2.8 works only for nonnegative sequences and is based on the ratio of successive terms in the sequence.

Theorem 5.2.8: Let a_n be a sequence of positive real numbers. Then

(i) $a_n \uparrow$ if and only if $\frac{a_{n+1}}{a_n} \geq 1, \ \forall n \in \mathbb{N}$.

(ii) $a_n \downarrow$ if and only if $\frac{a_{n+1}}{a_n} \leq 1, \ \forall n \in \mathbb{N}$.

Proof: Let a_n be a sequence of positive real numbers.

Proof of part (i): First, suppose that $\dfrac{a_{n+1}}{a_n} \geq 1, \ \forall n \in \mathbb{N}$. Then $a_{n+1} \geq a_n, \ \forall n \in \mathbb{N}$; therefore, $a_n \uparrow$ when $\dfrac{a_{n+1}}{a_n} \geq 1, \ \forall n \in \mathbb{N}$.

Conversely, suppose that $a_n \uparrow$. Then, $a_{n+1} \geq a_n, \ \forall n \in \mathbb{N}$, which means that $\dfrac{a_{n+1}}{a_n} \geq 1, \ \forall n \in \mathbb{N}$. Therefore, $\dfrac{a_{n+1}}{a_n} \geq 1, \ \forall n \in \mathbb{N}$ when $a_n \uparrow$.

Proof of part (ii): First, suppose that $\dfrac{a_{n+1}}{a_n} \leq 1, \ \forall n \in \mathbb{N}$. Then $a_{n+1} \leq a_n, \ \forall n \in \mathbb{N}$; therefore, $a_n \downarrow$ when $\dfrac{a_{n+1}}{a_n} \leq 1, \ \forall n \in \mathbb{N}$.

Conversely, suppose that $a_n \downarrow$. Then, $a_{n+1} \leq a_n, \ \forall n \in \mathbb{N}$, which means that $\dfrac{a_{n+1}}{a_n} \leq 1, \ \forall n \in \mathbb{N}$. Therefore, $\dfrac{a_{n+1}}{a_n} \leq 1, \ \forall n \in \mathbb{N}$ when $a_n \uparrow$.

■

Again, Theorem 5.2.8 applies only to nonnegative sequences. Furthermore, if the ratio $\dfrac{a_{n+1}}{a_n}$ is strictly greater than 1 (less than 1), the sequence will be strictly increasing (strictly decreasing). The second test for monotonicity is based on the difference of successive terms of the sequence and does not require the sequence to be a nonnegative sequence.

Theorem 5.2.9: Let a_n be a sequence of real numbers. Then

(i) $a_n \uparrow$ if and only if $a_{n+1} - a_n \geq 0, \ \forall n \in \mathbb{N}$.

(ii) $a_n \downarrow$ if and only if $a_{n+1} - a_n \leq 0, \ \forall n \in \mathbb{N}$.

Proof: The proof of Theorem 5.2.9 is left as an exercise.

■

Note that if the differences are strictly greater than 0, the sequence is strictly increasing, whereas if the differences are strictly less than 0, the sequence is strictly decreasing. The last test for monotonicity is based on the derivative of the general term in a sequence. Recall that a sequence a_n is actually shorthand for the function $a(n) = a_n$ defined on $\mathbb{N}$. Theorem 5.2.10 shows that when a sequence $a_n = a(n)$ is differentiable on $[1, \infty)$ and the derivative of $a(n)$ is positive on $[1, \infty)$ (negative on $[1, \infty)$), then the sequence a_n is increasing (decreasing).

Theorem 5.2.10: Let $a(n) = a_n$ be a sequence of real numbers. If the function $a(n)$ is differentiable on $[1, \infty)$, then

(i) a_n is a nondecreasing sequence if and only if $\frac{d}{dn}[a(n)] \geq 0, \forall\, n \in \mathbb{N}$.

(ii) a_n is a nonincreasing sequence if and only if $\frac{d}{dn}[a(n)] \leq 0, \forall\, n \in \mathbb{N}$.

Proof: The proof of Theorem 5.2.10 is left as an exercise.

■

Theorem 5.2.10 also implies that if the derivative is strictly greater than 0, then the sequence is strictly increasing; similarly, when the derivative is strictly less than 0, then the sequence will be strictly decreasing. The following two examples illustrate the use of the monotonicity tests of Theorems 5.2.8, 5.2.9, and 5.2.10.

Example 5.2.14: Show that each of the following sequences is a monotone sequence:

a. $a_n = \dfrac{n-1}{n+1}$

b. $b_n = \dfrac{2^n}{n!}$

c. $c_n = ne^{-n}$

Solutions:

a. Let $a_n = \dfrac{n-1}{n+1}$. Let $n \in \mathbb{N}$ be ABF, and consider the difference $a_{n+1} - a_n$.

$$a_{n+1} - a_n = \frac{(n+1)-1}{(n+1)+1} - \frac{n-1}{n+1} = \frac{n(n+1)-(n-1)(n+2)}{(n+2)(n+1)}$$

$$= \frac{n^2+n-n^2-n+2}{(n+2)(n+1)} = \frac{2}{(n+2)(n+1)} \geq 0$$

Thus, $a_{n+1} - a_n \geq 0$, $\forall\, n \in \mathbb{N}$. Hence, by Theorem 5.2.9 part (iii), $a_n \uparrow$. Also, since $\dfrac{2}{(n+2)(n+1)} > 0$ for every $n \in \mathbb{N}$, it follows that $a_n = \dfrac{n-1}{n+1}$ is strictly increasing.

b. Let $b_n = \dfrac{2^n}{n!}$. Let $n \in \mathbb{N}$ be ABF and consider the ratio $\dfrac{b_{n+1}}{b_n}$.

$$\frac{b_{n+1}}{b_n} = \frac{\frac{2^{n+1}}{(n+1)!}}{\frac{2^n}{n!}} = \frac{2}{(n+1)} \leq 1, \ \forall\, n \in \mathbb{N}$$

Thus, $\dfrac{b_{n+1}}{b_n} \leq 1$, $\forall\, n \in \mathbb{N}$ and therefore by Theorem 5.2.9 part (ii), $b_n \downarrow$.

Since $b1 = b_2$ it follows that $b_n = \dfrac{2^n}{n!}$ is not strictly decreasing. However, for $n \geq 2$, it is true that $\dfrac{b_{n+1}}{b_n} < 1$, and therefore b_n is strictly decreasing for $n \geq 2$.

c. Let $c_n = c(n) = ne^{-n}$. Since $c(n)$ is differentiable on $\mathbb{R}$, consider the derivative of $c(n)$:

$$\frac{d}{dn}\left[ne^{-n}\right] = e^{-n} - ne^{-n} = e^{-n}[1-n] \leq 0, \ \forall\, n \in \mathbb{N}$$

Thus, $\dfrac{d}{dn}[c(n)] \leq 0$, $\forall\, n \in \mathbb{N}$ and therefore by Theorem 5.2.10 part (ii), $c_n \downarrow$.

Determining whether c_n is strictly decreasing will involve a bit of analysis. Clearly, for $n > 1$, $e^{-n}[1-n] < 0$. Thus, the only terms where c_n could be constant are c_1 and c_2. But

$$\frac{c_2}{c_1} = \frac{2e^{-2}}{e^{-1}} = 2e^{-1} < 1$$

Thus, $c_1 > c_2$, and therefore c_n is strictly decreasing.

Example 5.2.15: For each of the following sequences, determine whether the sequence is monotonic. If it is monotonic, determine whether it is a nonincreasing or a nondecreasing sequence.

a. $a_n = \dfrac{(-1)^n}{n}$.

b. $b_n = \dfrac{n}{3^n}$.

c. $c_n = e^{-n}$

Solutions:

a. Let $a_n = \dfrac{(-1)^n}{n}$. Let $n \in \mathbb{N}$ be ABF. Since a_n can take on negative values, only the difference and derivative tests can be applied to this sequence. Consider the difference $a_{n+1} - a_n$:

$$a_{n+1} - a_n = \frac{(-1)^{n+1}}{n+1} - \frac{(-1)^n}{n} = \frac{(-1)^{n+1}n - (-1)^n(n+1)}{n(n+1)}$$

$$= \begin{cases} \frac{2n+1}{n(n+1)} > 0 & \text{if } n \text{ is odd} \\ \frac{-2n-1}{n(n+1)} < 0 & \text{if } n \text{ is even} \end{cases}$$

Thus, since the sign of $a_{n+1} - a_n$ depends on the value of n, it follows that a_n is not a monotone sequence. Note that the first five terms of a_n are $-1, \frac{1}{2}, -\frac{1}{3}, \frac{1}{4}, -\frac{1}{5}$, which clearly shows that a_n is not monotonic.

b. The solution to part (b) is left as an exercise.

c. The solution to part (c) is left as an exercise.

Now, knowing which of the three tests for monotonicity to use on a particular sequence can be difficult. However, given the three approaches outlined in Theorems 5.2.8–5.2.10, if one of the approaches does not work, simply try another one. For example, if the sequence is differentiable, Theorem 5.2.10 applies and might be a good first approach to try; likewise, when the sequence is nonnegative, then Theorem 5.2.9 applies and it might be a good approach. Also, nonnegative sequences involving powers or factorial terms are often best tested for monotonicity with the ratio test, and sequences involving linear or rational functions are often best tested for monotonicity with the difference test.

The following theorem provides several results concerning the monotonicity of a sequence. In fact, Theorem 5.2.11 shows that the monotonicity of a sequence is preserved under the addition of a constant, multiplication by a constant, and the addition and multiplication of sequences having the same monotonicity.

Theorem 5.2.11: Let a_n and b_n be a monotonic sequences of real numbers. Then

(i) $c + a_n$ is also a monotonic sequence, $\forall\, c \in \mathbb{R}$.

(ii) ca_n is also a monotonic sequence, $\forall\, c \in \mathbb{R}$.

(iii) If $a_n \uparrow$ and $b_n \uparrow$, then $(a_n + b_n) \uparrow$.

(iv) If $a_n \downarrow$ and $b_n \downarrow$, then $(a_n + b_n) \downarrow$.

(v) If a_n and b_n are nonnegative nondecreasing sequences, then $a_n b_n \uparrow$.

(vi) If a_n and b_n are nonnegative nonincreasing sequences, then $a_n b_n \downarrow$.

Proof: Let a_n and b_n be monotonic sequences of real numbers.

Proof of part (i): Let $c \in \mathbb{R}$ be ABF. Now, since a_n is a monotonic sequence, either $a_n \uparrow$ or $a_n \downarrow$.

Case 1: Suppose that $a_n \uparrow$. Then, $a_{n+1} \geq a_n$, $\forall\, n \in \mathbb{N}$, and hence $c + a_{n+1} \geq c + a_n$, $\forall n \in \mathbb{N}$. Therefore, $c + a_n \uparrow$ whenever $a_n \uparrow$.

Case 2: Suppose that $a_n \downarrow$. Then, $a_{n+1} \leq a_n$, $\forall\, n \in \mathbb{N}$ and hence, $c + a_{n+1} \leq c + a_n$, $\forall n \in \mathbb{N}$. Therefore, $c + a_n \downarrow$ whenever $a_n \downarrow$.

Proof of part (ii): The proof of part (ii) is left as an exercise.

Proof of part (iii): Let a_n and b_n be nondecreasing sequences. Then

$$a_{n+1} \geq a_n, \ \forall\, n \in \mathbb{N}$$

$$b_{n+1} \geq b_n, \ \forall\, n \in \mathbb{N}$$

Let $n \in \mathbb{N}$ be ABF and consider $a_n + b_n$:

$$a_n + b_n \underbrace{\leq a_{n+1} + b_n}_{a_n \leq a_{n+1}} \underbrace{\leq a_{n+1} + b_{n+1}}_{b_n \leq b_{n+1}}$$

Thus, $a_n + b_n \leq a_{n+1} + b_{n+1}$, $\forall\, n \in \mathbb{N}$; therefore, $a_n + b_n \uparrow$ whenever a_n and b_n are nondecreasing sequences.

Proof of part (iv): The proof of part (iv) is left as an exercise.

Proof of part (v): Let a_n and b_n be nonnegative nondecreasing sequences. Then

$$a_{n+1} \geq a_n, \ \forall\, n \in \mathbb{N}$$

$$b_{n+1} \geq b_n, \ \forall\, n \in \mathbb{N}$$

Let $n \in \mathbb{N}$ be ABF and consider $a_n b_n$.

$$a_n b_n \underbrace{\leq a_{n+1} b_n}_{a_n \leq a_{n+1}} \underbrace{\leq a_{n+1} b_{n+1}}_{b_n \leq b_{n+1}}$$

Thus, $a_n b_n \leq a_{n+1} b_{n+1}$, $\forall\, n \in \mathbb{N}$, and therefore $a_n b_n \uparrow$ whenever a_n and b_n are nonnegative nondecreasing sequences.

Proof of part (vi): The proof of part (vi) is left as an exercise.

■

Recall that every convergent sequence is bounded and thus, when a sequence a_n converges, there exists a real number M such that $|a_n| \leq M$, $\forall\, n \in \mathbb{N}$. Moreover, this means that $-M \leq a_n \leq M$, $\forall\, n \in \mathbb{N}$. Thus, a convergent sequence a_n is bounded above by M and below by $-M$. The following definition introduces several different types bounds for a set or real numbers or a sequence of real numbers.

Definition 5.2.4: Let X be a nonempty subset of $\mathbb{R}$.

a. A real number M satisfying $x \leq M$ for all $x \in X$ is called an *upper bound* of X. When there exists an upper bound for the set X, the set X is said to be *bounded from above.*

b. A real number m satisfying $x \geq m$ for all $x \in X$ is called a *lower bound* of X. When there exists a lower bound for the set X, the set X is said to be *bounded from below.*

c. An upper bound M is said to be a *least upper bound* or *supremum* if for any other upper bound $M^\star$, $M \leq M^\star$.

d. A lower bound m is said to be a *greatest lower bound* or *infinum* if for any other lower bound $m^\star$, $m \geq m^\star$.

The least upper bound is abbreviated by l.u.b.; the supremum by sup; the greatest lower bound, by g.l.b.; and the infinum, by inf. For example, if m is the infinum of a set X, this will be denoted by $\inf X = m$; similarly, the supremum, least upper bound, and greatest lower bound of X will be denoted by $\sup X$, l.u.b. X, and g.l.b X, respectively.

Example 5.2.16: Let $X = \left\{x_n : x_n = \dfrac{n}{n+1}\right\}$. Determine the l.u.b (infinum) and g.l.b. (supremum) of X.

Solutions:

Note that $X = \left\{\dfrac{1}{2}, \dfrac{2}{3}, \dfrac{3}{4}, \dfrac{4}{5}, \ldots\right\}$ so that $x_n \geq \dfrac{1}{2}$, for all $n \in \mathbb{N}$. Thus, $m = 1/2$ is a lower bound for X. Clearly, since $x_1 = \dfrac{1}{2} \in X$, there does not exist a lower bound greater than m. Therefore, m is the g.l.b., or infinum, of X; that is, inf $X = 1/2$.

For the upper bound, note that $x_n \leq 1$, $\forall\, n \in \mathbb{N}$. Thus, $M = 1$ is an upper bound for X. To see that $M = 1$ is the l.u.b., or supremum, of X, suppose that there exists an upper bound $M^\star$ such that $M^\star < 1$, say, $M^\star$. Then, $M^\star = 1 - \epsilon$ for some value of $\epsilon > 0$. Consider x_n. Since $M^\star$ is an upper bound for x_n, it follows that $x_n \leq 1 - \epsilon$, $\forall\, n \in \mathbb{N}$.

Let N be the smallest integer greater than $\frac{1}{\epsilon} - 1$. Then, since $x_n \uparrow$ and $N > \frac{1}{\epsilon} - 1$,

$$x_N = \frac{N}{N+1} > \frac{\frac{1}{\epsilon} - 1}{\frac{1}{\epsilon} - 1 + 1} = 1 - \epsilon = M^\star$$

contradicting $M^\star$ being an upper bound for x_n. Therefore, $M = 1$ is the l.u.b. for X. (i.e., sup $X = 1$).

Note that since $+\infty$ and $-\infty$ are not real numbers, the definition of a bound precludes $\pm\infty$ from being an upper or lower bound for a set X. When a set does not have an upper or lower bound, it is said to be *unbounded.* Also, when a real number M is an upper bound for X, then so is every real number greater than M. Thus, there is no unique upper bound for a set X. Similarly, if m is a lower bound for X, then so is every real number less than m and hence, there is not a unique lower bound for a set X. Now, while upper and lower bounds are not unique, the following theorem shows that the supremum and infinum of a set are indeed unique.

Theorem 5.2.12: Let X be a subset of $\mathbb{R}$. If

(i) $s = \sup X$, then s is unique.

(ii) $i = \inf X$, then i is unique.

Proof (Uniqueness Proof):

Proof or part (i): Let X be a subset of $\mathbb{R}$, and suppose that the supremum of X is not unique. Let s_1 and s_2 be two different supremums of X. Then, s_1 and s_2 are both upper bounds, and hence

(1) $s_1 \leq s_2$ since s_1 is a supremum and s_2 is an upper bound.

(2) $s_2 \leq s_1$ since s_2 is a supremum and s_1 is an upper bound.

Thus, $s_1 \leq s_2$ and $s_1 \geq s_2$, which means that $s_1 = s_2$, contradicting $s_1 \neq s_2$. Therefore, the supremum of X is unique.

Proof or part (ii): The proof of part (ii) is left as an exercise.

■

The following theorem shows that if a monotonic sequence of real numbers is bounded, then it also converges. Furthermore, the proof of this theorem reveals that the limit of a bounded monotonic sequence is the supremum of the sequence for nondecreasing sequences and the infinum for nonincreasing sequences.

Theorem 5.2.13: Let a_n be a monotonic sequence of real numbers.

(i) If a_n is a nondecreasing sequence, then a_n converges if and only if it is bounded.

(ii) If a_n is a nonincreasing sequence, then a_n converges if and only if it is bounded.

Proof (Biconditional Proof): Let a_n be a monotonic sequence of real numbers.

Proof of part (i): Suppose that $a_n \uparrow$ and that a_n converges. Since a_n converges, it follows from Lemma 5.2.1 that a_n is also bounded. Therefore, every convergent nondecreasing sequence is bounded.

Conversely, suppose that a_n is bounded, and let $\epsilon > 0$ be ABF. Define $A = \{a_n : n \in \mathbb{N}\}$. Since a_n is a bounded sequence, it follows that the set A is also bounded. Thus, there exists $s \in \mathbb{R}$ such that $s = \sup A$.

Now, since s is the supremum of A, it follows that $s-\epsilon$ is not an upper bound for A and $s+\epsilon$ is an upper bound for a_n. Furthermore, since $a_n \uparrow$, there exists $N \in \mathbb{N}$ such that $s-\epsilon < a_n$ whenever $n \geq N$. Thus, whenever $n \geq N$, it follows that $s - \epsilon < a_n < s + \epsilon$. Equivalently, $|a_n - s| < \epsilon$ whenever $n \geq N$, and hence $s = \lim_{n \to \infty} a_n$. Thus, every bounded nondecreasing sequence is a convergent sequence.

Therefore, a monotonic nondecreasing sequence converges if and only if it is bounded.

Proof of part (ii): The proof of part (ii) is left as an exercise.

■

As a result of Theorem 5.2.13, a monotonic sequence a_n can be tested for convergence by simply determining whether the sequence is bounded. Also, once it is shown that a monotonic sequence a_n is bounded, it follows from the proof of Theorem 5.2.13 that $\lim_{n \to \infty} a_n = \sup \{a_n : n \in \mathbb{N}\}$ for nondecreasing sequences and $\lim_{n\to\infty} a_n = \inf\{a_n : n \in \mathbb{N}\}$ for nonincreasing sequences. An algorithm that can be used to show that a monotone sequence converges is given below.

Algorithm for Showing a Monotone Sequence Converges: Let a_n be a sequence. To prove that a_n is a convergent monotonic sequence

1. Show that a_n is monotone and determine whether $a_n \uparrow$ or $a_n \downarrow$.
2. Show that a_n is bounded.
3. Find $\sup\{a_n : n \in \mathbb{N}\} = s$ for nondecreasing sequences; for nonincreasing sequences find $\inf\{a_n : n \in \mathbb{N}\} = i$.
4. Conclude $\lim_{n\to\infty} a_n = s$ for nondecreasing sequences and $\lim_{n\to\infty} a_n = i$ for nonincreasing sequences.
5. Clean up and rewrite the scratchwork into a clear and concise proof of the theorem. Make sure that each step of the proof makes sense and is clearly justified.
6. Read the proof over carefully and make any necessary corrections.

Example 5.2.17: Let $a_1 = 1$ and for $n \geq 2$, let $a_n = \sqrt{2 + \sqrt{a_{n-1}}}$. Show that a_n is monotone and bounded by 2.

Solution: First, show that a_n is monotone. This recursively defined sequence can be shown to be a nondecreasing sequence with mathematical induction.

Let $a_1 = 1$ and for $n \geq 2$ let $a_n = \sqrt{2 + \sqrt{a_{n-1}}}$. Define $\mathcal{P}_n := a_{n+1} \geq a_n$.

For $n = 1$, it follows that

$$a_2 = \sqrt{2 + \sqrt{1}} = \sqrt{3} \geq 1 = a_1$$

Therefore, $\mathcal{P}_1$ is true.

Suppose that $\mathcal{P}_k$ is true for some ABF $k \in \mathbb{N}$. This means that $a_{k+1} \geq a_k$. Now, if $\mathcal{P}_{k+1}$ is true, then $a_{k+2} \geq a_{k+1}$. Consider a_{k+2}:

$$a_{k+2} = \sqrt{2 + \sqrt{a_{k+1}}} \underbrace{\geq \sqrt{2 + \sqrt{a_k}}}_{\text{By } \mathcal{P}_k} = a_{k+1}$$

Thus, $\mathcal{P}_{k+1}$ is true whenever $\mathcal{P}_k$ is true, and therefore a_n is a nondecreasing sequence.

Now, a proof by contradiction will be used to show that $a_n \leq 2$ for all $n \in \mathbb{N}$. Suppose that $\exists\ N \in \mathbb{N}$ such that $a_N > 2$. Since $a_n \uparrow$, WLOG let N be the smallest value of n such that $a_N > 2$ and $a_{N-1} \leq 2$. Then

$$2 < a_n = \sqrt{2 + \sqrt{a_{N-1}}}$$

which means that

$$2 < \sqrt{2 + \sqrt{a_{N-1}}} \iff 4 < 2 + \sqrt{a_{N-1}} \iff 2 < \sqrt{a_{N-1}}$$

which means that $a_{N-1} > 4$, which contradicts $a_{N-1} \leq 2$.

Therefore, $a_n \leq 2$ for all $n \in \mathbb{N}$.

Finally, since a_n is nondecreasing and is bounded, it follows that a_n converges. Furthermore, the limit of a_n requires only the determination of the supremum of $\{a_n : n \in \mathbb{N}\}$. Since $\sup\{a_n : n \in \mathbb{N}\} = 2$ it follows that $\lim\limits_{n\to\infty} a_n = 2$.

Example 5.2.18: Let $a_n = \dfrac{e^n}{1+2e^n}$. Show that a_n converges to $\dfrac{1}{2}$.

Solution: Let $a_n = \dfrac{e^n}{1+2e^n}$. Since a_n is a differentiable function in n on $\mathbb{R}$, the derivative test can be used to determine whether a_n is a monotonic sequence. Consider $\dfrac{d}{dn}[a_n]$:

$$\frac{d}{dn}[a_n] = \frac{d}{dn}\left[\frac{e^n}{1+2e^n}\right] = \frac{e^n}{(1+2e^n)^2}$$

Clearly, $\dfrac{e^n}{(1+2e^n)^2} > 0$, $\forall\ n \in \mathbb{N}$ and therefore, $a_n \uparrow$. Furthermore

$$a_n = \frac{e^n}{1+2e^n} = \frac{e^n}{e^n(e^{-n}+2)} = \frac{1}{e^{-n}+2} \leq \frac{1}{2},\ \forall\ n \in \mathbb{N}$$

Thus, $a_n \uparrow$ and a_n is bounded, and hence, by Theorem 5.2.13 it follows that a_n converges.

Finally, note that $\forall\ n \in \mathbb{N}$:

$$\frac{e^n}{1+2e^n} = \frac{1}{e^{-n}+2} \geq \frac{1}{2+\frac{1}{n}}$$

Since $\dfrac{1}{n} \geq e^{-n}$, $\forall\ n \in \mathbb{N}$, it follows that

$$\frac{1}{2+\frac{1}{n}} \leq \frac{e^n}{1+2e^n} \leq \frac{1}{2}$$

Now, since $\lim_{n\to\infty} \frac{1}{2+\frac{1}{n}} = \frac{1}{2}$, by the squeeze theorem it follows that $a_n \to \frac{1}{2}$.

5.2.3 Cauchy Sequences

Note that the convergence of a monotonic sequence can be determined without knowing the limit; that is, Theorem 5.2.13 shows that every bounded monotonic sequence converges. However, for many monotonic sequences the limit, either $\sup\{a_n : n \in \mathbb{N}\}$ or $\inf\{a_n : n \in \mathbb{N}\}$, may be very difficult to determine. Now, for nonmonotonic sequences, using the definition of convergence to prove that the sequence converges requires knowledge of the limit of the sequence. When a sequence a_n converges to a known limit a, then this means that for every $\epsilon > 0$, there exists $N \in \mathbb{N}$ such that $|a_n - a| < \epsilon$ whenever $n \geq N$ and thus, from some point on, all terms in the tail end of the sequence become approximately equal to the limit a; moreover, this also means that from some point on, all terms in the sequence must be nearly equal. Sequences for which a_n and a_m are nearly equal from some point on are called *Cauchy sequences.* The definition of a Cauchy sequence is given below; Cauchy sequences are named after the French mathematician Augustin Cauchy (1789–1857).

Definition 5.2.5: A real-valued sequence a_n is said to be a *Cauchy sequence* if and only if for every $\epsilon > 0$, there exists $N \in \mathbb{N}$ such that $|a_n - a_m| < \epsilon$ whenever $n, m \geq N$.

Note that the definition of a Cauchy sequence depends only on the terms of the sequence in question and does not require knowledge of the limit of the sequence. In Theorem 5.2.14 it will be shown that all Cauchy sequences of real numbers converge; therefore, showing that a sequence of real numbers is a Cauchy sequence also proves that it is a convergent sequence. First, an algorithm for proving that a sequence a_n is a Cauchy sequence is given below.

Algorithm for Showing a Sequence is a Cauchy Sequence: Let a_n be a sequence of real numbers. To show that a_n is a Cauchy sequence

1. Let $\epsilon > 0$ be ABF.
2. Let $m, n \in \mathbb{N}$ and consider $|a_n - a_m|$.
3. Determine how to make $|a_n - a_m|$ arbitrarily small. In other words, determine the value of $N \in \mathbb{N}$ so that $|a_n - a_m| < \epsilon$ whenever $n, m \geq N$.
4. Conclude a_n is a Cauchy sequence.
5. Clean up and rewrite the scratchwork into a clear and concise proof of the theorem. Make sure that each step of the proof makes sense and is clearly justified.
6. Read the proof over carefully and make any necessary corrections.

Example 5.2.19: Let $a_n = 2 + \frac{1}{n}$. Show that a_n is a Cauchy sequence.

Solution:

1. Let $\epsilon > 0$ be ABF.
2. Let $m, n \in \mathbb{N}$. Without loss of generality (WLOG), assume $m > n$. Consider $|a_n - a_m|$.

$$\left|2 + \frac{1}{n} - \left(2 + \frac{1}{m}\right)\right| = \left|\frac{1}{n} - \frac{1}{m}\right| = \left|\frac{m-n}{mn}\right| \leq \frac{m+n}{mn}$$

   Now, since $m > n$, it follows that

$$\left|2 + \frac{1}{n} - \left(2 + \frac{1}{m}\right)\right| \leq \frac{m+n}{mn} \leq \frac{2m}{mn} = \frac{2}{n}$$

3. Let N be the smallest natural number greater than $\frac{2}{\epsilon}$. Then, for $n \geq N$

$$\left|2 + \frac{1}{n} - \left(2 + \frac{1}{m}\right)\right| \leq \frac{2}{n} < \epsilon$$

4. Therefore, $a_n = 2 + \frac{1}{n}$ is a Cauchy sequence.

Note that a proof showing that a sequence is a Cauchy sequence is very similar to a convergence proof; both proofs hinge on finding a value of $N \in \mathbb{N}$ for every $\epsilon > 0$. In a convergence proof the value of N must be found so that $|a_n - a| < \epsilon$ whenever $n \geq N$, while in proving a_n is a Cauchy sequence the value of N must be found so that $|a_n - a_m| < \epsilon$ whenever $n \geq N$.

Example 5.2.20: Let $a_n = e^{-n}$. Show that a_n is a Cauchy sequence.

Solution: Let $\epsilon > 0$ be ABF and let $m, n \in \mathbb{N}$. WLOG, assume $m > n$ and consider $|e^{-n} - e^{-m}|$.

$$|e^{-n} - e^{-m}| = |e^{-n}(1 - e^{-m+n})|$$

Now, since $m > n$, it follows that $-m + n < 0$ and $0 < e^{-m+n} < 1$. Since $0 < e^{-m+n} < 1$, it follows that $0 < 1 - e^{-m+n} < 1$. Thus

$$|e^{-n} - e^{-m}| = |e^{-n}(1 - e^{-m+n})| \leq |e^{-n}| = e^{-n}$$

Let N be the smallest natural number greater than $-\ln(\epsilon)$. Then, for $n \geq N$, it follows that

$$|e^{-n} - e^{-m}| \leq e^{-n} < \epsilon$$

Therefore, $a_n = e^{-n}$ is a Cauchy sequence.

The following theorem shows that every convergent sequence is a Cauchy sequence.

Theorem 5.2.14: Let a_n be a sequence of real numbers. If $a_n \to a$, then a_n is a Cauchy sequence.

Proof: Let $\epsilon > 0$ be ABF. Since $a_n \to a$, there exists $N \in \mathbb{N}$ such that $|a_n - a| < \frac{\epsilon}{2}$ whenever $n \geq N$.

Consider $|a_n - a_m|$:

$$|a_n - a_m| = |a_n - a + a - a_m| \leq |a_n - a| + |a_m - a|$$

Thus, for $n, m \geq N$, it follows that

$$|a_n - a_m| \leq |a_n - a| + |a_m - a| < \underbrace{\epsilon/2}_{\text{Since } n \geq N} + \underbrace{\epsilon/2}_{\text{Since } m \geq N} = \epsilon$$

Therefore a_n is a Cauchy sequence whenever a_n is a convergent sequence.

■

Theorem 5.2.15, which is stated below without proof, is the converse of Theorem 5.2.14; a proof of Theorem 5.2.15 is given in *Elementary Analysis: The Theory of Calculus* by Kenneth Ross (2003). In particular, Theorem 5.2.15 shows that if a sequence of real numbers is a Cauchy sequence, then it is also a convergent sequence. Thus, when a sequence is shown to be a Cauchy sequence, then that sequence has also been shown to be convergent. However, the actual limit of the sequence cannot be found by simply showing that a sequence is a Cauchy sequence. Thus, once the sequence is shown to be convergent, further analysis of the sequence will be required in order to determine its limit.

Theorem 5.2.15: Let a_n be a sequence of real numbers. If a_n is a Cauchy sequence, then a_n converges.

Proof: The proof of this theorem is a construction proof and requires knowledge of subsequences, which have not been discussed in this text.

■

Now, Theorems 5.2.14 and 5.2.15 can be combined into the following bi-conditional theorem, which states that a real-valued sequence a_n converges if and only if it is a Cauchy sequence. Hence, convergent sequences of real numbers are Cauchy sequences, and Cauchy sequences of real numbers converge.

Theorem 5.2.16: A sequence of real numbers a_n is a Cauchy sequence if and only if it converges.

Example 5.2.21: Prove the following result. If a_n is a Cauchy sequence of real numbers, then a_n is bounded.

> ***Proof:*** Suppose that a_n is a Cauchy sequence of real numbers. Then, by Theorem 5.2.18, a_n is a convergent sequence, and hence a_n is bounded.

Now, every convergent sequence of real numbers is a Cauchy sequence, Cauchy sequences are bounded, and Cauchy sequences of real numbers are convergent sequences. Another important property that is related to Cauchy sequences is defined below.

Definition 5.2.6: A set X is said to be *complete* if and only if each of the Cauchy sequences in X converges to an element of X.

An example of a complete set is $\mathbb{R}$. Specifically, since every Cauchy sequence of real numbers is a convergent sequence and converges to a real limit, it follows that $\mathbb{R}$ is complete. The following example shows that $\mathbb{Q}$ is not complete.

Example 5.2.22: Prove that $\mathbb{Q}$ is not complete.

Solution: $\mathbb{Q}$ can be shown to be incomplete using a proof by contradiction.

> ***Proof (by Contradiction):*** Suppose that $\mathbb{Q}$ is complete. Consider the sequence $a_n = \left(1 + \frac{1}{n}\right)^n$. Since $\mathbb{Q}$ is closed under addition and multiplication, it follows that a_n is a rational number for every value of $n \in \mathbb{N}$.
>
> Now, since $\left(1 + \frac{1}{n}\right)^n$ is a convergent sequence (see Table 5.2.1), it follows that a_n is a Cauchy sequence. However, $\lim_{n\to\infty} a_n = e \notin \mathbb{Q}$, which contradicts the assumption that $\mathbb{Q}$ is complete.
>
> Therefore, $\mathbb{Q}$ is not complete.

■

5.3 Limits of Functions

In the previous section, the limiting behavior of a special type of real-valued function, a sequence, was studied for large values of n. In this section, the behavior of real-valued functions will be studied, particularly, the limiting behavior of a function at a point $x_0 \in \mathbb{R}$. Now, in a first course on calculus, one of the first topics that is studied is the limiting behavior of function f at a point x_0. Also, in the traditional calculus course, the derivative of a real-valued function is introduced following a discussion of the limit of a function. Furthermore, the definition of the derivative of a function f at a point x_0 is based on a special type of limiting behavior. Finally, following a discussion of the derivative and differentiation, the concept of the definite integral is introduced, which is also based on a special type of limiting behavior. Thus, limits play an important role in the foundations of calculus.

Let f be a real-valued function with domain $\mathcal{D}$. Now, the limit of f at a point x_0 will be based on how f behaves near the point x_0. In particular, if f is nearly constant for x values near the point x_0, then f will have a limit at the point x_0, even if $x_0 \notin \mathcal{D}$. It turns out that whether in fact $x_0 \in \mathcal{D}$ is irrelevant with regard to the limiting behavior of f near x_0. For this reason, the definition of the limit of a real-valued function will be based on the behavior of the function in the neighborhood of x_0 only. The definition of a *neighborhood* and a *deleted neighborhood* are given below.

Definition 5.3.1: Let $x_0 \in \mathbb{R}$, and let $\epsilon > 0$. The open interval $(x_0 - \epsilon, x_0 + \epsilon)$ is called a *neighborhood* of x_0 with radius ϵ and will be denoted by $N_\epsilon(x_0)$.

Definition 5.3.2: Let $x_0 \in \mathbb{R}$ and $\epsilon > 0$. The set $\{x \in \mathbb{R} : 0 < |x - x_0| < \epsilon\}$ is called a *deleted neighborhood* of x_0 of radius ϵ.

For example, the interval $(0.5, 1.5)$ is a neighborhood of x_0 of radius $\epsilon = 0.5$, and the set $\{x \in \mathbb{R} : 0 < |x-1| < 0.5\} = (0.5, 1) \cup (1, 1.5)$ is a deleted neighborhood of x_0 of radius $\epsilon = 0.5$.

Example 5.3.1: Determine

a. A neighborhood of $x_0 = 10$ of radius $\epsilon = 0.1$.
b. A deleted neighborhood of $x_0 = 10$ of radius $\epsilon = 0.05$.
c. A deleted neighborhood of $x_0 = 10$ of radius $\epsilon = 0.01$.

Solutions:

a. $(9.9, 10.1)$ is a neighborhood of $x_0 = 10$ of radius $\epsilon = 0.1$.
b. The set $\{x \in \mathbb{R} : 0 < |x - 10| < 0.05\} = (9.95, 10) \cup (10, 10.05)$ is a deleted neighborhood of x_0 of radius $\epsilon = 0.05$.

c. The set $\{x \in \mathbb{R} : 0 < |x - x_0| < 0.01\} = (x_0 - 0.1, x_0) \cup (x_0, x_0 + 0.01)$ is a deleted neighborhood of x_0 of radius $\epsilon = 0.01$.

Note that the point x_0 is in every one of its neighborhoods; however, the point x_0 is not in any of its deleted neighborhoods. Since the point x_0 may not be in the domain of the function f, the definition of the limit of a real-valued function f at a point x_0 is based on a deleted neighborhood of x_0, rather than a neighborhood of x_0. The ϵ-δ definition of the limit of a real-valued function f is given below.

Definition 5.3.3: A real-valued function f, defined on a domain $\mathcal{D}$ that contains a deleted neighborhood of x_0, is said to have *limit* L as x approaches x_0 if and only if for every $\epsilon > 0$, there exists a $\delta > 0$ such that $|f(x) - L| < \epsilon$ whenever $|x - x_0| < \delta$. When the limit of a function f exists at the point x_0, this will be denoted by $\lim_{x \to x_0} f(x) = L$.

Note that this definition is analogous the definition for the limit of a sequence of real numbers as $n \to \infty$. Note the similarity of the following two definitions:

a. **Limit of a Sequence:** A real-valued sequence a_n has limit a if and only if $\forall\ \epsilon > 0, \exists\ N \in \mathbb{N}$ such that $|a_n - a| < \epsilon$ whenever $n \geq N$.

b. **Limit of a Function:** A real-valued function f has limit L if and only if $\forall\ \epsilon > 0, \exists\ \delta > 0$ such that $|f(x) - L| < \epsilon$ whenever $|x - x_0| < \delta$.

Thus, when working with the limit of a real-valued function f, the statements $N \in \mathbb{N}$ and $n \geq N$ that are used in the limit of a sequence are simply replaced by $\delta > 0$ and $|x - x_0| < \delta$. For this reason, many of the theorems concerning the limit of a real-valued function will have proofs similar to the proofs of the analogous results for the real-valued sequences.

Now, it is important to note that a function f might have a limit as x approaches x_0, even though the function is not defined at the point x_0. For example, consider the function $f(x) = \dfrac{x-1}{x^2-1}$. Then $f(x)$ has the limit $\dfrac{1}{2}$ as x approaches 1, however $x = 1$ is not in the domain of f. The plot of $f(x) = \dfrac{x-1}{x^2-1}$ is given in Figure 5.3.1 shows that $f(x)$ does approach a limit as x approaches 1.

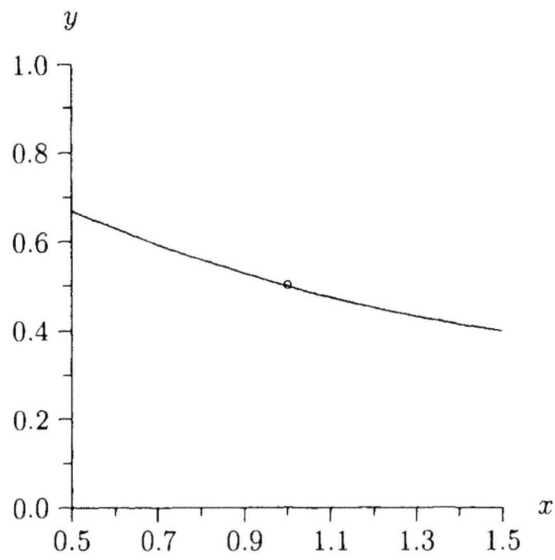

Figure 5.3.1 A plot of $f(x)=\frac{x-1}{x^2-1}$.

In order to show that a real-valued function has a limit L as x approaches x_0, or equivalently $\lim_{x \to x_0} f(x) = L$, it must shown that for an arbitrary value of $\epsilon > 0$, there is a δ such that $|f(x) - L| < \epsilon$ whenever $|x - x_0| < \delta$. A proof of this nature is called an ϵ-δ proof. Moreover, proving $\lim_{x \to x_0} f(x) = L$ with an ϵ-δ proof is analogous to proving that a sequence converges using an ϵ-N proof. An algorithm for proving $\lim_{x \to x_0} f(x) = L$ with an ϵ-δ proof is given below.

An Algorithm for an ϵ-δ Limit Proof: Let f be a real-valued function. To prove that $\lim_{x \to x_0} f(x) = L$

1. Let $\epsilon > 0$ be ABF.
2. Consider $|f(x) - L|$.
3. Determine how to make $|f(x) - L|$ arbitrarily small for the value of x in a deleted neighborhood of x_0.
4. From steps 2 and 3, determine the value of δ so that $|f(x) - L| < \epsilon$ whenever $|x - x_0| < \delta$.
5. Conclude $\lim_{x \to x_0} f(x) = L$.
6. Clean up and rewrite the scratchwork into a clear and concise proof of the theorem. Make sure that each step of the proof makes sense and is clearly justified.
7. Read the proof over carefully and make any necessary corrections.

Note that the value of δ is dependent on the value of ϵ and δ is simply the radius of a deleted neighborhood of x_0 for which $|f(x) - L| < \epsilon$. Also, if a smaller value of ϵ was chosen, then a smaller value of δ will generally be required. Thus, in order to require that $f(x)$ is closer to its limiting value L at the point x_0, x must be closer to x_0. An ϵ-δ proof is illustrated in the following example.

Example 5.3.2: Prove that $\lim\limits_{x \to 2} x^2 + 4x = 12$.

Solution (Scratchwork):

Step 1: Let $\epsilon > 0$ be arbitrary but fixed.

Step 2: Consider $|f(x) - L|$:

$$|f(x) - L| = |x^2 + 4x - 12| = |(x+6)(x-2)| = |x+6| \cdot |x-2|$$

Step 3: A value of δ must be found so that whenever $|x - 2| < \delta$ it will follow that $|x^2 + 4x - 12| < \epsilon$. Now

$$|x^2 + 4x - 12| = |x+6| \cdot |x-2|$$

Thus, if an upper bound can be placed on $|x + 6|$ when x is near 2, then it will be possible to relate $|x^2 + 4x - 12|$ to $|x - 2|$. Consider x values in neighborhood of 2 of radius 1 (i.e., x values such that $|x - 2| < 1$). Then, for $|x - 2| < 1$ it follows that

$$|x + 6| = |x - 2 + 8| \leq |x - 2| + 8 < 9$$

Thus, $|x + 6| < 9$ whenever $|x - 2| < 1$, and therefore

$$|x^2 + 4x - 12| = |x+6| \cdot |x-2| < 9 \cdot |x-2|$$

Step 4: Now, let $\delta = \min\left(1, \frac{\epsilon}{9}\right)$. Then whenever $|x - 2| < \delta$,

$$|x^2 + 4x - 12| = |x+6| \cdot |x-2| < 9 \cdot |x-2| < 9 \cdot \delta < 9 \cdot \frac{\epsilon}{9} = \epsilon$$

Therefore, $\lim\limits_{x \to 2} x^2 + 4x = 12$.

This scratchwork is written up in a formal ϵ-δ proof below.

Proof: Let $\epsilon > 0$ be arbitrary but fixed. Consider $|f(x) - L|$:

$$|f(x) - L| = |x^2 + 4x - 12| = |x+6| \cdot |x-2|$$

Now, for $|x - 2| < 1$, it follows that

$$|x + 6| = |x - 2 + 8| \leq |x - 2| + 8 < 9$$

Thus, $|x + 6| < 9$ whenever $|x - 2| < 1$, and therefore

$$|x^2 + 4x - 12| = |x + 6| \cdot |x - 2| < 9 \cdot |x - 2|$$

Let $\delta = \min(1, \epsilon/9)$. Then, whenever $|x - 2| < \delta$, it follows that

$$|x^2 + 4x - 12| = |x + 6| \cdot |x - 2| < 9 \cdot |x - 2| < 9 \cdot \delta < 9 \cdot \frac{\epsilon}{9} = \epsilon$$

Therefore, $\lim_{x \to 2} x^2 + 4x = 12$.

■

Note that in the proof above, $|x+6|$ needed to be bounded. A bound was found by considering a neighborhood of 2 of radius 1. The choice of the radius of the neighborhood in this proof was not special, and in fact, any radius could have been used in place of the value of 1. For example, if a radius of 2 were used in place of 1, the bound would have become $|x + 6| < 10$; similarly, if a radius of 0.25 were used in lieu of 1, the bound would have become $|x + 6| < 8.25$. It was important to bound $|x + 6|$, but the actual value of the bound, which depends on the radius of the neighborhood, was not important.

Example 5.3.3: Prove that $\lim_{x \to x_0} x^2 = x_0^2$ for all $x_0 \in \mathbb{R}$.

Solution:

Proof: Let $\epsilon > 0$, and let $x_0 \in \mathbb{R}$ be arbitrary but fixed. Consider $|x^2 - x_0^2|$:

$$|x^2 - x_0^2| = |x - x_0| \cdot |x + x_0|$$

For $|x - x_0| < 1$, it follows that

$$|x + x_0| = |x - x_0 + 2x_0| \leq |x - x_0| + 2|x_0| < 1 + 2|x_0|$$

Thus, whenever $|x - x_0| < 1$, it follows that

$$|x^2 - x_0^2| = |x - x_0| \cdot |x + x_0| \leq (1 + 2|x_0|) \cdot |x - x_0|$$

Let $\delta = \min\left(1, \dfrac{\epsilon}{1 + 2|x_0|}\right)$. Then, whenever $|x - x_0| < \delta$, it follows that

$$|x^2 - x_0^2| \leq (1 + 2|x_0|) \cdot |x - x_0| < (1 + 2|x_0|) \cdot \delta$$

$$< (1 + 2|x_0|) \cdot \frac{\epsilon}{1 + 2|x_0|} = \epsilon$$

Therefore, $\lim_{x \to x_0} x^2 = x_0^2$.

■

Recall that the actual value of δ will depend on both the size of ϵ and the value of x_0. Specifically, for a fixed value of x_0, the smaller ϵ is chosen to be, the smaller δ will need to be. Also, for a fixed value of ϵ, the δ value for two different values of x_0 may differ. For instance, in the previous example the value of δ for a fixed value of ϵ was $\delta = \min\left(1, \frac{\epsilon}{1+2|x_0|}\right)$. Thus, for $x_0 = 1$, $\delta = \min\left(1, \frac{\epsilon}{3}\right)$ and for $x_0 = 2$, $\delta = \min\left(1, \frac{\epsilon}{5}\right)$. Clearly, the value of δ will depend on the values of both ϵ and x_0.

The following lemma shows that when f has a nonzero limit at the point x_0, then $f(x)$ can be bounded from below. This lemma will be needed in the proof of Theorem 5.3.1.

Lemma 5.3.1: Let f be a real-valued function defined on a domain $\mathcal{D}$ containing a deleted neighborhood of x_0. If $\lim_{x \to x_0} f(x) = L$ and $L \neq 0$, then there exists a $\delta > 0$ such that $|f(x)| > \frac{|L|}{2}$ whenever $|x - x_0| < \delta$.

Proof: Let f be a real-valued function with $\lim_{x \to x_0} f(x) = L$ and suppose $L \neq 0$. Then there exists $\delta > 0 \ni: |f(x) - L| < \frac{|L|}{2}$ whenever $|x - x_0| < \delta$.

Consider $|f(x)|$:

$$|f(x)| = |f(x) - L + L| = |L - (L - f(x))|$$

$$\geq |L| - |L - f(x)| = |L| - |f(x) - L|$$

Thus, whenever $|x - x_0| < \delta$, it follows that

$$|f(x)| \geq |L| - |f(x) - L| > |L| - \frac{|L|}{2} = \frac{|L|}{2}$$

■

The following theorem provides powerful results for determining the limit of the sum, product, and ratio of two real-valued functions. Even more importantly, the rules proved in Theorem 5.3.1 are often the key components in the proofs of several theorems presented later in this section and Section 5.4, also. Note that each result in Theorem 5.3.1 is analogous to a result proved in Chapter 4 for the sum, product, or ratio of two convergent sequences.

Theorem 5.3.1: Let f and g be functions defined on domains $\mathcal{D}_f$ and $\mathcal{D}_g$ containing deleted neighborhoods of x_0. If $\lim_{x \to x_0} f(x) = L_1$ and $\lim_{x \to x_0} g(x) = L_2$, then

(i) $\lim_{x \to x_0} \left[k \cdot f(x) + l\right] = k \cdot L_1 + l$ for all $k, l \in \mathbb{R}$.

(ii) $\lim_{x \to x_0} \left[f(x) + g(x)\right] = L_1 + L_2$.

(iii) $\lim_{x \to x_0} \left[f(x) \cdot g(x)\right] = L_1 \cdot L_2$.

(iv) $\lim_{x \to x_0} \left[\frac{1}{f(x)}\right] = \frac{1}{L_1}$ provided that $L_1 \neq 0$.

(v) $\lim_{x \to x_0} \left[\frac{f(x)}{g(x)}\right] = \frac{L_1}{L_2}$ provided that $L_2 \neq 0$.

Proof: Let $\epsilon > 0$ be ABF, and let f and g be functions defined on the domains $\mathcal{D}_f$ and $\mathcal{D}_g$ containing deleted neighborhoods of x_0. Suppose that $\lim_{x \to x_0} f(x) = L_1$ and $\lim_{x \to x_0} g(x) = L_2$, then

Proof of part (i): The proof of part (i) is left as an exercise.

Proof of part (ii): Since $\lim_{x \to x_0} f(x) = L_1$ and $\lim_{x \to x_0} g(x) = L_2$, there exists $\delta_1 > 0$ and $\delta_2 > 0$ such that

$$|f(x) - L_1| < \frac{\epsilon}{2} \quad \text{whenever} \quad |x - x_0| < \delta_1$$

$$|g(x) - L_2| < \frac{\epsilon}{2} \quad \text{whenever} \quad |x - x_0| < \delta_2$$

Consider $|f(x) + g(x) - (L_1 + L_2)|$:

$$|f(x) + g(x) - (L_1 + L_2)| = |f(x) - L_1 + g(x) - L_2|$$

$$\underbrace{\leq |f(x) - L_1| + |g(x) - L_2|}_{\text{By the triangle inequality}}$$

Let $\delta = \min(\delta_1, \delta_2)$. Then, whenever $|x - x_0| < \delta$, it follows that

$$|f(x) + g(x) - (L_1 + L_2)| = |f(x) - L_1| + |g(x) - L_2| < \frac{\epsilon}{2} + \frac{\epsilon}{2} = \epsilon$$

Therefore, $\lim_{x \to x_0} \left[f(x) + g(x) \right] = L_1 + L_2.$

Proof of part (iii): Since $\lim_{x \to x_0} f(x) = L_1$ and $\lim_{x \to x_0} g(x) = L_2$, there exists $\delta_1 > 0$ and $\delta_2 > 0$ such that

$$|f(x) - L_1| < \frac{\epsilon}{3(|L_2| + 1)} \quad \text{whenever} \quad |x - x_0| < \delta_1$$

$$|g(x) - L_2| < \frac{\epsilon}{3(|L_1| + 1)} \quad \text{whenever} \quad |x - x_0| < \delta_2$$

Also, there exist δ_3 and δ_4 such that

$$|f(x) - L_1| < \sqrt{\frac{\epsilon}{3}} \quad \text{whenever} \quad |x - x_0| < \delta_3$$

$$|g(x) - L_2| < \sqrt{\frac{\epsilon}{3}} \quad \text{whenever} \quad |x - x_0| < \delta_4$$

Consider $|f(x)g(x) - L_1 L_2|$. First, note that

$$f(x) = (f(x) - L_1) + L_1$$

and

$$g(x) = (g(x) - L_2) + L_2$$

Thus

$$\begin{aligned} f(x)g(x) &= [(f(x) - L_1) + L_1] \cdot [(g(x) - L_2) + L_2] \\ &= L_1(g(x) - L_2) + L_2(f(x) - L_1) \\ &\quad + (f(x) - L_1) \cdot (g(x) - L_2) + L_1 L_2 \end{aligned}$$

Hence

$$\begin{aligned} |f(x)g(x) - L_1 L_2| &= |L_1(g(x) - L_2) \\ &\quad + L_2(f(x) - L_1) + (f(x) - L_1) \cdot (g(x) - L_2)| \\ &\le |L_1| \cdot |g(x) - L_2| + |L_2| \cdot |f(x) - L_1| \\ &\quad + |f(x) - L_1| \cdot |g(x) - L_2| \end{aligned}$$

by the triangle inequality. Now, let $\delta = \min(\delta_1, \delta_2, \delta_3, \delta_4)$. Then, whenever $|x - x_0| < \delta$, it follows that

$$|f(x)g(x) - L_1L_2| \leq |L_1|\cdot|g(x) - L_2| + |L_2|\cdot|f(x) - L_1|$$

$$+ |f(x) - L_1|\cdot|g(x) - L_2|$$

$$< |L_1| \cdot \frac{\epsilon}{3(|L_1| + 1)} + |L_2| \cdot \frac{\epsilon}{3(|L_2| + 1)} + \sqrt{\frac{\epsilon}{3}} \cdot \sqrt{\frac{\epsilon}{3}}$$

$$< \frac{\epsilon}{3} + \frac{\epsilon}{3} + \frac{\epsilon}{3} = \epsilon$$

Therefore, $\lim_{x \to x_0} \left[f(x)\cdot g(x)\right] = L_1 \cdot L_2$.

Proof of part (iv): Suppose that $\lim_{x \to x_0} f(x) = L_1$ and $L_1 \neq 0$. Since $L_1 \neq 0$, there exists $\delta_1 > 0$ such that

$$|f(x) - L_1| < \epsilon \cdot \frac{|L_1|^2}{2}$$

whenever $|x - x_0| < \delta_1$, and by Lemma 5.3.1 there exists δ_2 such that $|f(x)| \geq \frac{|L_1|}{2}$ whenever $|x - x_0| < \delta_2$.

Let $\delta = \min(\delta_1, \delta_2)$. Then, whenever $|x - x_0| < \delta$, it follows that

$$\left|\frac{1}{f(x)} - \frac{1}{L_1}\right| = \frac{1}{|f(x)\cdot L_1|}\cdot|f(x) - L_1| < \frac{1}{\frac{|L_1|}{2}\cdot|L_1|}\cdot\epsilon\cdot\frac{|L_1|^2}{2} = \epsilon$$

Therefore, $\lim_{x \to x_0} \left[\frac{1}{f(x)}\right] = \frac{1}{L_1}$ whenever $\lim_{x \to x_0} f(x) = L_1$ and $L_1 \neq 0$.

Proof of part (v): Since

$$\frac{f(x)}{g(x)} = f(x)\cdot\frac{1}{g(x)}$$

and $L_2 \neq 0$, this result follows directly from the application of parts (iii) and (iv) of this theorem.

■

Note that Theorem 5.3.1 applies only when both limits exist. However, it is possible for the sum of two real-valued functions to have a limit at x_0 when neither of the individual functions has a limit at x_0. For example, let $f(x) = \frac{1}{x}$ and $g(x) = -\frac{1}{x}$. Then, $\lim_{x \to 0} \left[f(x) + g(x)\right] = \lim_{x \to 0} \left[\frac{1}{x} - \frac{1}{x}\right] = 0$, yet neither $\lim_{x \to 0} \frac{1}{x}$ nor $\lim_{x \to 0} -\frac{1}{x}$ exists. In this example, clearly the statement

$$\lim_{x \to 0} \left[\frac{1}{x} - \frac{1}{x}\right] = \lim_{x \to 0} \frac{1}{x} + \lim_{x \to 0} -\frac{1}{x}$$

would make no sense at all, since neither of the limits on the righthand side exists.

The following corollary to Theorem 5.3.1 shows that limits are linear functions. Thus, when the limits exist, then $\lim_{x \to x_0} \left[af(x) + bg(x)\right] = aL_1 + bL_2$ and $\lim_{x \to x_0} \left[f(x) - g(x)\right] = L_1 - L_2$, also.

Corollary to Theorem 5.3.1: Let f and g be functions defined on domains $\mathcal{D}_f$ and $\mathcal{D}_g$ containing deleted neighborhoods of x_0. If $a, b \in \mathbb{R}$, $\lim_{x \to x_0} f(x) = L_1$, and $\lim_{x \to x_0} g(x) = L_2$, then

(i) $\lim_{x \to x_0} \left[af(x) + bg(x)\right] = aL_1 + bL_2$.

(ii) $\lim_{x \to x_0} \left[f(x) - g(x)\right] = L_1 - L_2$.

Proof: Let f and g be functions defined on a domain $\mathcal{D}$ containing a deleted neighborhood of x_0 and suppose that $\lim_{x \to x_0} f(x) = L_1$ and $\lim_{x \to x_0} g(x) = L_2$.

Proof of part (i): Part (i) follows directly from parts (i) and (ii) of Theorem 5.3.1.

Proof of part (ii): Part (ii) follows directly from part (i) with $a = 1$ and $b = -1$.

■

The next theorem will be used to prove that the limit of a polynomial $p(x)$ at the point x_0 is simply $p(x_0)$. In particular, the following theorem shows that $\lim_{x \to x_0} x^n = x_0^n$, for all natural numbers n.

Theorem 5.3.2: If $x_0 \in \mathbb{R}$, then $\lim_{x \to x_0} x^n = x_0^n,\ \forall\, n \in \mathbb{N}$.

Proof: The proof of Theorem 5.3.2 is left as an exercise.

■

Now, Theorem 5.3.3 shows that if $p(x)$ is a polynomial, the limit as x approaches x_0 of $p(x)$ can be found by simply evaluating the $p(x)$ at the value x_0. The corollary to Theorem 5.3.3 will show that a similar result holds for rational functions.

Theorem 5.3.3: If $p(x)$ is a polynomial, then $\lim_{x \to x_0} p(x) = p(x_0),\ \forall\, x \in \mathbb{R}$.

Proof: The proof of Theorem 5.3.3 is left as an exercise.

■

Corollary to Theorem 5.3.3: Let $p(x)$ and $q(x)$ be polynomials of degree n and m, respectively. If $q(x_0) \neq 0$, then $\lim_{x \to x_0} \dfrac{p(x)}{q(x)} = \dfrac{p(x_0)}{q(x_0)}$.

Proof: This result follows directly from Theorems 5.3.3 and 5.3.1 part (v).

■

Example 5.3.4: Let $f(x) = x^2 - 2x - 1$ and $g(x) = x + 1$. Determine

a. $\lim_{x \to 2} (f(x) + g(x))$.

b. $\lim_{x \to 1} f(x) \cdot g(x)$.

c. $\lim_{x \to 3} \dfrac{f(x)}{g(x)}$.

Solutions: Note that f and g are polynomials, and therefore by Theorem 5.3.3, it follows that

$$\lim_{x \to x_0} f(x) = x_0^2 - 2x_0 - 1 \quad \text{and} \quad \lim_{x \to x_0} g(x) = x_0 + 1$$

a. $\lim_{x \to 2} f(x) = -1$ and $\lim_{x \to 2} g(x) = 3$. Therefore

$$\lim_{x \to 2} [f(x) + g(x)] = \underbrace{\lim_{x \to 2} f(x) + \lim_{x \to 2} g(x)}_{\text{By Theorem 5.3.1 part (i)}} = -1 + 3 = 2$$

b. $\lim_{x \to 1} f(x) = -2$ and $\lim_{x \to 1} g(x) = 2$. Therefore

$$\lim_{x \to 1} f(x) \cdot g(x) = \underbrace{\lim_{x \to 1} f(x) \cdot \lim_{x \to 1} g(x)}_{\text{By Theorem 5.3.1 part (iii)}} = -2(2) = -4$$

c. $\lim_{x \to 3} f(x) = 2$ and $\lim_{x \to 3} g(x) = 4$. Therefore

$$\lim_{x \to 3} \frac{f(x)}{g(x)} = \underbrace{\frac{\lim_{x \to 3} f(x)}{\lim_{x \to 3} g(x)}}_{\text{By corollary to Theorem 5.3.3}} = \frac{2}{4} = 0.5$$

The squeeze theorem for real-valued functions is an important theorem that is often used when the function of interest is hard to work with. In particular, when the function of interest, say, $h(x)$, is trapped between the two functions $f(x)$ and $g(x)$ that have known and equal limits, then the squeeze theorem shows that $\lim_{x \to x_0} f(x) = \lim_{x \to x_0} h(x) = \lim_{x \to x_0} g(x)$. The squeeze theorem is given below.

Theorem 5.3.4 (The Squeeze Theorem): Let f, g, and h be functions defined on domains containing a deleted neighborhood of x_0 where $f(x) \leq h(x) \leq g(x)$. If $\lim_{x \to x_0} f(x) = \lim_{x \to x_0} g(x) = L$, then $\lim_{x \to x_0} h(x) = L$.

Proof: Let f, g, and h be functions defined on common domains containing a deleted neighborhood of x_0 where $f(x) \leq h(x) \leq g(x)$. Suppose that $\lim_{x \to x_0} f(x) = \lim_{x \to x_0} g(x) = L$, and let $\epsilon > 0$.

Then, since $\lim_{x \to x_0} f(x) = L$, there exists δ_1 such that

$$\underbrace{L - \epsilon < f(x) < L + \epsilon}_{|f(x) - L| < \epsilon} \qquad \text{whenever} \qquad |x - x_0| < \delta_1$$

and since $\lim_{x \to x_0} g(x) = L$, there exists δ_2 such that

$$\underbrace{L - \epsilon < g(x) < L + \epsilon}_{|g(x) - L| < \epsilon} \qquad \text{whenever} \qquad |x - x_0| < \delta_2$$

Furthermore, since there exists a deleted neighborhood of x_0 where $f(x) \leq h(x) \leq g(x)$, it follows that there exists $\delta_3 > 0$ such that $f(x) \leq h(x) \leq g(x)$ whenever $|x - x_0| < \delta_3$.

Let $\delta = \min(\delta_1, \delta_2, \delta_3)$. Then, whenever $|x - x_0| < \delta$, it follows that

$$L - \epsilon < f(x) \leq h(x) \leq g(x) < L + \epsilon$$

Hence, $L - \epsilon < h(x) < L + \epsilon$, or equivalently $|h(x) - L| < \epsilon$, whenever $|x - x_0| < \delta$. Therefore, $\lim\limits_{x \to x_0} h(x) = L$.

■

The squeeze theorem is now used to prove that $\lim\limits_{x \to 0} \dfrac{\sin(x)}{x} = 1$.

Example 5.3.5: Prove that $\lim\limits_{x \to 0} \dfrac{\sin(x)}{x} = 1$.

Solution: First, a plot of the functions $\dfrac{\sin(x)}{x}$ and $\cos(x)$, on a neighborhood of radius 1 of the point $x_0 = 0$, is given in Figure 5.3.2; the dashed line is $\dfrac{\sin(x)}{x}$ and the solid line is $\cos(x)$.

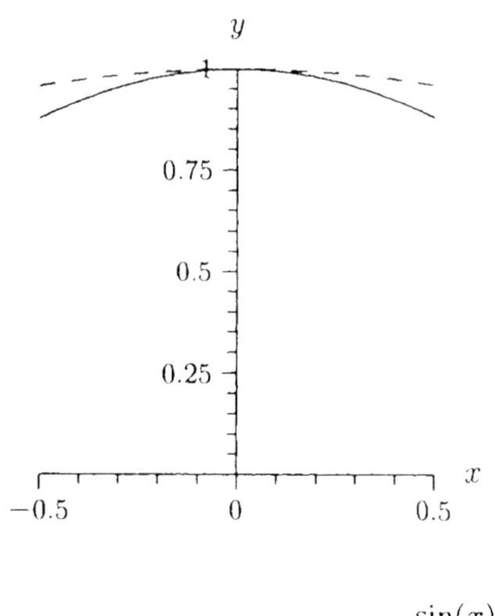

Figure 5.3.2 A plot of $\cos(x)$ and $\dfrac{\sin(x)}{x}$.

Note that $\cos(x) \leq \dfrac{\sin(x)}{x} \leq 1$ in this neighborhood, and $\lim\limits_{x \to 0} \cos(x) = 1$. Thus, by the squeeze theorem it follows that $\lim\limits_{x \to 0} \dfrac{\sin(x)}{x} = 1$.

Example 5.3.6: Let $f(x) = x^{1-x}$ on the domain $\mathcal{D} = [0, \infty)$. Prove that $\lim\limits_{x \to 0} x^{1-x} = 0$.

Solution: First, the functions x^{1-x}, $\sqrt{x}$, and x, on the interval $[0, 0.5]$, are plotted in Figure 5.3.3. The solid line represents x; the dashed line, $\sqrt{x}$; and the dotted line, x^{1-x}.

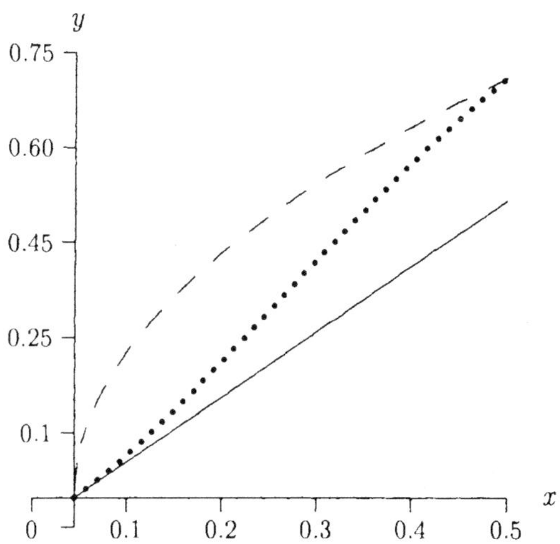

Figure 5.3.3 A plot of the functions x, $\sqrt{x}$, and x^{1-x}.

Note that $x \leq x^{1-x} \leq x^{.5}$ on $[0, .5]$ and $\lim\limits_{x \to 0} x = \lim\limits_{x \to 0} \sqrt{x} = 0$. Thus, by the squeeze theorem it follows that $\lim\limits_{x \to 0} x^{1-x} = 0$.

The following two theorems provide results for the limit of powers of f when $\lim\limits_{x \to x_0} f(x) = L$. Theorem 5.3.5 states that if $\lim\limits_{x \to x_0} f(x) = L$, then $\lim\limits_{x \to x_0} f(x)^n = L^n$, and Theorem 5.3.6 shows that when $L \geq 0$, it follows that $\lim\limits_{x \to x_0} \sqrt{f(x)} = \sqrt{L}$.

Theorem 5.3.5: Let f be a function defined on the domain $\mathcal{D}$ containing a deleted neighborhood of x_0. If $\lim\limits_{x \to x_0} f(x) = L$, then $\lim\limits_{x \to x_0} f(x)^n = L^n$, $\forall\, n \in \mathbb{N}$.

Proof: The proof of Theorem 5.3.5 is left as an exercise.

■

Theorem 5.3.6: Let f be a function defined on the domain $\mathcal{D}$ containing a deleted neighborhood of x_0. If $\lim\limits_{x \to x_0} f(x) = L \geq 0$, then $\lim\limits_{x \to x_0} \sqrt{f(x)} = \sqrt{L}$.

Proof: The proof of Theorem 5.3.6 is left as an exercise.

■

Example 5.3.7: Suppose that $\lim\limits_{x \to x_0} f(x) = 3$ and $\lim\limits_{x \to x_0} g(x) = 2$. Determine

a. $\lim\limits_{x \to x_0} \sqrt{3f(x) + 8g(x)}$.

b. $\lim\limits_{x \to x_0} \sqrt{f(x)g(x) + 10}$.

c. $\lim\limits_{x \to x_0} f(x)^2 \sqrt{g(x)}$.

Solutions: Suppose that $\lim\limits_{x \to x_0} f(x) = 3$ and $\lim\limits_{x \to x_0} g(x) = 2$.

a. By the corollary to Theorem 5.3.1, it follows that

$$\lim_{x \to x_0} [3f(x) + 8g(x)] = 3 \lim_{x \to x_0} f(x) + 8 \lim_{x \to x_0} g(x) = 3 \cdot 3 + 8 \cdot 2 = 25$$

and by Theorem 5.3.6, it follows that

$$\lim_{x \to x_0} \sqrt{3f(x) + 8g(x)} = \sqrt{\lim_{x \to x_0} 3f(x) + 8g(x)} = 5.$$

b. By Theorem 5.3.1 parts (ii) and (iii), it follows that

$$\lim_{x \to x_0} f(x)g(x) + 10 = 2(3) + 10 = 16$$

and by Theorem 5.3.6, it follows that

$$\lim_{x \to x_0} \sqrt{f(x)g(x) + 10} = \sqrt{2(3) + 10} = 4$$

c. By Theorem 5.3.5, it follows that

$$\lim_{x \to x_0} f(x)^2 = 3^2 = 9 \text{ and } \lim_{x \to x_0} \sqrt{g(x)} = \sqrt{2}$$

Thus, by Theorem 5.3.1 part (iii), it follows that

$$\lim_{x \to x_0} f(x)^2 \sqrt{g(x)} = 3^2\sqrt{2} = 9\sqrt{2}$$

The following theorem shows that if $\lim_{x \to x_0} f(x) = L$ and x_n is of any sequence of real numbers converging to x_0, then $\lim_{n \to \infty} f(x_n) = L$, also.

Theorem 5.3.7: Let f be a function with domain $\mathcal{D}$ containing a deleted neighborhood of x_0. If $\lim_{x \to x_0} f(x) = L$ and x_n is a sequence of real numbers contained in $\mathcal{D}$ with $x_n \to x_0$, then $\lim_{n \to \infty} f(x_n) = L$.

Proof: Let f be a function with domain $\mathcal{D}$ containing a deleted neighborhood of x_0 and $\lim_{x \to x_0} f(x) = L$. Let $\epsilon > 0$ be ABF, and let x_n be an arbitrary sequence converging to x_0 which is contained in $\mathcal{D}$.

Since $\lim_{x \to x_0} f(x) = L$, there exists δ such that $|f(x) - L| < \epsilon$ whenever $|x - x_0| < \delta$. Now, since $x_n \to x_0$, there exists $N \in \mathbb{N}$ such that $|x - x_0| < \delta$ whenever $n \geq N$.

Now, $x_n \in \mathcal{D}$ for each $n \in \mathbb{N}$; thus, $|f(x_n) - L| < \epsilon$ whenever $n \geq N$. Therefore, $\lim_{x \to x_0} f(x_n) = L$ for any sequence $x_n \to x_0$ that is contained in $\mathcal{D}$.

■

It is important to note that Theorem 5.3.7 does not say when x_n is a sequence of real numbers converging to the point x_0 and $\lim_{n \to \infty} f(x_n) = L$, then it follows that $\lim_{x \to x_0} f(x) = L$. Thus, it may be possible that a sequence x_n converges to x_0 and that $\lim_{n \to \infty} f(x_n) = L$; however, $\lim_{x \to x_0} f(x) \neq L$. For example, while $x_n = \frac{1}{n} \to 0$ and $\lim_{n \to \infty} \frac{|x_n|}{x_n} = 1$, $\lim_{x \to 0} \frac{|x|}{x}$ does not exist. Theorem 5.3.7 simply states that when $\lim_{x \to x_0} f(x) = L$, then $\lim_{n \to \infty} f(x_n) = L$ for every sequence of real numbers converging to x_0.

5.4 Continuity

When a real-valued function $f(x)$ has the property that $\lim_{x \to x_0} f(x) = f(x_0)$, then the function f is said to be *continuous* at the point x_0. In this section the continuity of real-valued functions will be studied. A definition of the continuity of a function f at a point $x = x_0$ is given below.

Definition 5.4.1: A function f is said to be *continuous* at a point x_0 if and only if

(i) $\lim_{x \to x_0} f(x)$ exists.

(ii) $x_0 \in \mathcal{D}$.

(iii) $\lim_{x \to x_0} f(x) = f(x_0)$.

Definition 5.4.2: A function f is said to be *discontinuous* at the point x_0 if and only if $x_0 \in \mathcal{D}$ and f is not continuous at x_0.

Recall that the limit of a real-valued function f, as x approaches x_0, does not depend on the value of $f(x_0)$. In particular, it might be the case that $f(x_0)$ does not even exist (i.e., $x_0 \notin \mathcal{D}$). In fact, there are five possibilities for the relationship between $\lim_{x \to x_0} f(x)$ and $f(x_0)$:

(1) $\lim_{x \to x_0} f(x)$ exists and equals $f(x_0)$.

(2) $\lim_{x \to x_0} f(x)$ exists and does not equal $f(x_0)$.

(3) $\lim_{x \to x_0} f(x)$ exists but $f(x_0)$ is undefined.

(4) $\lim_{x \to x_0} f(x)$ does not exist and $f(x_0)$ does exist.

(5) Neither $\lim_{x \to x_0} f(x)$ nor $f(x_0)$ exists.

Note that only in case (1), where $\lim_{x \to x_0} f(x)$ exists and equals $f(x_0)$, is the function f continuous at $x = x_0$. In each of the other cases, f is not continuous at $x = x_0$. The following example shows that the function $f(x) = \dfrac{|x|}{x}$ is not continuous at $x = 0$.

Example 5.4.1: Let $f(x) = \dfrac{|x|}{x}$. The plot of $f(x)$ on a deleted neighborhood of 0 is given in Figure 5.4.1.

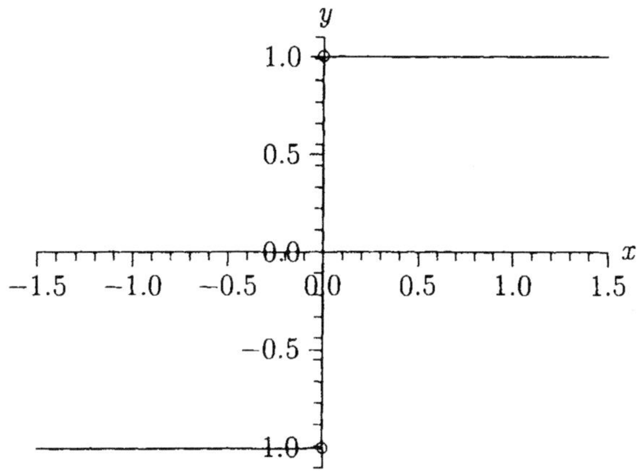

Figure 5.4.1 A plot of $f(x)=\dfrac{|x|}{x}$.

Now

$$f(x) = \frac{|x|}{x} = \begin{cases} -1 & x < 0 \\ \text{undefined} & x = 0 \\ 1 & x > 0 \end{cases}$$

and clearly from Figure 5.4.1 it can be seen that f is continuous everywhere except at $x = 0$.

An equivalent definition of continuity based on the ϵ-δ definition of the limit of $f(x)$ is given below.

Definition 5.4.3: A function f is said to be *continuous* at a point x_0 if and only if $x_0 \in \mathcal{D}$ and for every $\epsilon > 0$, $\exists\ \delta > 0$ such that $|f(x) - f(x_0)| < \epsilon$ whenever $|x - x_0| < \delta$.

Note that in Definition 5.4.3, the value of δ depends on two values, namely, the value of ϵ and the particular value of x_0. In general, the value of δ for a fixed value of ϵ must change as x moves from one point to another. The dependence of δ on a particular value of x is illustrated in the following example.

Example 5.4.2: If $f(x) = x^2$, then $\lim_{x \to 0} f(x) = 0$ and $\lim_{x \to 2} f(x) = 4$. Let $\epsilon = 1$ and determine the value of δ

a. So that $|x^2 - 0^2| < 1$ whenever $|x - 0| < \delta$.

b. So that $|x^2 - 2^2| < 1$ whenever $|x - 2| < \delta$.

Solution:

a. Note that when $|x^2 - 0| = x^2 < 1$, it follows that $|x| < \sqrt{1} = 1$. Now, let $\delta_1 = 1$. Then, $|x^2 - 0| < 1$ whenever $|x - 0| < 1$.
b. Note that $|x^2 - 4^2| = |x - 4| \cdot |x + 4|$. Now, for $|x - 4| < 1$, it follows that

$$|x - 4| < 1 \Longrightarrow -3 < x < 5 \Longrightarrow 1 < x + 4 < 9$$

Therefore, when $\delta \leq 1$, it follows that $|x + 4| \leq 9$ and therefore

$$|x^2 - 4^2| = |x - 4| \cdot |x + 4| \leq 9 \cdot |x - 4|$$

Thus, let $\delta = \min(1, 1/9) = 1/9$. Then, whenever $|x - 2| < \delta$, it follows that $|x^2 - 4| < 1$.

Note that the values of δ in parts (a) and (b) are different because the particular value of δ for $\epsilon = 1$ depends on the particular value of x_0 under consideration in the limit.

An algorithm for proving that f is continuous at the point $x = x_0$ with an ϵ-δ proof is given below.

An Algorithm for an ϵ-δ Continuity Proof: Let $f(x)$ be a real-valued function. To prove that f is continuous at the point x_0

1. Let $\epsilon > 0$ be ABF.
2. Consider $|f(x) - f(x_0)|$.
3. Determine how to make $|f(x) - f(x_0)|$ arbitrarily small for x near x_0.
4. From steps 2 and 3, determine the value of δ so that $|f(x) - f(x_0)| < \epsilon$ whenever $|x - x_0| < \delta$.
5. Conclude $\lim\limits_{x \to x_0} f(x) = f(x_0)$ and therefore, f is continuous at $x = x_0$.
6. Clean up and rewrite the scratchwork in a clear and concise proof of the theorem. Make sure that each step of the proof makes sense and is clearly justified.
7. Read the proof over carefully and make any necessary corrections.

Note that the algorithm given above is simply the ϵ-δ algorithm for showing that the limit of the function f at $x = x_0$ is $L = f(x_0)$. The ϵ-δ algorithm is used to show that $f(x) = x^2$ is continuous at every point x_0 in Example 5.4.3.

Example 5.4.3: Let $f(x) = x^2$ for $x \in \mathbb{R}$. Prove that f is continuous at $x = x_0$, $\forall\, x_0 \in \mathbb{R}$.

Solution:

Proof: Let $\epsilon > 0$ and $x_0 \in \mathbb{R}$ be ABF. Consider $|x^2 - x_0^2|$.

$$|x^2 - x_0^2| = |x - x_0| \cdot |x + x_0|$$

Now, whenever $|x - x_0| < 1$, it follows that $|x + x_0| < 2|x_0| + 1$. Therefore

$$|x^2 - x_0^2| = |x - x_0| \cdot |x + x_0| < |x - x_0| \cdot (2|x_0| + 1)$$

Now, let $\delta = \min\left(\dfrac{\epsilon}{2|x_0| + 1}, 1\right)$. Then, whenever $|x - x_0| < \delta$, it follows that

$$|x^2 - x_0^2| = |x - x_0| \cdot |x + x_0| < |x - x_0| \cdot (2|x_0| + 1)$$

$$< \delta(2|x_0| + 1) < \frac{\epsilon}{2|x_0| + 1}(2|x_0| + 1) = \epsilon$$

Therefore, by Definition 5.4.3, $f(x) = x^2$ is continuous at $x = x_0$.

■

In the previous example, since the same proof works for every value of $x_0 \in \mathbb{R}$, it follows that $f(x) = x^2$ is continuous over the entire real line. When a function f is continuous at every point in an interval, then the function is said to be continuous on the interval.

Definition 5.4.4: A function f is said to be *continuous over an interval I* if and only if f is continuous at each point $x \in I$.

Note that the form of the interval I in Definition 5.4.4 is not specified. This means that a function f can be continuous on an open interval, a closed interval, or a half-open interval. For example, $f(x) = \dfrac{1}{x}$ is continuous over the interval $(0, \infty)$ but is not continuous on the interval $[0, \infty)$.

Example 5.4.4: Let $f(x) = x^2 + 4x - 1$. Show that f is continuous on $\mathbb{R}$.

Solution: Let $x_0 \in \mathbb{R}$ be ABF. By Theorem 5.3.3 it follows that

$$\lim_{x \to x_0} \left[x^2 + 4x - 1\right] = x_0^2 - 4x_0 - 1 = f(x_0)$$

Now, since $x_0 \in \mathbb{R}$ was arbitrary, it follows that $\lim_{x \to x_0} f(x) = f(x_0)$ for all $x_0 \in \mathbb{R}$ and therefore, $f(x) = x^2 + 4x - 1$ is continuous on $\mathbb{R}$.

The following two theorems are direct results of Theorem 5.3.3 and its corollary and show that the polynomials and rational functions are continuous functions on $\mathbb{R}$.

Theorem 5.4.1: If $p(x)$ is a polynomial in x, then $p(x)$ is continuous for all $x \in \mathbb{R}$.

Proof: Theorem 5.4.1 follows directly from Theorem 5.3.3.

■

Theorem 5.4.2: Let $p(x)$ and $q(x)$ be polynomials, and let $x_0 \in \mathbb{R}$. If $q(x_0) \neq 0$, then $\dfrac{p(x)}{q(x)}$ is continuous at $x = x_0$.

Proof: Theorem 5.4.2 follows directly from the corollary to Theorem 5.3.3

■

The next theorem shows that when a function f is continuous, then so are the functions $af(x) + b$ and $|f(x)|$. In proving Theorem 5.4.3, the proof of the continuity of $af(x) + b$ is based on the limit results of Section 5.3, while the continuity of $|f(x)|$ is proved using an ϵ-δ approach.

Theorem 5.4.3: Let f be a real-valued function, and let $x_0 \in \mathcal{D}$. If f is continuous at $x = x_0$, then

(i) $a \cdot f(x) + b$ is continuous function at $x = x_0$, $\forall\ a, b \in \mathbb{R}$.

(ii) $|f(x)|$ is continuous at $x = x_0$.

Proof: Let f be a real-valued function that is continuous at $x = x_0$.

Proof of part (i): Let $a, b \in \mathbb{R}$ be ABF and define $g(x) := a \cdot f(x) + b$. Then $\mathcal{D}_g = \mathcal{D}_f$ and $g(x_0) = a \cdot f(x_0) + b$. Now

$$\lim_{x \to x_0} g(x) = \lim_{x \to x_0} (a \cdot f(x) + b) = a \cdot \lim_{x \to c} f(x) + b$$

$$= a \cdot f(c) + b = g(x_0)$$

Hence, $\lim_{x \to x_0} g(x) = g(x_0)$; therefore, $g(x) = a \cdot f(x) + b$ is continuous at $x = x_0$, $\forall\ a, b \in \mathbb{R}$.

Proof of part (ii): Let $\epsilon > 0$ be ABF. Since $\lim_{x \to x_0} f(x) = f(x_0)$ there exists a $\delta > 0$ such that $|f(x) - f(x_0)| < \epsilon$ whenever $|x - x_0| < \delta$.

Consider $\Big| |f(x)| - |f(x_0)| \Big|$:

$$\Big| |f(x)| - |f(x_0)| \Big| \underbrace{\leq |f(x) - f(x_0)|}_{\text{By Theorem 4.2.14}}$$

Thus, whenever $|x - x_0| < \delta$, it follows that

$$\Big| |f(x)| - |f(x_0)| \Big| \leq |f(x) - f(x_0)| < \epsilon$$

Therefore, $|f(x)|$ is continuous at $x = x_0$.

■

The following theorem shows that when two functions f and g are continuous at $x = x_0$, then so are the sum, product, and ratio of f and g. Furthermore, the proof of each part of this theorem follows directly from the limit theorems of Section 5.3.

Theorem 5.4.4: Let f and g be real-valued functions with common domain $\mathcal{D}$. If f and g are continuous at the point $x = x_0 \in \mathcal{D}$, then

(i) $f + g$ is continuous at $x = x_0$.

(ii) $af + bg$ is continuous at $x = x_0$, $\forall\ a, b \in \mathbb{R}$.

(iii) $f \cdot g$ is continuous at $x = x_0$.

(iv) $\dfrac{f}{g}$ is continuous at $x = x_0$, provided that $g(x_0) \neq 0$.

Proof: Let f and g be real-valued functions with common domain $\mathcal{D}$, and suppose that f and g are continuous at the point $x = x_0 \in \mathcal{D}$.

Proof of part (i): Let $s(x) = f(x) + g(x)$. Then, $\mathcal{D}_s = \mathcal{D}$ and $s(x_0) = f(x_0) + g(x_0)$, and since f and g are continuous at $x = x_0$, it follows that $\lim_{x \to x_0} f(x) = f(x_0)$ and $\lim_{x \to x_0} g(x) = g(x_0)$.

Consider $\lim_{x \to x_0} s(x)$:

$$\lim_{x \to x_0} s(x) = \lim_{x \to x_0} [f(x) + g(x)] = \underbrace{\lim_{x \to x_0} f(x) + \lim_{x \to x_0} g(x)}_{\text{by Theorem 5.3.1(i)}}$$

$$= f(x_0) + g(x_0) = s(x_0)$$

Therefore, $s(x) = f(x) + g(x)$ is continuous at $x = x_0$ whenever f and g are continuous at $x = x_0$.

Proof of part (ii): The proof of part (ii) is left as an exercise.

Proof of part (iii): The proof of part (iii) is left as an exercise.

Proof of part (iv): The proof of part (iv) is left as an exercise.

∎

The continuity of the functions $f(x)^2$, $\sqrt{f(x)}$, and $f(x)^n$ follow directly from previous results. In particular, Theorem 5.4.4 can be used to show that $f(x)^2$ is continuous, Theorem 5.3.6 can be used to show that $\sqrt{f(x)}$ is continuous, and Theorem 5.3.5 can be used to show that $f(x)^n$ is continuous whenever the function f is continuous.

Theorem 5.4.5: Let f be a real-valued function with domain $\mathcal{D}$. If f is continuous at $x = x_0$, then

(i) $f(x)^2$ is continuous at $x = x_0$.
(ii) $\sqrt{f(x)}$ is continuous at $x = x_0$, provided that $f(x_0) > 0$.
(iii) $f(x)^n$ is continuous at $x = x_0$, $\forall\, n \in \mathbb{N}$.

Proof: Let f be a real-valued function with domain $\mathcal{D}$, and suppose that f is continuous at $x = x_0$.

Proof of part(i): Part (i) follows directly from Theorem 5.4.4 part (iii) with $g(x) = f(x)$.

Proof of part(ii): Part (ii) follows directly from Theorem 5.3.6.

Proof of part(i): Part (iii) follows directly from Theorem 5.3.5.

∎

The previous three theorems show that the continuity of one or more functions is a powerful property that is maintained even when the functions are combined using the basic arithmetic operations. For example, when $f(x)$ and $g(x)$ are continuous at the point $x = x_0$, then so are each of the following functions:

$$kf(x) + l,\ |f(x)|,\ f(x) + g(x),\ f(x)g(x),\ \frac{f(x)}{g(x)},\ f(x)^2,\ \sqrt{f(x)}$$

Furthermore, the next theorem shows that the composition of two continuous functions can also be a continuous function.

Theorem 5.4.6: Let f and g be real-valued functions with domains $\mathcal{D}_f$ and $\mathcal{D}_g$. If f is continuous at $x = g(x_0)$ and g is continuous at $x = x_0$, then the function $f \circ g(x) = f\,(g(x))$ is continuous at $x = x_0$.

> ***Proof:*** Let f and g be real-valued functions with domains $\mathcal{D}_f$ and $\mathcal{D}_g$, and suppose that f is continuous at $x = g(x_0)$ and g is continuous at $x = x_0$. Let $\epsilon > 0$ be ABF.
>
> Since f is continuous at $x = g(x_0)$ and g is continuous at $x = x_0$, $\exists\ \delta_1 > 0$ such that for $y \in \mathcal{D}_f$, $|f(y) - f\,((g(x_0))\,| < \epsilon$ whenever $|y - g(x_0)| < \delta_1$.
>
> Also, since g is a continuous function, $\exists\ \delta_2 > 0$ such that for $x \in \mathcal{D}_g$, $|g(x) - g(x_0)| < \delta_1$ whenever $|x - x_0| < \delta_2$.
>
> Thus, for $x_0 \in \mathcal{D}_{f\circ g}$ it follows that $|f(x) - f\,((g(x_0))\,| < \epsilon$ whenever $|g(x) - g(x_0)| < \delta_1$ which occurs whenever $|x - x_0| < \delta_2$. Hence, $|f(x) - f\,((g(x_0))\,| < \epsilon$ whenever $|x - x_0| < \delta_2$.
>
> Therefore, the function $f \circ g(x) = f\,(g(x))$ is continuous at $x = x_0$.
>
> ■

While the continuity is preserved under the ordinary arithmetic operations, the following example illustrates that the composition of two continuous functions may or may not preserve continuity.

Example 5.4.5: Let $f(x) = \sqrt{x}$ and $g(x) = \frac{x}{x+1}$.

a. Determine the domain of $f \circ g$.

b. Determine the domain of $g \circ f$.

c. Determine whether $f \circ g$ is continuous at $x = -2$.

d. Determine whether $g \circ f$ is continuous at $x = -2$.

Solutions: Let $f(x) = \sqrt{x}$ and $g(x) = \frac{x}{x+1}$. Then, the domains of f and g are $\mathcal{D}_f = [0, \infty)$ and $\mathcal{D}_g = (-\infty, -1) \cup (-1, \infty)$.

a. The domain of $f \circ g$ is $\mathcal{D}_{f \circ g} = (-\infty, -1) \cup [0, \infty)$

b. The domain of $g \circ f$ is $[0, \infty)$.

c. $f \circ g$ is continuous at $x = -2$ since (1) g is continuous at $x = -2$, (2) $g(-2) = 2 \in \mathcal{D}_f$, and (3) f is continuous at $x = 2$.

d. $g \circ f$ is not continuous at $x = -2$ since f is not continuous at $x = -2$ since $-2 \notin \mathcal{D}_f$.

Example 5.4.6: Let $f(x) = \sqrt{x-2}$ and $g(x) = \sqrt{x+1}$.

a. Determine the domain of $f \circ g$.

b. Determine the domain of $g \circ f$.

c. Determine whether $f \circ g$ is continuous at $x = -1$.

d. Determine whether $g \circ f$ is continuous at $x = -1$.

Solutions: The solutions to Example 5.4.6 are left as exercises.

The next theorem, Theorem 5.4.7, will show that the maximum and the minimum of two continuous functions are also both continuous functions. For example, if $f(x)$ and $g(x)$ are continuous at $x = x_0$, then $\max(f(x), g(x))$ and $\min(f(x), g(x))$ are also continuous at $x = x_0$. Before stating and proving Theorem 5.4.7, the following lemma, which will be used in the proof of the Theorem 5.4.7, will be proved.

Lemma 5.4.1: Let f and g be real-valued functions with domains $\mathcal{D}_f$ and $\mathcal{D}_g$. If x is in both $\mathcal{D}_f$ and $\mathcal{D}_g$, then

(i) $\max\Big(f(x),\, g(x)\Big) = \frac{1}{2}\Big[f(x) + g(x)\Big] + \frac{1}{2}|f(x) - g(x)|$

(ii) $\min\Big(f(x),\, g(x)\Big) = -\max\Big(-f(x),\, -g(x)\Big)$

Proof: Let f and g be real-valued functions with domains $\mathcal{D}_f$ and $\mathcal{D}_g$. Let x be an arbitrary point in both $\mathcal{D}_f$ and $\mathcal{D}_g$.

Proof of part (i): Let $M(x) = \max\Big(f(x),\, g(x)\Big)$. Now, either $f(x) \geq g(x)$ or $g(x) < f(x)$.

Case 1: Suppose that $f(x) \geq g(x)$, and consider $M(x)$. Since $f(x) \geq g(x)$ it follows that $M(x) = \max\Big(f(x), g(x)\Big) = f(x)$ and $|f(x) - g(x)| = f(x) - g(x)$. Thus

$$\begin{aligned} &\frac{1}{2}\Big(f(x) + g(x)\Big) + \frac{1}{2}|f(x) - g(x)| \\ &= \frac{1}{2}\Big(f(x) + g(x)\Big) + \frac{1}{2}\Big[f(x) - g(x)\Big] \\ &= f(x) \end{aligned}$$

Therefore, when $f(x) \geq g(x)$, it follows that

$$M(x) = \max\Big(f(x), g(x)\Big) = \frac{1}{2}\Big(f(x) + g(x)\Big) + \frac{1}{2}|f(x) - g(x)|$$

Case 2: Suppose that $f(x) < g(x)$, and consider $M(x)$. Since $f(x) < g(x)$ it follows that, $M(x) = \max\Big(f(x), g(x)\Big) = g(x)$ and $|f(x) - g(x)| = g(x) - f(x)$. Thus

$$\begin{aligned} &\frac{1}{2}\Big(f(x) + g(x)\Big) + \frac{1}{2}|f(x) - g(x)| \\ &= \frac{1}{2}\Big(f(x) + g(x)\Big) + \frac{1}{2}\Big(g(x) - f(x)\Big) \\ &= g(x) \end{aligned}$$

Therefore, when $f(x) < g(x)$, it follows that

$$M(x) = \max\Big(f(x), g(x)\Big) = \frac{1}{2}\Big(f(x) + g(x)\Big) + \frac{1}{2}|f(x) - g(x)|$$

Hence, for all x in $\mathcal{D}_f$ and $\mathcal{D}_g$, it follows that

$$M(x) = \max\Big(f(x), g(x)\Big) = \frac{1}{2}\Big(f(x) + g(x)\Big) + \frac{1}{2}|f(x) - g(x)|$$

Proof of part (ii): The proof of part (ii) is left as an exercise.

■

Theorem 5.4.7 shows that $\max(f(x), g(x))$ and $\min(f(x), g(x))$ are continuous functions whenever f and g are continuous functions.

Theorem 5.4.7: Let f and g be functions that are continuous at the point $x = x_0$. Then

(i) $\max\left[f(x),\, g(x)\right]$ is continuous at $x = x_0$.

(ii) $\min\left[f(x),\, g(x)\right]$ is continuous at $x = x_0$.

Proof: Let f and g be functions that are continuous at the point $x = x_0$.

Proof of part (i): First, note that by Lemma 5.4.7

$$\max\left[f(x), g(x)\right] = \frac{1}{2}\left[f(x) + g(x)\right] + \frac{1}{2}|f(x) - g(x)|$$

Since f and g are continuous at $x = x_0$, by Theorem 5.4.4 parts (i) and (ii), it follows that $f + g$ and $f - g$ are continuous at $x = x_0$.

Also, since $f + g$ and $f - g$ are continuous at $x = x_0$, by Theorem 5.4.4 part (ii), it follows that $\frac{1}{2}(f + g)$ is continuous at $x = x_0$, and by Theorem 5.4.3 part (ii) and Theorem 5.4.4 part (ii), it follows that $\frac{1}{2}|f - g|$ is continuous at $x = x_0$.

Hence, by Theorem 5.4.4(i), it follows that

$$\max\left[f(x), g(x)\right] = \frac{1}{2}\left[f(x) + g(x)\right] + \frac{1}{2}|f(x) - g(x)|$$

is continuous at $x = x_0$.

Therefore, $\max\Big(f(x),\, g(x)\Big) = \frac{1}{2}\Big(f(x) + g(x)\Big) + \frac{1}{2}|f(x) - g(x)|$ is continuous at $x = x_0$ whenever f and g are continuous at $x = x_0$.

Proof of part (ii): The proof of part (ii) is left as an exercise.

■

Examples of the maximum and minimum of two continuous functions are given in Example 5.4.7. Note that the continuity of the maximum and minimum functions is clearly illustrated in Figures 5.4.2 and 5.4.3.

Example 5.4.7: Let $f(x) = \sin(x)$ and $g(x) = \cos(x)$. Since $\cos(x)$ and $\sin(x)$ are continuous on $\mathbb{R}$, it follows that $\max(\cos(x), \sin(x))$ is continuous on $\mathbb{R}$ as is $\min(\cos(x), \sin(x))$. The plots of $\max\big(\cos(x), \sin(x)\big)$ and $\min\big(\cos(x), \sin(x)\big)$ are given in Figures 5.4.2 and 5.4.3; Figure 5.4.2 displays $\max(\cos(x), \sin(x))$, and Figure 5.4.3 displays $\min(\cos(x), \sin(x))$.

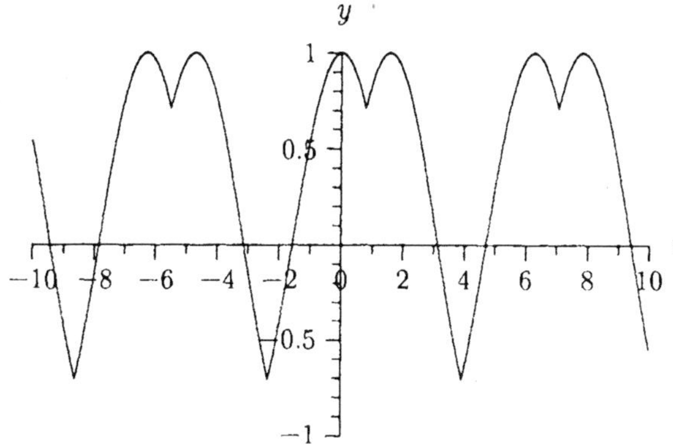

Figure 5.4.2 Plot of max(sin(x),cos(x)).

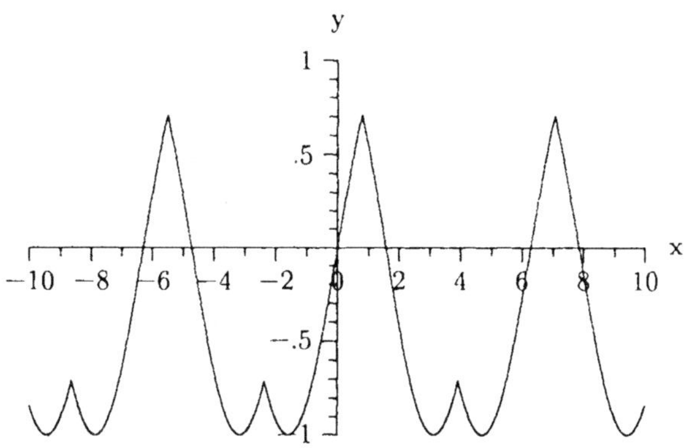

Figure 5.4.3 Plot of min(sin(x), cos(x)).

When a function f is continuous on a closed interval $[a, b]$, it can be shown that for any value y lying between $f(a)$ and $f(b)$, there is a point $c \in [a, b]$ such that $f(c) = y$. This result is known as the the *Intermediate Value Theorem* and is one of the most important results concerning continuity. This theorem is often used to show that there exists, or there does not exist, a solution to an equation of the form $f(x) = 0$ on a closed interval $[a, b]$. The Intermediate

Value Theorem is presented below without proof; for a proof of this theorem, see *Elementary Analysis: The Theory of Calculus* by K. A. Ross (2003).

Theorem 5.4.8 (The Intermediate Value Theorem): Let $f(x)$ be a real-valued function which is continuous on the closed interval $[a, b]$. If y is any number between $f(a)$ and $f(b)$, then there exists at least one number $c \in [a, b]$ such that $f(c) = y$.

The Intermediate Value Theorem can be used to determine whether there exists a solution to the equation $f(x) = y$ when $f(x)$ is a continuous function on a closed interval $[a, b]$. Specifically, given a function f that is continuous on $[a, b]$ and an equation $f(x) = y$, there will always exist a solution to $f(x) = y$ when y lies between $f(a)$ and $f(b)$. Furthermore, if there exist points x_1 and x_2 in $[a, b]$ such that y lies between $f(x_1)$ and $f(x_2)$, then there is also a solution to the equation $f(x) = y$ that is in $[a, b]$. However, when y does not lie between $f(a)$ and $f(b)$, the Intermediate Value Theorem does not imply that there is no solution to $f(x) = y$ in the interval $[a, b]$. For example, if $f(x) = \sin(x)$ and the equation being solved is $f(x) = 0.5$ on $[0, \pi]$, then since $\sin(0) = \sin(\pi) = 0$, checking the endpoints does not reveal that there is actually a solution to this equation, namely, $x = \dfrac{\pi}{6}$ in the interval $[0, \pi]$.

Example 5.4.8: Determine whether there is a solution to each of the following equations:

a. Let $f(x) = x^2 - 2x$ on the interval $[1, 3]$. Is it clear that there is a solution to the equation $f(x) = 2$ in $[1, 3]$?

b. Let $g(x) = e^{-x}$ on the interval $[0, 3]$. Is it clear that there is a solution to the equation $g(x) = 0.5$ in $[0, 3]$?

c. Let $c(x) = \cos(x)$ on the interval $\left[\dfrac{\pi}{4}, \dfrac{\pi}{2}\right]$. Is it clear that there is a solution to the equation $c(x) = 0.85$ in $[0, \pi]$?.

Solutions:

a. Since $f(1) = -1$ and $f(3) = 3$, it follows from the Intermediate Value Theorem that there does exist $c \in [1, 3]$ such that $f(c) = 2$.

b. Since $g(0) = 1$ and $g(3) = e^{-3} = 0.0498$, it follows from the Intermediate Value Theorem that there does exist $c \in [0, 3]$ such that $g(c) = 0.5$.

c. Since $c\left(\dfrac{\pi}{4}\right) = \dfrac{\sqrt{2}}{2} < 0.71$ and $c\left(\dfrac{\pi}{2}\right) = 0$, it is not clear from the Intermediate Value Theorem whether there exists a value in the interval $\left[\dfrac{\pi}{4}, \dfrac{\pi}{2}\right]$ such that $c(x) = 0.85$.

Corollary to Theorem 5.4.8: Let f be a real-valued function which is continuous on the closed interval $[a, b]$. If $f(a) \cdot f(b) < 0$, then there exists at least one number $c \in [a, b]$ such that $f(c) = 0$.

> ***Proof:*** This corollary follows directly from the Intermediate Value Theorem since $f(a) \cdot f(b) < 0$ only when $f(a) > 0$ and $f(b) < 0$ or vice versa.
>
> ■

5.5 Derivatives

Another important application of limits is found in the derivative of a function. Recall from calculus that the derivative of a function at a point $x = x_0$ is the slope of the line tangent to the graph of the function f at the point $(x_0, f(x_0))$. Furthermore, the derivative of a real-valued function contains information on where the function is increasing, decreasing, or constant, as well as many other important properties related to the behavior of the function. A definition of the derivative of a function at a point x_0 is given below.

Definition 5.5.1: Let f be a function with domain $\mathcal{D}$ containing an interval I. The *derivative* of a function f at the point $x_0 \in I$ is said to exist if and only if

$$\lim_{x \to x_0} \frac{f(x) - f(x_0)}{x - x_0}$$

exists, and in this case the derivative of f at the point $x = x_0$ will be denoted by $f'(x_0)$.

Definition 5.5.2: A function f is said to be *differentiable* at a point x_0 if and only if the derivative of f exists at the point x_0; f is said to be *differentiable* on an interval I if and only if the derivative of f exists at every point in I.

The derivative of a function f at a point x_0 is also sometimes denoted by $\frac{d}{dx}[f(x)]_{x=x_0}$, $D_x[f(x)]_{x=x_0}$, or $\dot{f}(x_0)$. The derivative of a real-valued function f, at the point x_0, measures the rate of change of the function f in a neighborhood of x_0. In fact, the derivative of f, at $x = x_0$, is often called the *instantaneous rate of change* of f at the point $(x_0, f(x_0))$. Two alternative, and equivalent, definitions of the derivative are given below.

Definition 5.5.3: Let f be a function with domain $\mathcal{D}$ containing a neighborhood of x_0. The *derivative* of a function f at the point $x_0 \in \mathcal{D}$ is said to exist if and only if

$$\lim_{h \to 0} \frac{f(x_0 + h) - f(x_0)}{h}$$

exists. The derivative of f at the point $x = x_0$ will be denoted by $f'(x_0)$.

Definition 5.5.4: Let f be a function with domain $\mathcal{D}$ containing a neighborhood of x_0. The *derivative* of a function f at the point $x_0 \in \mathcal{D}$ is said to exist if and only if for every $\epsilon > 0$, there exists $\delta > 0$ such that

$$\left|\frac{f(x) - f(x_0)}{x - x_0} - f'(x_0)\right| < \epsilon$$

whenever $|x - x_0| < \delta$. The derivative of f at the point $x = x_0$ will be denoted by $f'(x_0)$.

Note that among the three definitions of the derivative of f, only Definition 5.5.4 requires that $f'(x_0)$ be known. Thus, in order to prove that f has a derivative at $x = x_0$ using Definition 5.5.4, the actual value of the derivative must be known. On the other hand, Definitions 5.5.1 and 5.5.3 can be used to determine the derivative of a function f at a point x_0 without knowing the derivative, $f'(x_0)$. For this reason, Definitions 5.5.1 and 5.5.3 can be used to derive the general form of the derivative of a function f, while Definition 5.5.4 cannot. The following four examples illustrate how Definitions 5.5.1 and 5.5.3 can be used to determine the generic form of the derivative of a function f.

Example 5.5.1: Let $f(x) = x^2 + 2x$. Determine the derivative of $f(x)$ at the point $x = x_0$ using

a. Definition 5.5.1.

b. Definition 5.5.3.

Solutions:

a. Consider $\lim\limits_{x \to x_0} \dfrac{f(x) - f(x_0)}{x - x_0}$:

$$\lim_{x \to x_0} \frac{f(x) - f(x_0)}{x - x_0} = \lim_{x \to x_0} \frac{x^2 + 2x - x_0^2 - 2x_0}{x - x_0}$$

$$\lim_{x \to x_0} \frac{(x - x_0)(x + x_0) + 2(x - x_0)}{x - x_0} = \lim_{x \to x_0} x + x_0 + 2$$
$$= 2x_0 + 2$$

Therefore, $f'(x_0) = 2x_0 + 2$.

b. Consider $\lim_{h \to 0} \dfrac{f(x_0+h) - f(x_0)}{h}$:

$$\lim_{h \to 0} \frac{f(x_0+h) - f(x_0)}{h} = \lim_{h \to 0} \frac{(x_0+h)^2 + 2(x_0+h) - x_0^2 - 2x_0}{h}$$

$$= \lim_{h \to 0} \frac{x_0^2 + 2x_0 h + h^2 + 2x_0 + 2h - x_0^2 - 2x_0}{h}$$

$$cr = \lim_{h \to 0} \frac{2x_0 h + h^2 + 2h}{h}$$

$$= \lim_{h \to 0} \Big(2x_0 + h + 2\Big) = 2x_0 + 2$$

Therefore, $f'(x_0) = 2x_0 + 2$.

Example 5.5.2: Let $f(x) = \sqrt{x}$. Determine the derivative of $f(x)$ at the point $x = x_0$ using

a. Definition 5.5.1.

b. Definition 5.5.3.

Solutions:

a. Consider $\lim_{h \to 0} \dfrac{f(x_0+h) - f(x_0)}{h}$:

$$\lim_{h \to 0} \frac{f(x_0+h) - f(x_0)}{h} = \lim_{h \to 0} \frac{\sqrt{x_0+h} - \sqrt{x_0}}{h}$$

$$= \lim_{h \to 0} \left[\frac{\sqrt{x_0+h} - \sqrt{x_0}}{h} \cdot \frac{\sqrt{x_0+h} + \sqrt{x_0}}{\sqrt{x_0+h} + \sqrt{x_0}}\right]$$

$$= \lim_{h \to 0} \frac{x_0 + h - x_0}{h\left[\sqrt{x_0+h} + \sqrt{x_0}\right]}$$

$$= \lim_{h \to 0} \frac{1}{\sqrt{x_0+h} + \sqrt{x_0}} = \frac{1}{2\sqrt{x_0}}$$

Therefore, $f'(x_0) = \dfrac{1}{2\sqrt{x_0}}$.

b. The solution to part (b) is left as an exercise.

Example 5.5.3: Let $f(x) = \frac{1}{x^2}$. Determine the derivative of $f(x)$ at the point $x = x_0$.

Solution: Using Definition 5.5.3, note that

$$\lim_{h \to 0} \frac{f(x_0 + h) - f(x_0)}{h} = \lim_{h \to 0} \frac{\frac{1}{(x_0+h)^2} - \frac{1}{x_0^2}}{h}$$

$$= \lim_{h \to 0} \frac{x_0^2 - (x_0 + h)^2}{hx_0^2(x_0 + h)^2}$$

$$= \lim_{h \to 0} \frac{x_0^2 - x_0^2 - 2x_o h - h^2}{h\left[x_0^2(x_0 + h)^2\right]} = \lim_{h \to 0} \frac{-2x_0 - h}{\left[x_0^2(x_0 + h)^2\right]}$$

$$= \frac{-2x_0}{x_0^4} = -\frac{2}{x_0^3}$$

Therefore, $f'(x_0) = -\frac{2}{x_0^3}$.

Example 5.5.4: Let $f(x) = \sin(x)$. Determine the derivative of $f(x)$ at the point $x = x_0$.

Solution: Using Definition 5.5.3, note that

$$\lim_{h \to 0} \frac{f(x_0 + h) - f(x_0)}{h} = \lim_{h \to 0} \frac{\sin(x_0 + h) - \sin(x_0)}{h}$$

$$= \lim_{h \to 0} \frac{\sin(x_0)\cos(h) + \sin(h)\cos(x_0) - \sin(x_0)}{h}$$

$$= \lim_{h \to 0} \left[\sin(x_0) \cdot \frac{\cos(h) - 1}{h} + \cos(x_0) \cdot \frac{\sin(h)}{h}\right]$$

Now, in Example 5.3.4 it was shown that $\lim_{h \to 0} \frac{\sin(h)}{h} = 1$.

Consider $\lim\limits_{h \to 0} \dfrac{\cos(h) - 1}{h}$:

$$\lim_{h \to 0} \frac{\cos(h) - 1}{h} = \lim_{h \to 0} \left[\frac{\cos(h) - 1}{h} \cdot \frac{\cos(h) + 1}{\cos(h) + 1}\right]$$

$$= \lim_{h \to 0} \frac{\cos^2(h) - 1}{h(\cos(h) + 1)}$$

$$= \lim_{h \to 0} \frac{-\sin^2(h)}{h(\cos(h) + 1)}$$

$$= \lim_{h \to 0} \left[\frac{-\sin(h)}{h} \cdot \frac{\sin(h)}{\cos(h) + 1}\right]$$

Now, since $\lim\limits_{h \to 0} \dfrac{\sin(h)}{h} = 1$ and $\lim\limits_{h \to 0} \dfrac{\sin(h)}{\cos(h) + 1} = 0$, it follows that

$$\lim_{h \to 0} \frac{\cos(h) - 1}{h} = \lim_{h \to 0} \frac{-\sin(h)}{h} \cdot \lim_{h \to 0} \frac{\sin(h)}{\cos(h) + 1} = -1 \cdot 0 = 0$$

Thus

$$\lim_{h \to 0} \frac{\sin(x_0 + h) - \sin(x_0)}{h}$$

$$= \lim_{h \to 0} \left[\sin(x_0) \cdot \frac{\cos(h) - 1}{h} + \cos(x_0) \cdot \frac{\sin(h)}{h}\right]$$

$$= \sin(x_0) \lim_{h \to 0} \frac{\cos(h) - 1}{h} + \cos(x_0) \lim_{h \to 0} \frac{\sin(h)}{h}$$

$$= \sin(x_0) \cdot 0 + \cos(x_0) \cdot 1 = \cos(x_0)$$

Therefore, $\dfrac{d}{dx}\Big[\sin(x)\Big]_{x=x_0} = \cos(x_0)$.

Theorem 5.5.1 shows that when a function is differentiable at $x = x_0$, then that function is also continuous at $x = x_0$. Furthermore, the contrapositive of the theorem shows that when a function is not continuous at $x = x_0$, it is not differentiable at $x = x_0$, either.

Theorem 5.5.1: Let f be a real-valued function. If f is differentiable at $x = x_0$, then f is continuous at $x = x_0$.

Proof: Let f be a real-valued function that is differentiable at x_0. Since f is differentiable at x_0, it follows that

$$f'(x_0) = \lim_{x \to x_0} \frac{f(x) - f(x_0)}{x - x_0}$$

Consider $\lim_{x \to x_0} [f(x) - f(x_0)]$:

$$\lim_{x \to x_0} [f(x) - f(x_0)] = \lim_{x \to x_0} \left[\frac{x - x_0}{x - x_0} \cdot [f(x) - f(x_0)]\right]$$

$$= \underbrace{\lim_{x \to x_0} [(x - x_0)] \lim_{x \to x_0} \left[\frac{f(x) - f(x_0)}{x - x_0}\right]}_{\text{By Theorem 5.3.1(iii)}}$$

$$= 0 \cdot f'(x_0) = 0$$

Hence, $\lim_{x \to x_0} [f(x) - f(x_0)] = 0$. Thus, $\lim_{x \to x_0} f(x) = f(x_0)$. Therefore, f is continuous at $x = x_0$.

■

Note that the converse of Theorem 5.5.1 is not true. Specifically, there exist functions that are continuous at a point x_0, yet the function is not differentiable at x_0. An example of a function that is continuous on $\mathbb{R}$ that is not differentiable everywhere on $\mathbb{R}$ is given in the following example.

Example 5.5.5: Let $f(x) = |x|$. The plot of $f(x) = |x|$ is given in Figure 5.5.1.

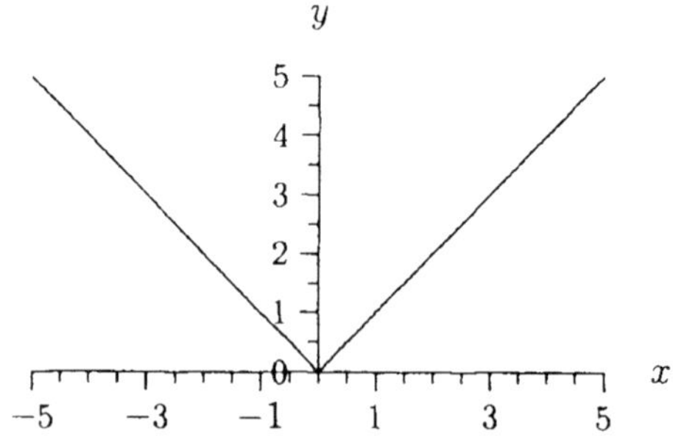

Figure 5.5.1 A plot of $f(x)=|x|$.

By Theorem 5.4.3(ii), $f(x) = |x|$ is continuous on $\mathbb{R}$, and hence f is continuous at $x = 0$. Consider $\lim\limits_{h \to 0} \dfrac{f(x_0 + h) - f(x_0)}{h}$ for $x_0 = 0$:

$$\lim_{h \to 0} \frac{f(0+h) - f(0)}{h} = \lim_{h \to 0} \frac{|0+h| - |0|}{h} = \lim_{h \to 0} \frac{|h|}{h}$$

This limit does not exist since $\lim\limits_{h \to 0^-} \dfrac{|h|}{h} = -1$ and $\lim\limits_{h \to 0^+} \dfrac{|h|}{h} = 1$.

Therefore, $f(x) = |x|$ is not differentiable at $x_0 = 0$.

The next theorem provides useful results for determining the derivatives of the following functions $mf(x) + b$, $f(x) + g(x)$, $f(x) \cdot g(x)$, and $\dfrac{f(x)}{g(x)}$ at a point $x = x_0$ whenever the functions f and g are differentiable at x_0.

Theorem 5.5.2: Let f and g be real-valued functions. If f and g are differentiable at the point $x = x_0$, then

(i) $\dfrac{d}{dx}[mf(x) + b]_{x=x_0} = mf'(x_0)$.

(ii) $\dfrac{d}{dx}[f(x) + g(x)]_{x=x_0} = f'(x_0) + g'(x_0)$.

(iii) $\dfrac{d}{dx}[f(x) - g(x)]_{x=x_0} = f'(x_0) - g'(x_0)$.

(iv) $\dfrac{d}{dx}[f(x)g(x)]_{x=x_0} = f'(x_0)g(x_0) + f(x_0)g'(x_0)$.

(v) $\dfrac{d}{dx}\left[\dfrac{f(x)}{g(x)}\right]_{x=x_0} = \dfrac{f'(x_0)g(x_0) - f(x_0)g'(x_0)}{g(x_0)^2}$, provided $g(x_0) \neq 0$.

Proof: Let f and g be real-valued functions, and suppose that f and g are differentiable at the point $x = x_0$.

Proof of part (i): From Definition 5.5.3, it follows that the derivative of $mf(x) + b$ at $x = x_0$ is

$$= \lim_{h \to 0} \frac{mf(x_0 + h) + b - mf(x_0) - b}{h}$$

$$= m \cdot \lim_{h \to 0} \frac{f(x_0 + h) - f(x_0)}{h} = mf'(x_0)$$

Therefore, $\dfrac{d}{dx}[mf(x) + b]_{x=x_0} = mf'(x_0)$.

Proof of part (ii): From Definition 5.5.3, it follows that the derivative of $f(x) + g(x)$ at $x = x_0$ is

$$= \lim_{h \to 0} \frac{f(x_0 + h) + g(x_0 + h) - f(x_0) - g(x_0)}{h}$$

$$= \underbrace{\lim_{h \to 0} \frac{f(x_0 + h) - f(x_0)}{h} + \lim_{h \to 0} \frac{g(x_0 + h) - g(x_0)}{h}}_{\text{By Theorem 5.3.1(i)}}$$

$$= f'(x_0) + g'(x_0)$$

Therefore

$$\frac{d}{dx} [f(x) + g(x)]_{x=x_0} = f'(x_0) + g'(x_0)$$

whenever f and g are differentiable at $x = x_0$.

Proof of part (iii): The proof of part (iii) is left as an exercise.

Proof of part (iv): From Definition 5.5.3, it follows that the derivative of $f(x)g(x)$ at $x = x_0$ is $\lim_{h \to 0} \dfrac{f(x_0 + h)g(x_0 + h) - f(x_0)g(x_0)}{h}$.

Now

$$f(x_0 + h)g(x_0 + h) - f(x_0)g(x_0)$$

$$= f(x_0 + h)g(x_0) + f(x_0 + h)g(x_0) - f(x_0 + h)g(x_0) - f(x_0)g(x_0)$$

Thus $\dfrac{d}{dx} [f(x)g(x)]_{x=x_0}$

$$= \lim_{h \to 0} \left[\frac{f(x_0 + h)g(x_0 + h) - f(x_0 + h)g(x_0)}{h} \right.$$

$$\left. + \frac{f(x_0 + h)g(x_0) - f(x_0)g(x_0)}{h} \right]$$

$$= \lim_{h \to 0} \frac{f(x_0 + h)\left[g(x_0 + h) - g(x_0)\right] + g(x_0)\left[f(x_0 + h) - f(x_0)\right]}{h}$$

Now, since f and g are differentiable at $x = x_0$, from Theorem 5.5.2 it follows that f and g are continuous at $x = x_0$, and thus $\lim_{h \to 0} f(x_0 + h) = f(x_0)$ and $\lim_{h \to 0} g(x_0 + h) = g(x_0)$. Hence,

$$\lim_{h \to 0} \frac{f(x_0 + h)g(x_0 + h) - f(x_0)g(x_0)}{h}$$

$$= \lim_{h \to 0} f(x_0 + h) \lim_{h \to 0} \frac{g(x_0 + h) - g(x_0)}{h} + g(x_0) \lim_{h \to 0} \frac{f(x_0 + h) - f(x_0)}{h}$$

by Theorem 5.3.1 parts (i) and (iii). Thus

$$\lim_{h \to 0} \frac{f(x_0 + h)g(x_0 + h) - f(x_0)g(x_0)}{h} = f(x_0)g'(x_0) + g(x_0)f'(x_0)$$

Therefore

$$\frac{d}{dx} [f(x)g(x)]_{x=x_0} = f'(x_0)g(x_0) + f(x_0)g'(x_0)$$

whenever f and g are differentiable at $x = x_0$.

Proof of part (v): Suppose that $g(x_0) \neq 0$. From Definition 5.5.3, it follows that the derivative of $\dfrac{f(x)}{g(x)}$ at $x = x_0$ is

$$= \lim_{h \to 0} \frac{\frac{f(x_0+h)}{g(x_0+h)} - \frac{f(x_0)}{g(x_0)}}{h}$$

$$= \lim_{h \to 0} \frac{f(x_0 + h)g(x_0) - f(x_0)g(x_0 + h)}{g(x_0 + h)g(x_0)h}$$

$$= \lim_{h \to 0} \frac{f(x_0 + h)g(x_0) \overbrace{-f(x_0)g(x_0) + f(x_0)g(x_0)}^{0} -f(x_0)g(x_0 + h)}{g(x_0 + h)g(x_0)h}$$

$$= \lim_{h \to 0} \left[\frac{f(x_0 + h) - f(x_0)}{g(x_0 + h)h} - f(x_0)\frac{g(x_0 + h) - g(x_0)}{g(x_0 + h)g(x_0)h}\right]$$

Now, since f and g are differentiable at $x = x_0$, it follows that f and g are also continuous at $x = x_0$. Thus, $\lim_{h \to 0} g(x_0 + h) = g(x_0)$, and since $g(x_0) \neq 0$, it follows that $\lim\limits_{h \to 0} \dfrac{1}{g(x_0+h)} = \dfrac{1}{g(x_0)}$. Hence, by Theorem 5.3.1 parts (i) and (iii) it follows that

$$\lim_{h \to 0} \frac{\frac{f(x_0+h)}{g(x_0+h)} - \frac{f(x_0)}{g(x_0)}}{h} = \lim_{h \to 0} \frac{1}{g(x_0+h)} \cdot \lim_{h \to 0} \frac{f(x_0+h) - f(x_0)}{h}$$

$$- \frac{f(x_0)}{g(x_0)} \lim_{h \to 0} \frac{1}{g(x_0+h)} \cdot \lim_{h \to 0} \frac{g(x_0+h) - g(x_0)}{h}$$

Thus, $\dfrac{d}{dx}\left[\dfrac{f(x)}{g(x)}\right]_{x=x_0}$

$$= \lim_{h \to 0} \frac{\frac{f(x_0+h)}{g(x_0+h)} - \frac{f(x_0)}{g(x_0)}}{h}$$

$$= \frac{1}{g(x_0)} \cdot f'(x_0) + \frac{f(x_0)}{g(x_0)} \cdot \frac{1}{g(x_0)} \cdot g'(x_0)$$

$$= \frac{g(x_0)f'(x_0) - f(x_0)g'(x_0)}{g(x_0)^2}$$

Therefore

$$\frac{d}{dx}\left[\frac{f(x)}{g(x)}\right]_{x=x_0} = \frac{f'(x_0)g(x_0) - f(x_0)g'(x_0)}{g(x_0)^2}$$

whenever f and g are differentiable at $x = x_0$ and $g(x_0) \neq 0$.

■

Example 5.5.6: Let $f(x) = x^2$ and $g(x) = \sin x$. Determine

a. $\dfrac{d}{dx}\Big[f(x) + g(x)\Big]_{x=x_0}$

b. $\dfrac{d}{dx}\Big[f(x)g(x)\Big]_{x=x_0}$

c. $\dfrac{d}{dx}\left[\dfrac{f(x)}{g(x)}\right]_{x=x_0}$

Solutions:

a. Using Theorem 5.5.3(ii),

$$\frac{d}{dx}\Big[f(x) + g(x)\Big]_{x=x_0} = 2x_0 + \cos x_0$$

b. Using Theorem 5.5.3(iii),

$$\frac{d}{dx}\Big[f(x)g(x)\Big]_{x=x_0} = 2x_0 \sin x_0 + x_0^2 \cos x_0$$

c. Using Theorem 5.5.3(iv),

$$\frac{d}{dx}\left[\frac{f(x)}{g(x)}\right]_{x=x_0} = \frac{2x_0 \sin x_0 - x_0^2 \cos x_0}{\sin x_0{}^2}$$

EXERCISES

5.1 For each of the following sequences, find the smallest value of N such that $|a_n - a| < 0.1,\ \forall\ n \geq N$:

a. $a_n = \dfrac{n}{n+5}$ and $a = 1$

b. $a_n = \dfrac{3n+11}{2n+6}$ and $a = \dfrac{3}{2}$

c. $a_n = \dfrac{1}{n^2}$ and $a = 0$.

d. $a_n = 1 - \dfrac{(-1)^n}{n^2}$ and $a = 1$.

5.2 Repeat Exercise 5.1 for $|a_n - a| < 0.01$.

5.3 Find a value of M that bounds each of the sequences in Exercise 5.1.

5.4 Suppose that $\lim_{n\to\infty} x_n = 2$, $\lim_{n\to\infty} y_n = 3$, and $y_n > 0,\ \forall\ n \in \mathbb{N}$. Determine

a. $\lim_{n\to\infty} (x_n + y_n)$.

b. $\lim_{n\to\infty} \dfrac{x_n + 2y_n}{y_n - 1}$.

c. $\lim_{n\to\infty} x_n^2 + 5y_n^2$.

d. $\lim_{n\to\infty} \sqrt{6y_n - 3x_n}$.

5.5 Prove each of the following theorems:

a. **Theorem:** If a_n and b_n are sequences of real numbers with $\lim_{n\to\infty} a_n = a$ and $\lim_{n\to\infty} b_n = b$, then

$$\lim_{n\to\infty} (r \cdot a_n + s \cdot b_n) = r \cdot a + s \cdot b,\ \forall\ r, s \in \mathbb{R}$$

b. **Theorem:** If a_n and b_n are sequences of real numbers with $\lim_{n\to\infty} a_n = a$ and $\lim_{n\to\infty} b_n = b$, then $\lim_{n\to\infty} (a_n + b_n)^2 = (a+b)^2$.

c. **Theorem:** If a_n and b_n are sequences of real numbers with $\lim_{n\to\infty} a_n = a$ and $\lim_{n\to\infty} b_n = b$, then $\lim_{n\to\infty} \left(a_n^2 + b_n^2\right) = a^2 + b^2$.

d. **Theorem:** If a_n and b_n are sequences of real numbers with $\lim_{n\to\infty} a_n = a$ and $\lim_{n\to\infty} a_n + b_n = c$, then $\lim_{n\to\infty} b_n = c - a$.

e. **Theorem:** If a_n and b_n are sequences of real numbers with $\lim_{n\to\infty} a_n = a$, $a \neq 0$, and $\lim_{n\to\infty} \frac{b_n}{a_n} = c$, then $\lim_{n\to\infty} b_n = ca$.

5.6 Let a_n be a bounded sequence and suppose that b_n is a sequence of real numbers such that $\lim_{n\to\infty} b_n = 0$. Prove that $\lim_{n\to\infty} (a_n b_n) = 0$.

5.7 Determine the following limits:

a. $\lim_{n\to\infty} \left(1 + \frac{2}{n}\right)^{3n}$

b. $\lim_{n\to\infty} \frac{\sqrt{n^2+1} + \sqrt{n} - \sqrt{n^2-1}}{n+1}$

c. $\lim_{n\to\infty} \frac{17n^4 + 12n^3 - 3n + 1000}{12n^4 - 11}$

5.8 Prove each of the following theorems using an ϵ-N proof:

a. **Theorem:** If a_n is a sequence with $\lim_{n\to\infty} a_n = a$, then $\lim_{n\to\infty} |a_n| = |a|$.

b. **Theorem:** If a_n is a sequence with $\lim_{n\to\infty} a_n = a$, then $\lim_{n\to\infty} a_n^2 = a^2$.

c. **Theorem:** If a_n is a sequence of nonnegative real numbers with $\lim_{n\to\infty} a_n = a$, then $\lim_{n\to\infty} \sqrt{a_n} = \sqrt{a}$.

d. **Theorem:** If a_n is a convergent sequence with $\lim_{n\to\infty} a_n = a$, then $\frac{1}{n}\sum_{i=1}^{n} a_n \to a$.

e. **Theorem:** If a_n is a convergent sequence with $\lim_{n\to\infty} a_n = a$, then $\exists\, N_k \in \mathbb{N}$ such that $|a_n - a| < \frac{1}{k}$ whenever $n \geq N_k$, $\forall\, k \in \mathbb{N}$.

f. **Theorem:** If a_n is a convergent sequence with $\lim_{n\to\infty} a_n = a$, then $\forall\, \epsilon > 0$, $\exists\, n \in \mathbb{N}$ such that $\left|\frac{a_n + a_m}{2} - a\right| < \epsilon$ whenever $n, m \geq N$.

5.9 Prove that if a_n is a sequence of nonnegative real numbers with $a_n \to a$, then $a_n^k \to a^k$, $\forall\, k \in \mathbb{N}$.

5.10 Use the squeeze theorem to determine the limit for each of the following sequences:

a. $a_n = \dfrac{\sin(n^2)}{n}$

b. $b_n = \dfrac{1 - \cos(n)}{n+1}$

c. $c_n = \left(1 + \dfrac{1}{n+1}\right)^n$

d. $c_n = \left(\dfrac{n-1}{n+1}\right)^{\sin(n)}$

5.11 Prove that each of the following sequences is a monotone sequence:

a. $a_n = \sqrt{n^2 + n} - n$

b. $d_n = \left(1 + \dfrac{1}{n+2}\right)^2$

c. $c_n = \dfrac{n}{\sqrt{n^2+1}}$

d. $b_1 = \sqrt{2}, \quad b_2 = \sqrt{2 + \sqrt{2}}, \quad b_{n+1} = \sqrt{2 + b_n}$

e. $e_n = \dfrac{1}{e^n + 1}$

5.12 Prove that each of the sequences in Exercise 5.11 converges to a limit.

5.13 Let a_n be a sequence of real numbers. Assuming that a_n is differentiable with respect to n, prove that

a. a_n is a nondecreasing sequence if and only if $\dfrac{d}{dn}[a_n] \geq 0, \ \forall\, n \in \mathbb{N}$.

b. a_n is a nonincreasing sequence if and only if $\dfrac{d}{dn}[a_n] \leq 0, \ \forall\, n \in \mathbb{N}$.

5.14 Let $a_n^{(i)} \to a^{(i)}, \ \forall\, i \in \mathbb{N}$. Use mathematical induction to prove each of the following theorems:

a. **Theorem:** $\displaystyle\lim_{n\to\infty} \sum_{i=1}^{k+1} a_n^{(i)} = \sum_{i=1}^{k+1} a^{(i)}, \ \forall\, k \in \mathbb{N}.$

b. **Theorem:** $\displaystyle\lim_{n\to\infty} \prod_{i=1}^{k+1} a_n^{(i)} = \prod_{i=1}^{k+1} a^{(i)}, \ \forall\, k \in \mathbb{N}.$

5.15 Let a_n be a Cauchy sequence of real numbers. Prove that

a. a_n is bounded.

b. There exists a real number a such that $\lim_{n\to\infty} a_n = a$.

c. If a_n and b_n are Cauchy sequences, then $a_n + b_n$ is a Cauchy sequence.

d. If a_n and b_n are Cauchy sequences, then $a_n b_n$ is a Cauchy sequence.

5.16 Let a_n and b_n be monotonic sequences of real numbers. Prove that

a. ca_n is a monotone sequence $\forall\ c \in \mathbb{R}$.

b. $a_n \downarrow$ and $b_n \downarrow$, then $(a_n + b_n) \downarrow$.

c. If a_n and b_n are nonnegative nonincreasing sequences, then $a_n b_n \downarrow$.

5.17 Let $a_n = \sqrt{4n^2 + 8n} - 2n$. Then

a. Show that $a_n \uparrow$.

b. Show that a_n is bounded by 4.

c. Determine $\lim_{n\to\infty} \sqrt{4n^2 + 8n} - 2n$.

5.18 Let $a_n = \left(1 + \frac{1}{n}\right)^n$. Prove that a_n is a nondecreasing sequence that is bounded above by 3.

5.19 Let $b_n = \left(1 + \frac{1}{n}\right)^{n+1}$. Prove that b_n is a nonincreasing sequence that is bounded below by 2.

5.20 Let a_n be a nonincreasing sequence of positive real numbers and define $\gamma_n = \dfrac{a_1 + a_2 + a_3 + \cdots + a_n}{n}$:

a. Show that $a_1 \geq \gamma_n \geq a_n$, $\forall\ n \in \mathbb{N}$.

b. Show that γ_{n+1} can be written as $\gamma_{n+1} = \dfrac{n\gamma_n + a_{n+1}}{n+1}$.

c. Show that γ_n is a nonincreasing sequence.

d. Let $a_1 = 4$ and show that $\gamma_n \to \gamma$ for some value of $0 \leq \gamma \leq 4$.

5.21 Prove that if X is a subset of $\mathbb{R}$ and $i = \inf X$, then i is unique.

5.22 Prove that if a_n is a nonincreasing sequence of real numbers, then a_n converges if and only if it is bounded from below.

5.23 Determine the following limits:

a. $\lim\limits_{x \to 2} \dfrac{x^2 + x}{x^2 - 1}$

b. $\lim\limits_{x \to -1} \dfrac{x^2 + x}{x^2 - 1}$

c. $\lim\limits_{x \to -3} \dfrac{x^2 + 4x + 3}{x + 3}$

5.24 Prove each of the following theorems using an ϵ-δ proof:

a. **Theorem:** If $f(x)$ is a real-valued function with $\lim\limits_{x \to a} f(x) = L$, then $\lim\limits_{x \to a} kf(x) + l = kL + l$, $\forall\, k, l \in \mathbb{R}$.

b. **Theorem:** If $f(x)$ is a real-valued function with $\lim\limits_{x \to a} f(x) = L$, then $\lim\limits_{x \to a} |f(x)| = |L|$.

c. **Theorem:** If $f(x)$ is a real-valued function with $\lim\limits_{x \to a} f(x) = L$, then $\lim\limits_{x \to a} f(x)^2 = L^2$.

d. **Theorem:** If $f(x)$ is a nonnegative real-valued function with $\lim\limits_{x \to a} f(x) = L$, then $\lim\limits_{x \to a} \sqrt{f(x)} = \sqrt{L}$.

5.25 Prove each of the following theorems:

a. **Theorem:** If $a \in \mathbb{R}$, then $\lim\limits_{x \to a} x^n = a^n$, $\forall\, n \in \mathbb{N}$.

b. **Theorem:** If $\lim\limits_{x \to a} f(x) = L$, then $\lim\limits_{x \to a} f(x)^n = L^n$, $\forall\, n \in \mathbb{N}$.

c. **Theorem:** If $p(x)$ is a polynomial, then $\lim\limits_{x \to a} p(x) = p(a)$, $\forall\, a \in \mathbb{N}$.

5.26 For $i \in \mathbb{N}$, let $f_i(x)$ be functions defined on a common domain $\mathcal{D}$. Given that $\lim\limits_{x \to c} f_i(x) = L_i$ for all $i \in \mathbb{N}$, prove that

a. $\lim\limits_{x \to c} \sum\limits_{i=1}^{n+1} f_i(x) = \sum\limits_{i=1}^{n+1} L_i$, $\forall\, n \in \mathbb{N}$.

b. $\lim\limits_{x \to c} \prod\limits_{i=1}^{m+1} f_i(x) = \prod\limits_{i=1}^{n+1} L_i$, $\forall\, n \in \mathbb{N}$.

5.27 Let f and g be real-valued functions with common domain $\mathcal{D}$. Assuming f and g to be continuous at $x = x_0$, prove that

a. $af + bg$ is continuous at $x = x_0$, $\forall\, a, b \in \mathbb{R}$.

b. fg is continuous at $x = x_0$.

c. $\dfrac{f}{g}$ is continuous at $x = x_0$, provided that $g(x_0) \neq 0$.

5.28 Prove each of the continuity results in Exercise 5.21 with an ϵ-δ proof.

5.29 Let f and g be real-valued functions with domains $\mathcal{D}_f$ and $\mathcal{D}_G$, respectively. Prove that

a. If $x \in \mathcal{D}_f$ and $x \in \mathcal{D}_g$, then

$$\min\left(f(x), g(x)\right) = -\max\left(-f(x), -g(x)\right)$$

b. If f and g are continuous at $x = x_0$, then $\min\left(f(x), g(x)\right)$ is continuous at $x = x_0$.

5.30 Let f be a function with domain $(-\infty, \infty)$, having the property that $f(x+h) = f(x) \cdot f(h)$ for all $x, h \in \mathbb{R}$ and $f(0) \neq 0$. Then

a. Show that$f(0) = 1$.

b. Determine $f'(x) = \lim\limits_{h \to 0} \dfrac{f(x+h) - f(x)}{h}$.

c. Show that $f(x) = e^{\lambda x}$ where $\lambda = f'(0)$.

5.31 Prove that if $f(x)$ is bounded (i.e., $|f(x)| \leq M$ for some constant M) and $g(x)$ approaches 0 as x approaches c, then $\lim\limits_{x \to c} f(x) \cdot g(x) = 0$.

5.32 Prove that $\lim\limits_{x \to 0} \left(x \cdot \sin \dfrac{1}{x}\right) = 0$.

5.33 Let $f_i(x)$ be continuous function on a domain $\mathcal{D}$, $\forall\, i \in \mathbb{N}$. Prove that

a. $\sum\limits_{i=1}^{n+1} f_i(x)$ is a continuous function on $\mathcal{D}$, $\forall\, n \in \mathbb{N}$.

b. $\prod\limits_{i=1}^{n+1} f_i(x)$ is a continuous function on $\mathcal{D}$, $\forall\, n \in \mathbb{N}$.

c. $\max(f_i(x))$ is a continuous function on $\mathcal{D}$.

d. $\min(f_i(x))$ is a continuous function on $\mathcal{D}$.

5.34 Let $f(x) = mx + b$. Prove that for any $\epsilon > 0$, $\delta = \dfrac{\epsilon}{1 + |m|}$ is a positive number such that $|f(x) - f(c)| < \epsilon$ whenever $|x - c| < \delta$, proving that $f(x)$ is continuous at $x = c$.

5.35 Let f and g be continuous at $x = x_0$. Prove that

a. $\max\left(f(x), g(x)\right)^2$ is continuous at $x = x_0$.
b. $\max\left(|f(x)|, |g(x)|\right)$ is continuous at $x = x_0$.
c. $\max\left(\min\left(f(x)^2, g(x)\right), f(x) + g(x)\right)$ is continuous at $x = x_0$.

5.36 Determine where each of the following functions is continuous in $\mathbb{R}$—also, for each isolated point where a function is not defined, determine whether the function can be defined there so as to make it continuous:

a. $f(x) = \dfrac{x}{(x+3)^2}$
b. $g(x) = \dfrac{2x-4}{x^2-4}$
c. $h(x) = \dfrac{1}{1-|x|}$

5.37 Let $f(x)$ be continuous at $x = x_0$ and $g(x)$ continuous on an interval containing $f(x_0)$. Prove that $g \circ f$ is continuous at $x = x_0$.

5.38 Let f be a continuous function with domain $[0, 1]$. Prove that if for all $x \in [0, 1]$, $0 \leq f(x) \leq 1$, then $\exists\, \psi \in [0, 1]$ such that $f(\psi) = \psi$.

5.39 Use the Intermediate Value Theorem to show that the following equations have at least one real solution in the specified interval:

a. $x^2 - 5 = 0$ on $[2, 3]$
b. $x^3 + x + 1 = 0$ on $[-1, 0]$
c. $x^3 - 3x^2 + 1 = 0$ on $[0, 1]$
d. $\cos x = x$ on $\left[0, \frac{\pi}{2}\right]$
e. $x^3 \cos(x) + 1 = x^2$ on $[-\pi, \pi]$

5.40 Prove that if $f(x)$ is a continuous function on $\mathbb{R}$ with $f(c) > 0$, then there exists $\delta > 0$ such that $f(x) > 0$ whenever $|x - c| < \delta$.

5.41 Use an ϵ-δ proof to show that $k \cdot f(x) + lg(x)$ is continuous at $x = c$ whenever $f(x)$ and $g(x)$ are continuous at $x = c$.

5.42 Use Definition 5.5.3 to find the derivative of each of the following functions:

a. $f(x) = x^2 + 4x$
b. $g(x) = \dfrac{1}{x^2}$

c. $h(x) = \dfrac{3x+4}{2x-1}$ for $x \neq \dfrac{1}{2}$

d. $p(x) = x^n$ for $n \in \mathbb{N}$

e. $s(x) = \sin(x)$

5.43 Use Definition 5.5.1 to find the derivative of each of the functions given in Exercise 5.4.2.

5.44 Show that the function

$$f(x) = \begin{cases} x^2 + x + 1 & \text{if } x \geq 1 \\ 4x - 1 & \text{if } x < 1 \end{cases}$$

is continuous at $x = 1$ but not differentiable at $x = 1$.

5.45 Let f be a real-valued function defined on $\mathbb{R}$, and suppose that for all $x, y \in \mathbb{R}$ with $x \neq y$ that $|f(x) - f(y)| < (x-y)^2$. Prove that there exists a real number c such that $f'(x) = 0$, $\forall x \in \mathbb{R}$.

5.46 Let $f(x)$ be differentiable on the interval $[a, b]$. Prove that

a. If $f(x)$ is a monotone nondecreasing, then $f'(x) \geq 0$, $\forall\, x \in [a, b]$.

b. If $f(x)$ is a monotone nonincreasing function, then $f'(x) \leq 0$ for all $x \in [a, b]$.

Chapter 6

The Foundations of Algebra

The logical foundations of most areas of modern mathematics, including mathematics encompassing algebra and calculus, are based on the field of mathematics known as *set theory.* Unlike the development of any other area of mathematics, the development of set theory was not based on the need to solve some physical or earthly problem. Moreover, set theory is a relatively new addition to modern mathematics and has changed the direction of mathematics. Georg Cantor (1845–1918) is credited with first introducing the ideas of sets and set theory in late nineteenth century, but not without some controversy. Cantor is also credited with recognizing that infinite sets can have different sizes. In particular, Cantor proved that the set of rational numbers contains fewer elements that the set of real numbers; for more information on Georg Cantor and the development of set theory, see *GEORG CANTOR: His Mathematics and Philosophy of the Infinite* by J. W. Dauben (1979). Other mathematicians with key contributions to the development of set theory include Bernhard Bolzano (1741–1848), Richard Dedekind (1831–1916), Ernst Zermelo (1871–1953), Bertrand Russell, George Boole, and Kurt Gödel.

6.1 Introduction to Sets

The study of modern mathematics requires that a student be well prepared in the area of set theory. For example, the theory of calculus as well as the theory of algebra have modern foundations built on set theory, and even the field of statistics is built on a foundation starting with set theory. The first use of the term *set* was due to Bolzano, and the definition of a set and *element* of a set are given below.

Definition 6.1.1: A well-defined collection of objects is called a *set.* The collection of all objects of interest is called the *universal set* or the *universe* and is denoted by Ω. An object in a set is called an *element.*

Definition 6.1.2: A set is said to be a *well-defined set* if and only if there is a method of determining whether a particular element is in the set.

The importance of using only well-defined sets was first illustrated by Bertrand Russell in 1901 with the following example. Let R be the set of all sets that are not members of themselves (i.e., $R = \{A : A \notin A\}$). Russell's set R leads to the following paradox, known as "Russell's paradox":

Is $R \in R$? If $R \in R$, then $R \notin R$. However, if $R \notin R$, then $R \in R$.

Clearly, a paradox of this nature is undesirable and hence only well-defined sets will be considered in this section. An example of a well-defined set is

$$A = \{\text{odd natural numbers less than } 20\}$$

and an example of a set that is not well defined is

$$B = \{\text{some odd natural numbers}\}$$

Clearly an element is in A if and only if it is one of the numbers 1, 3, 5, 7, 9, 11, 13, 15, 17, or 19, and hence A is a well-defined set. On the other hand, there is no way of knowing from the definition of the set B whether a particular odd natural number is in B. Hence B is not a well-defined set.

Example 6.1.1: The set of prime numbers less than 80 is a well-defined set with

$$\mathcal{P}_{80} = \{2, 3, 5, 7, 11, 13, 17, 19, 23, 29, 31, 33, 41, 43, 47, 53, 59, 61, 67, 71, 73, 79\}$$

The elements of $\mathcal{P}_{80}$ are 2, 3, 5, 7, 11, 13, 17, 19, 23, 29, 31, 33, 41, 43, 47, 53, 59, 61, 67, 71, 73, and 79.

Note that there is nothing in the definition of a set that requires that the elements of a set be numbers. Now, when x is an element of a set A, this will denoted by $x \in A$, and when x is not an element of the set A this will be denoted by $x \notin A$. Also, there are several different ways to represent a set, including roster notation, set-builder notation, and interval notation. A set is listed using roster notation by listing the elements of the set, separated by commas, and enclosed in curly set braces; the use of the set bracket notation $\{\ \}$ is due to Cantor in an article published in *Mathematische Annalen* (Cantor 1895). Roster notation is frequently used with small sets; for example, the set consisting of the prime numbers less than 80 was listed in roster notation in Example 6.1.1.

Definition 6.1.3: A set A is said to be a *finite set* if and only if the number of elements in A is a natural number. A set that is not finite is called an *infinite set.*

For example, $A = \{-3, -1, 2, 4, 11, 112\}$ is a finite set and $\mathbb{Z}$, the set of integers, is an infinite set. Finite sets can often be listed using roster notation, while infinite sets can sometimes be listed using roster notation by listing enough elements to show the pattern of elements in the set and then indicating that same pattern continues on indefinitely. For example, $\mathbb{N}= \{1, 2, 3, \ldots\}$. The first few elements illustrate the pattern, and the ellipsis (i.e., the three dots $\ldots$) indicate that the same pattern continues on indefinitely.

Example 6.1.2: Determine whether each of the following sets is finite or infinite:

a. $\mathcal{S} :=$ the set of real-valued solutions to the equation $x^4 - 5x^2 + 4 = 0$.

b. $\mathbb{Z}_O :=$ is the set of odd integers.

c. $\mathcal{P} :=$ is the set of prime numbers.

Solutions:

a. $\mathcal{S}$ is finite since $\mathcal{S} = \{-2, -1, 1, 2\}$.

b. $\mathbb{Z}_O$ is infinite since $\mathbb{Z}_O = \{0, \pm 1, \pm 2, \pm 3, \ldots\}$

c. $\mathcal{P}$ is infinite by Theorem 4.3.16.

With infinite sets, and in some cases finite sets, it is often impossible to list the set using roster notation. When this is the case, a set is usually written using mathematical notation for describing a set and its elements known as set-builder notation. A set is represented using set-builder notation by using mathematical notation to describe the set by listing the properties that its elements must satisfy. Thus, set-builder notation is a simply way of using mathematical notation to represent the set of elements having a particular property called the *defining property*. For example, the set of the prime numbers less than 100 can be expressed as using set-builder notation as $\mathcal{P}_{100} = \{p : p \text{ is a prime number and } p < 100\}$, and the set of rational numbers can be written using set-builder notation as

$$\mathcal{Q} = \left\{r : r = \frac{a}{b} \text{ where } a, b \in \mathbb{Z} \text{ and } b \neq 0\right\}$$

Example 6.1.3: Let $\mathcal{F}$ be the set of Fibonacci numbers. Express $\mathcal{F}$ using set-builder notation.

Solution: The set of Fibonacci numbers can be expressed as

$$\mathcal{F} = \{F_n : F_1 = F_2 = 1, \text{ and } F_{n+2} = F_{n+1} + F_n \text{ for } n \in \mathbb{N}\}$$

Finally, recall that an interval is an infinite collection of real numbers forming a continuum. The ends of an interval can be open or closed. That is, the end of an interval is open when the end does not include its endpoint and closed when the end of the interval does contain the endpoint. For example, the set $\{x \in \mathbb{R} : 0 < x < 10\}$ is the open interval $(0, 10)$, and the set $\{x \in \mathbb{R} : 0 \leq x \leq 10\}$ is the closed interval $[0, 10]$. Recall further that an interval can be half-open/half-closed such as the interval $[0, 10)$.

Example 6.1.4: Several examples of sets and the different methods used to represent them are given below:

a. $A = \{\text{dog, cat, cow, pig, horse, chicken, fish}\}$ is a set listed using roster notation.

b. $B = \{0, \pm 1, \pm 2, \pm 3, \pm 4, \ldots\}$ is the set of integers listed using roster notation.

c. $I = [0, \infty)$ is the set of nonnegative real numbers listed using interval notation.

d. $C = \{f : f \text{ is a real-valued continuous function}\}$ is the set of real-valued continuous functions, listed using set-builder notation.

e. $E = \{x \in \mathbb{R} : x^2+1 < 0\}$ is the set of values of the function $f(x) = x^2+1$, where $f(x) < 0$, listed using set-builder notation.

Note that set E in Example 6.1.4 contains no elements since $x^2 + 1 > 0$ for all $x \in \mathbb{R}$. Thus, the set E is empty and hence is called an *empty set*. Empty sets are commonly encountered in set theory, and the mathematical definition of an empty set is given below.

Definition 6.1.4: The set containing no elements is called the *empty set* and is denoted by $\emptyset$.

Note that the empty set may also be represented by $\{\ \}$. It is important to note that the set $\{\emptyset\}$ is not the same as $\emptyset$, since $\{\emptyset\}$ does contain one element, namely, the empty set; however $\emptyset$ has no elements. In other words, $\emptyset \in \{\emptyset\}$ and thus, the set $\{\emptyset\}$ is not empty.

Now, the two basic reference sets in set theory are Ω, the universe, and $\emptyset$, the empty set. However, most set theory is based on subcollections of elements in Ω. Subcollections of Ω are called *subsets*, and several important definitions that are used to relate subsets are given below.

Definition 6.1.5: A subcollection of elements of a universe Ω is called a *subset* of Ω. When a set A is a subset of Ω, this will be denoted by $A \subset \Omega$.

Definition 6.1.6: Let A and B be subsets of Ω. The set A is said to be a *subset* of the set B if and only if every element of A is also an element of B. When A is a subset of B, this will be denoted by $A \subset B$.

Definition 6.1.7: A set A is said to be a *proper subset* of a set B if and only if A is a subset that is strictly contained in B. When A is the subset of B that contains all the elements of B, then A is called an *improper subset* of B.

The subset relationship between two sets is analogous to the ordering of real numbers. In particular, $A \subset B$ is the set version of $a \leq b$ for numbers. In many presentations on set theory the special notation $A \subset B$ is used to denote that A is a proper subset of B, while $A \subseteq B$ is used to denote that

A may be an improper subset of B. However, since this distinction will not affect any of the results presented in this chapter, no distinction will be made between proper and improper subsets. Thus, the notation $A \subset B$ is meant to imply that A might be either a proper or improper subset of B.

Example 6.1.5: Let the sets A, B, and C be defined as follows:

$$A = \{z \in \mathbb{Z} : z = 2k \text{ for some } k \in \mathbb{Z}\}$$

$$B = \{z \in \mathbb{Z} : z = 3k \text{ for some } k \in \mathbb{Z}\}$$

$$C = \{z \in \mathbb{Z} : z = 6k \text{ for some } k \in \mathbb{Z}\}$$

Then, $C \subset A$ and $C \subset B$. In fact, C is a proper subset of both A and B, while B does contain many even numbers $B \not\subset A$.

Note that the empty set is a proper subset of any nonempty set. Also, any proper subset of a set B must necessarily exclude at least one member of B. For example, if the set A is a proper subset of a set B, then $\forall\ x \in A$, it follows that $x \in B$; however, there must be elements in B that are not in A. For example, the set of even integers $\mathbb{Z}_E$ is a proper subset of the integers since every element of $\mathbb{Z}_E$ is in $\mathbb{Z}$ but not every integer is in $\mathbb{Z}_E$. Similarly, the odd integers $\mathbb{Z}_O$ is a proper subset of $\mathbb{Z}$. However, the set formed by combining the even and odd integers is an improper subset of $\mathbb{Z}$ since it contains every possible integer and hence, is exactly $\mathbb{Z}$.

Now, suppose that A and B are subsets of Ω, A is a subset of B, and B is a subset of A, also. In this case, since every element of A is also an element of B and vice versa, it follows that A and B must have exactly the same elements. When two sets have exactly the same elements, these sets are said to be *equal sets*. The mathematical definition of equal sets is given below.

Definition 6.1.8: Let A and B be subsets of Ω. The sets A and B are said to be *equal sets* if and only if $A \subset B$ and $B \subset A$.

The definition of equal sets given above is equivalent to saying that two sets are equal if and only if they have exactly the same elements. For example, the set of even integers and the set of integer multiples of 2 are clearly equal. In Section 6.1.3 an algorithm for proving that two sets are equal is given.

Example 6.1.6: Let $A = [0, 10]$, $B = \{1, 2, 3, 4, 5, 6, 7, 8, 9, 10\}$, and let the set $C = \{x \in \mathbb{R} : x^2 + 1 < 10\}$. Determine which of the sets A, B, and C are subsets of each other.

Solution: The solution to Example 6.1.6 is left as a exercise.

The next theorem shows that the empty set is a subset of every subset of Ω. Furthermore, the empty set is a proper subset of every nonempty subset of Ω.

Theorem 6.1.1: Let A be a subset of Ω. Then, $\emptyset \subset A$.

> ***Proof:*** Let A be a subset of Ω. Note that $\emptyset$ is a subset of A if and only if every element in $\emptyset$ is also in A. Since there are no elements in $\emptyset$, this is true by default.
>
> ■

6.1.1 Set Algebra

The algebra used in set theory is somewhat analogous to simple arithmetic. In particular, new sets are often of created by combining the elements of the sets, which is analogous to addition of numbers, or by taking the elements that are in one set but not another, which is analogous to subtraction of numbers; there is even a method for creating a new set analogous to multiplication (i.e., the Cartesian product of sets), which will not be discussed in this text. Other set algebraic ways of creating new sets include forming a set by taking the elements common to two sets or the elements that are in the universal set but are not in the original set. In particular, the basic set operations that are used to create new sets are *union*, *intersection*, and *complementation*. The definitions of the sets resulting from the application of these operators are given below.

Definition 6.1.9: Let A and B be subsets of Ω. The *union* of the sets A and B is defined to be $\{x \in \Omega : x \in A \text{ or } x \in B\}$ and is denoted by $A \cup B$.

Definition 6.1.10: Let A and B be subsets of Ω. The *intersection* of the sets A and B is defined to be $\{x \in \Omega : x \in A \text{ and } x \in B\}$ and is denoted by $A \cap B$.

Definition 6.1.11: Let A be a subset of Ω. The *complement* of the set A is defined to be $\{x \in \Omega : x \notin A\}$ and is denoted by A^c.

Note that $\cup$ and $\cap$ are binary operators whose arguments are sets; complementation is a unary operator whose argument is a single set. The symbols $\cup$ and $\cap$ were introduced by Giuseppe Peano (1858–1932) in 1888. In fact, most of the set notation currently used in set theory is due to Cantor, Peano, and Ernst Schröder (1841–1902) and was introduced between 1880 and 1920. The three set operations union, intersection, and complementation are often used in forming new sets. For example, given two sets A and B, several examples of new sets that can be created using these set operations are listed below:

$$A \cap B^c, \quad (A \cap B) \cup (A \cap B^c), \quad (A^c \cup B^c) \cap (A \cup B), \quad (A \cup B)^c, \quad (A \cap B)^c$$

Example 6.1.7: Let $\Omega = \mathbb{R}$, $A = (0, 10)$, $B = \{x : |x - 1| < 5\}$, and $C = \mathbb{Q}$. Determine

a. $A \cup B$
b. $A \cap B$
c. A^c
d. C^c

Solutions: Let $\Omega = \mathbb{R}$, $A = (0, 10)$, $B = \{x : |x - 1| < 5\}$, and $C = \mathbb{Q}$.

a. First, note that B is the interval $(-4, 6)$. Now, by definition, $A \cup B$ contains all the points that are in either A or B or both A and B. Thus

$$A \cup B = (0, 10) \cup (-4, 6) = (-4, 10)$$

b. By definition, $A \cap B$ contains only those points that are in both A and B. Thus

$$A \cap B = (0, 10) \cap (-4, 6) = (0, 6)$$

c. By definition, A^c contains all the points that are in Ω that are not in the set A. Thus

$$A^c = (0, 10)^c = (-\infty, 0] \cup [10, \infty)$$

d. By definition, C^c contains all of the points that are in Ω that are not in the set C. Thus, $C^c = \mathbb{Q}^c$, which is the set of irrational numbers ($\mathbb{I}$).

The following properties of sets will be taken as axioms (i.e, as self-evidently true propositions). If A, B, and C are subsets of Ω, then

Set Axiom 1: $A \cup B = B \cup A$ (commutative property of unions)

Set Axiom 2: $A \cap B = B \cap A$ (commutative property of intersections)

Set Axiom 3: $(A \cup B) \cup C = A \cup (B \cup C)$ (associative property of unions)

Set Axiom 4: $(A \cap B) \cap C = A \cap (B \cap C)$ (associative property of intersections)

Note that these four axioms are analogous to the commutative and associative axioms for addition and multiplication given in Chapter 4.

Definition 6.1.12: Let A and B be subsets of Ω. The sets A and B are said to be *disjoint* if and only if $A \cap B = \emptyset$.

An example of two disjoint sets is $A = \{1, 3, 5\}$ and $B = \{2, 4, 6, 8\}$. Two sets that are always disjoint are A and A^c. Note that when two sets A and B are disjoint, then they have no elements in common and there is no nonempty subset of A that is a subset of B, and vice versa. Other commonly encountered disjoint sets include the set of odd integers and the set of even integers and the set of nonnegative real numbers $[0, \infty)$ and the negative real numbers $(-\infty, 0)$. A final example of a common use of disjoint sets is the deleted neighborhood discussed in Chapter 5. In particular, a deleted neighborhood of x_0 of radius ϵ is $\{x \in \mathbb{R} : 0 < |x - x_0| < \epsilon\}$, which can be written as the disjoint union $(x_0 - \epsilon, x_0) \cup (x_0, x_0 + \epsilon)$.

Example 6.1.8: Let $\Omega = \mathbb{Z}$, $A = \{x \in \mathbb{Z} : 3|x\}$, $B = \{x \in \mathbb{Z} : 2|x\}$, and $C = \mathbb{Z}_O$. Determine

a. $A \cup B$

c. $A \cap B$

c. $B \cap C$

Solutions:

a. $A \cup B = \{x \in \mathbb{Z} : x \text{ is divisible by 3 or is even}\} = \{0, \pm 2, \pm 3, \pm 4, \ldots\}$.

b. $A \cap B = \{x \in \mathbb{Z} : 3|x \text{ and } 2|x\} = \{x \in \mathbb{Z} : x = 6k \text{ for } k \in \mathbb{Z}\}$.

c. $B \cap C = \{x \in \mathbb{Z} : x \text{ is even and } x \text{ is odd}\} = \emptyset$. Hence, B and C are disjoint sets.

6.1.2 Element Chasing Proofs

A common method of proof that is used in set theory is the method of the *element chasing proof.* An element chasing proof is often used to show that two sets are equal or to show that one set is a subset of the other. An algorithm for showing that a set A is a subset of a set B using an element chasing proof is given below.

Algorithm for Showing A is a Subset of B: Let A and B be subsets of Ω. To prove that A is a subset of B

1. Let $x \in A$ be arbitrary but fixed.
2. Using a sequence of logical arguments, show that $x \in B$. This is known as "chasing x from the set A to the set B."
3. Conclude that $A \subset B$.

Note that showing that $A \subset B$ simply amounts to taking an arbitrary element x in A and then "chasing" it to the set B with a sequence of logical arguments. The following two theorems will be proved with an element chasing proof. The first theorem shows that set containment is a transitive relation, and the second theorem provides a result concerning the relation between the complements of A and B when $A \subset B$.

Theorem 6.1.2: Let A, B, and C be subsets of Ω. If $A \subset B$ and $B \subset C$, then $A \subset C$ (i.e., $A \subset B \subset C$).

> ***Proof:*** Let A, B, and C be subsets of Ω, and suppose that $A \subset B$ and $B \subset C$. Now, to show that $A \subset C$, it must be shown that every element in A is also in C.
>
> Let $x \in A$ be ABF. Now, since $A \subset B$, it follows that $x \in B$. Furthermore, since $B \subset C$, it follows that $x \in C$, also. Thus, when $x \in A$ it follows that $x \in C$, and therefore $A \subset C$.

■

Theorem 6.1.3: Let $A, B \subset \Omega$. If $A \subset B$, then $B^c \subset A^c$.

> ***Proof:*** Let $A, B \subset \Omega$ and suppose that $A \subset B$.
>
> Let $x \in B^c$. Then, $x \notin B$, which means that $x \notin A$ since $A \subset B$. Hence, $x \in A^c$ and therefore, $B^c \subset A^c$ whenever $A \subset B$.

■

Now, Definition 6.1.8 states that the sets A and B are equal sets if and only if $A \subset B$ and $B \subset A$. Proving that two sets are equal requires proofs of both $A \subset B$ and $B \subset A$. A proof of equality is often called a *dual-containment proof.* An algorithm for proving that two sets are equal with an element chasing proof is given below.

Algorithm for Showing that Two Sets Are Equal: Let A and B be subsets of Ω. To prove that the sets A and B are equal

1. Prove that $A \subset B$; that is, let $x \in A$ be arbitrary but fixed, and show that $x \in B$.
2. Prove that $B \subset A$. To do this, let $x \in B$ be arbitrary but fixed, and show that $x \in A$.
3. Steps 1 and 2 prove that $A \subset B$ and $B \subset A$ (i.e., dual-containment), and therefore it follows that $A = B$.

It is very important to remember that a proof of set equality is a two-step proof (i.e., dual containment) that first requires a proof of $A \subset B$ followed by a proof of $B \subset A$. The following theorem, which shows that $(A^c)^c = A$, will be used to illustrate an element chasing proof of set equality.

Theorem 6.1.4: Let A be a subset of Ω. Then, $(A^c)^c = A$.

> ***Proof:*** Let A be a subset of Ω. To show that $(A^c)^c = A$, it must be shown that $(A^c)^c \subset A$ and also that $A \subset (A^c)^c$.
>
> First, let $x \in (A^c)^c$ be ABF. Since $x \in (A^c)^c$, it follows from the definition of complementation that $x \notin A^c$. Now, since $x \notin A^c$, it follows that $x \in A$, and hence $(A^c)^c \subset A$.
>
> Conversely, let $x \in A$ be ABF. Then clearly $x \notin A^c$, which means that $x \in (A^c)^c$, and hence $A \subset (A^c)^c$.
>
> Thus, $(A^c)^c \subset A$ and $A \subset (A^c)^c$, and therefore $(A^c)^c = A$.
>
> ■

Note that the result $(A^c)^c = A$ is analogous to the result for the double negation of a statement given in Chapter 2 (i.e., $\neg(\neg P) = P$) and is also analogous to the double negative of a number (i.e., $-(-a) = a$). The remaining theorems presented in this section provide useful relationships and tools for working with the union, intersection, and complements of sets.

Theorem 6.1.5: Let A, B be subsets of Ω. If $A \subset B$, then $A \cup B = B$.

> ***Proof:*** Let $A, B \subset \Omega$ with $A \subset B$. To show that $A \cup B = B$, it must be shown that $A \cup B \subset B$ and $B \subset A \cup B$, also.
>
> First, let $x \in A \cup B$ be ABF. Since $x \in A \cup B$, it follows from the definition of union that $x \in A$ or $x \in B$.
>
> **Case 1:** If $x \in A$, then $x \in B$ since $A \subset B$, and hence $A \cup B \subset B$.
>
> **Case 2:** If $x \notin A$, then x must be in B and again $A \cup B \subset B$.
>
> Therefore, in either case, $A \cup B \subset B$.
>
> Conversely, let $x \in B$ be ABF. Then, clearly $x \in A$ or $x \in B$, which means that $x \in A \cup B$. Hence, $B \subset A \cup B$.
>
> Thus, $A \cup B \subset B$ and $B \subset A \cup B$, and therefore $A \cup B = B$ whenever $A \subset B$.
>
> ■

The following results follow directly from Theorem 6.1.5. In particular, the corollary to Theorem 6.1.5 shows that the union of any set A with the universal set is the universal set, and the union of A with the empty set is A.

Corollary to Theorem 6.1.5: Let A be a subset of Ω. Then

(i) $A \cup \Omega = \Omega$.

(ii) $\emptyset \cup A = A$.

Proof: Both parts of this corollary follows directly from Theorem 6.1.5 since $A \subset \Omega$ and $\emptyset \subset A$.

■

Theorem 6.1.6: Let A, B be subsets of Ω. If $A \subset B$, then $A \cap B = A$.

Proof: Let A, B be subsets of Ω with $A \subset B$.

First, let $x \in A \cap B$ be ABF. Since $x \in A \cap B$, it follows that $x \in A$ and $x \in B$, and hence $x \in A$. Thus, $A \cap B \subset A$.

Conversely, let $x \in A$ be ABF. Then, $x \in B$ since $A \subset B$, and hence $x \in A \cap B$. Therefore, $A \subset A \cap B$.

Thus, $A \cap B \subset B$ and $B \subset A \cap B$, and therefore $A \cap B = B$ whenever $A \subset B$.

■

Corollary to Theorem 6.1.6: Let A be a subset of Ω. Then

(i) $A \cap \Omega = A$.

(ii) $\emptyset \cap A = \emptyset$.

Proof: Both parts of this corollary follows directly from Theorem 6.1.6 since $A \subset \Omega$ and $\emptyset \subset A$.

■

Definition 6.1.13: Two sets A and B are said to form a *partition* of Ω if and only if $A \cap B = \emptyset$ and $A \cup B = \Omega$.

The following theorem shows that the sets A and A^c form a partition of the universe. In particular, part (i) of this theorem shows that Ω is the union of the sets A and A^c, and part (ii) shows that A and A^c are always disjoint sets.

Theorem 6.1.7: Let A be a subset of Ω. Then

(i) $A^c \cup A = \Omega$

(ii) $A^c \cap A = \emptyset$

Proof: Let A be a subset of Ω.

Proof of part (i): Clearly, $A^c \cup A \subset \Omega$ since A and A^c are subsets of Ω. Now, all that remains to be shown is that $\Omega \subset A^c \cup A$. Let $x \in \Omega$. Then, $x \in A$ or $x \notin A$. Thus, $x \in A \cup A^c$, and hence $\Omega \subset A^c \cup A$.

Therefore, by dual containment $A^c \cup A = \Omega$.

Proof of part (ii): Let $x \in A^c \cap A$. Then, $x \in A^c$ and $x \in A$, which is impossible. Hence, there are no values of $x \in \Omega$ such that $x \in A^c \cap A$, and therefore $A^c \cap A = \emptyset$.

■

Several basic relationships between the sets A, B, $A \cup B$, and $A \cap B$ are given in Theorem 6.1.8. In particular, Theorem 6.1.8 shows that the union of two sets is a set that is at least as large as either set and that the intersection of two sets is no larger than either of the two sets.

Theorem 6.1.8: Let A, B be subsets of Ω. Then

(i) $A \cap B \subset A$.

(ii) $A \cap B \subset B$.

(iii) $A \subset A \cup B$.

(iv) $B \subset A \cup B$.

(v) $A \cap B \subset A \cup B$.

Proof: Let A, B be subsets of Ω.

Proof of part (i): Let $x \in A \cap B$ be ABF. Since $x \in A \cap B$, it follows that $x \in A$ and $x \in B$. Thus, $x \in A$, and hence $A \cap B \subset A$.

Proof of parts (ii)–(v): The proofs of parts (ii)–(v) are left as exercises.

■

Note that $\emptyset$ satisfies the necessary condition for an identity under the binary operator $\cup$, which is $\emptyset \cup A = A \cup \emptyset = A$ for every subset A of Ω. Thus, part (ii) of the corollary to Theorem 6.1.6 shows that $\emptyset$ is the identity element for the binary operator $\cup$. Furthermore, since $A \subset A \cup B$ for any set B, it follows that there is no nonempty subset B such that $A \cup B = \emptyset$; hence, there are no inverses for the nonempty sets under the binary operator $\cup$. Similarly, Ω satisfies the necessary condition for an identity under the binary operator $\cap$, which is $\Omega \cap A = A \cap \Omega = A$ for every subset of A of Ω. Thus, part (ii) of the corollary to Theorem 6.1.6 shows that Ω is the identity element for the operator $\cap$. However, since $A \cap B \subset A$ for every set B, it follows that there is no subset B of Ω such that $A \cap B = \Omega$ when A is a proper subset of Ω; hence, there are no inverses for any of the proper subsets of Ω under the binary operator $\cap$.

Example 6.1.9: Under what conditions will it be true that $A \cap B = A \cup B$?

Solution: First, by Theorem 6.1.8 part (v), $A \cap B \subset A \cup B$. Thus, a necessary condition for $A \cap B = A \cup B$ is that $A \cup B \subset A \cap B$. Now, since the union of two sets is at least as large as either set and the intersection is no larger than either individual set, the only way that $A \cup B \subset A \cap B$ is for $A = B$. Thus, $A \cap B = A \cup B$ only when $A = B$.

The next theorem is known as *DeMorgan's laws for sets* and provides two very important results concerning the complementation of the union and intersection of two sets. Note that DeMorgan's laws for sets is analogous to DeMorgan's laws for statements, which was presented in Chapter 2.

Theorem 6.1.9 (DeMorgan's Laws): Let A, B be subsets of Ω. Then

(i) $(A \cup B)^c = A^c \cap B^c$.

(ii) $(A \cap B)^c = A^c \cup B^c$.

Proof: Let A, B be subsets of Ω.

Proof of part (i): First, let $x \in (A \cup B)^c$. Then, since $x \in (A \cup B)^c$, it follows that $x \notin (A \cup B)$.

Thus, $x \notin A$ and $x \notin B$, which means that $x \in A^c$ and $x \in B^c$, and hence $x \in A^c \cap B^c$. Therefore, $(A \cup B)^c \subset A^c \cap B^c$.

Conversely, let $x \in A^c \cap B^c$. Then, since $x \in A^c \cap B^c$, it follows that $x \notin A$ and $x \notin B$.

Thus, x is not in A and x is not in B which means x is not in A or B. Hence, $x \notin (A \cup B)$. Therefore, it follows that $x \in (A \cup B)^c$, and hence $A^c \cap B^c \subset (A \cup B)^c$.

Therefore, $(A \cup B)^c \subset A^c \cap B^c$ and $A^c \cap B^c \subset (A \cup B)^c$, and hence $(A \cup B)^c = A^c \cap B^c$.

Proof of part (ii): The proof of part (ii) is left as an exercise.

■

Theorem 6.1.10 shows that the union and intersection operators can be distributed over each other. The distributive laws for distributing union over intersection and vice versa are analogous to the distributive properties for the conjunction and disjunction operators of Chapter 2, and also for the ordinary arithmetic operators of multiplication and addition. For example, the distributive law for distributing multiplication over addition with numbers is

$$a \times (b + c) = a \times b + a \times c$$

The analogous result for distributing intersection over union for sets is

$$A \cap (B \cup C) = (A \cap B) \cup (A \cap C)$$

which is given in the following theorem.

Theorem 6.1.10 (The Distributive Properties): Let A, B, C be subsets of Ω. Then

(i) $A \cup (B \cap C) = (A \cup B) \cap (A \cup C)$.
(ii) $A \cap (B \cup C) = (A \cap B) \cup (A \cap C)$.

Proof: Let A, B, C be subsets of Ω.

Proof of part (i): First, let $x \in A \cup (B \cap C)$. Since $x \in A \cup (B \cap C)$, it follows that $x \in A$ or $x \in B \cap C$. Thus, either $x \in A$ or $x \notin A$.

Case 1: Suppose that $x \in A$. Then, $x \in A \cup B$ and $x \in A \cup C$, and hence

$$x \in (A \cup B) \cap (A \cup C)$$

Therefore, $A \cup (B \cap C) \subset (A \cup B) \cap (A \cup C)$ whenever $x \in A$.

Case 2: Suppose that $x \notin A$, then $a \in B \cap C$. Thus, $x \in B$ and $x \in C$, and it follows that $x \in A \cup B$ and $x \in A \cup C$, also. Thus

$$x \in (A \cup B) \cap (A \cup C)$$

Therefore, $A \cup (B \cup C) \subset (A \cup B) \cap (A \cup C)$ whenever $x \in B \cap C$.

Thus, in either case $A \cup (B \cap C) \subset (A \cup B) \cap (A \cup C)$.

Conversely, let $x \in (A \cup B) \cap (A \cup C)$. Since $x \in (A \cup B) \cap (A \cup C)$, it follows that $x \in (A \cup B)$ and $x \in (A \cup C)$. Now, $x \in A$ or $x \notin A$.

Case 1: Suppose that $x \in A$. Then, $x \in A \cup (B \cap C)$, and it follows that $(A \cup B) \cap (A \cup C) \subset A \cup (B \cap C)$ whenever $x \in A$.

Case 2: Suppose that $x \notin A$. Since $x \notin A$, $x \in (A \cup B)$, and $x \in (A \cup C)$, it follows that $x \in B$ and $x \in C$. Thus, $x \in B \cap C$, and hence $x \in A \cup (B \cap C)$. Therefore, $(A \cup B) \cap (A \cup C) \subset A \cup (B \cap C)$ whenever $x \notin A$.

Thus, in either case $(A \cup B) \cap (A \cup C) \subset A \cup (B \cap C)$, and therefore $A \cup (B \cap C) = (A \cup B) \cap (A \cup C)$.

Proof of part (ii): The proof of part (ii) is left as an exercise.

■

An interesting result that follows from the distributive law is that a set A can always be partitioned using any other subset B of Ω. In fact, the following corollary to Theorem 6.1.10 shows that the sets $A \cap B$ and $A \cap B^c$ form a partition of the set A, for any $B \subset \Omega$.

Corollary to Theorem 6.1.10: Let A be a subset of Ω. If B is any subset of Ω, then $A = (A \cap B) \cup (A \cap B^c)$.

Proof: Let A be a subset of Ω, and let B be an ABF subset of Ω. Then

$$A = \underbrace{A \cap \Omega}_{\text{By corollary to Theorem 6.1.6(i)}}$$

$$= \underbrace{A \cap (B \cup B^c)}_{\text{By theorem 6.1.7(i)}}$$

$$= \underbrace{(A \cap B) \cup (A \cap B^c)}_{\text{By Theorem 6.1.10(ii)}}$$

■

Example 6.1.10: Let $A = [0, 1)$, $B = (-1, 1)$, and $C = (-2, 1]$. Determine

a. A^c, B^c, and C^c

b. $A \cap B \cap C$

c. $A \cup B \cup C$

d. $(A \cap B) \cup (A^c \cap C^c)$

e. $(A \cap B) \cup C \cup B^c$

f. $A^c \cap B^c \cap C^c$

Solutions: Let $A = [0, 1)$ $B = (-1, 1)$, and $C = (-2, 1]$.

a. The complements of A, B, and C are

$$A^c = (-\infty, 0) \cup [1, \infty)$$

$$B^c = (-\infty, -1] \cup [1, \infty)$$

$$C^c = (-\infty, -2] \cup (1, \infty)$$

b. $A \cap B \cap C = [0, 1) \cap (-1, 1) \cap (-2, 1) = [0, 1)$

c. $A \cup B \cup C = (-2, 1]$

d. $(A \cap B) \cup (A^c \cap C^c)$

$$= \Big([0, 1) \cap (-1, 1)\Big) \cup \Big((-\infty, -2] \cup (1, \infty)\Big)$$

$$= [0, 1) \cup (-\infty, -2] \cup (1, \infty)$$

$$= (-\infty, -2] \cup [0, 1) \cup (1, \infty)$$

e. $(A \cap B) \cup C \cup B^c$

$$= \underbrace{\Big([0, 1) \cap (-1, 1)\Big)}_{[0,1)} \cup (-\infty, -1] \cup [1, \infty) = (-\infty, -1] \cup [0, \infty)$$

f. $A^c \cap B^c \cap C^c$

$$= \Big((-\infty, 0) \cup [1, \infty)\Big) \cap \Big((-\infty, -1] \cup [1, \infty)\Big) \cap \Big((-\infty, -2] \cup (1, \infty)\Big)$$

$$= (-\infty, -2] \cup (1, \infty).$$

Example 6.1.11: Let A, B, and C be subsets of Ω. Use set algebra to simplify the following expressions:

a. $(A \cap B) \cup (A \cap B^c)$.
b. $(A \cup B) \cap (A \cup B^c)$.
c. $(A^c \cap B^c) \cup (A^c \cap C^c)$.
d. $(A \cap B) \cup (A \cap B^c) \cup (A^c \cap B^c) \cup (A^c \cap B)$.

Solutions: Let A, B, and C be subsets of Ω. Note that

a. $(A \cap B) \cup (A \cap B^c)$

$$= \underbrace{A \cap (B \cup B^c)}_{\text{By Theorem 6.1.10(ii)}}$$

$$= \underbrace{A \cap \Omega}_{\text{By Theorem 6.1.7(i)}} = A$$

b. $(A \cup B) \cap (A \cup B^c)$

$$= \underbrace{A \cup (B \cap B^c)}_{\text{By Theorem 6.1.10(ii)}}$$

$$= \underbrace{A \cap \emptyset}_{\text{By Theorem 6.1.6(ii)}} = \emptyset$$

c. $(A^c \cap B^c) \cup (A^c \cap C^c)$

$$= \underbrace{A^c \cap (B^c \cup C^c)}_{\text{By Theorem 6.1.10(ii)}} = \underbrace{A^c \cap (B \cap C)^c}_{\text{By Theorem 6.1.9(ii)}}$$

$$= \underbrace{\Big(A \cup (B \cap C)\Big)^c}_{\text{By Theorem 6.1.9(ii)}}$$

d. $(A \cap B) \cup (A \cap B^c) \cup (A^c \cap B^c) \cup (A^c \cap B) = \Omega$. The details of part (d) are left as an exercise.

6.1.3 Unions and Intersections of Finite Collections of Sets

Recall that DeMorgan's laws for sets state that

$$(A \cup B)^c = A^c \cap B^c \text{ and } (A \cap B)^c = A^c \cup B^c$$

However, DeMorgan's laws do not state anything about the complements of the union of three (or more) sets $(A \cup B \cup C)^c$ nor the intersection of three or more sets. For example, what is $(A \cap B \cap C)^c$? Similarly, the distributive laws state only that

$$A \cap (B \cup C) = (A \cap B) \cup (A \cap C)$$

and

$$A \cup (B \cap C) = (A \cup B) \cap (A \cup C)$$

but state nothing about distributing a set A over the union or intersection of more than two sets. The purpose of this section is to extend the results of previous section to more general results that apply to a finite collection of sets. In particular, DeMorgan's laws and the Distributive laws are two of the key results of the previous section that will be generalized in this section. The following example shows that DeMorgan's law for unions is easily generalized from two sets to three sets.

Example 6.1.12: Let $A, B, C \subset \Omega$. Prove that $(A \cup B \cup C)^c = A^c \cap B^c \cap C^c$.

Solution: Let $A, B, C \subset \Omega$, and let $D = A \cup B$. Then

$$(A \cup B \cup C)^c = (D \cup C)^c = \underbrace{D^c \cap C^c}_{\text{By Theorem 6.1.9(i)}} = (A \cup B)^c \cap C^c$$

$$= \underbrace{(A^c \cap B^c) \cap C^c}_{\text{By Theorem 6.1.9(i)}} = A^c \cap B^c \cap C^c$$

Before further generalizations are made, the following definitions of the union and intersections of a finite number of sets are needed.

Definition 6.1.14: Let $A_1, A_2, \ldots, A_n$ be subsets of Ω. The *finite union* of these n sets is denoted by $\bigcup_{i=1}^{n} A_i$ and is defined to be

$$\bigcup_{i=1}^{n} A_i = \left\{x \in \Omega : x \in A_i \text{ for some } i \in \{1, 2, 3, \ldots, n\}\right\}$$

Definition 6.1.15: Let $A_1, A_2, \ldots, A_n$ be subsets of Ω. The *finite intersection* of the sets $A_1, A_2, \ldots, A_n$ is denoted by $\bigcap_{i=1}^{n} A_i$ and is defined to be

$$\bigcap_{i=1}^{n} A_i = \Big\{x \in \Omega \ : \ x \in A_i \text{ for every } i \in \{1, 2, 3, \ldots, n\}\Big\}$$

Note that an element x will be in the finite union of the sets $A_1, \ldots, A_n$ if and only if x is in at least one of the sets. Thus, it follows that x will be in $\cup_{i=1}^{n} A_i$ if and only if $\exists\, i \in \{1, 2, \ldots, n\}$ such that $x \in A_i$. Similarly, an element x will be in the finite intersection of the sets $A_1, \ldots, A_n$ if and only if x is in every single one of the sets. Thus, for the intersection of a finite number of sets, it follows that x will be in $\cap_{i=1}^{n} A_i$ if and only if $x \in A_i$, $\forall\, i \in \{1, 2, \ldots, n\}$.

Also, note that the mathematical expressions for finite unions and finite intersections use mathematical shorthand that is analogous to summation notation. For example,

$$\bigcup_{i=1}^{n} A_i$$

is used to represent a finite union $A_1 \cup A_2 \cdots \cup A_n$, while $\sum_{i=1}^{n} a_i$ is used to represent the finite sum $a_1 + \cdots + a_n$.

Example 6.1.13: Let $A_i = [0, 1/n]$ and $B_i = [1 - 1/n, 1 + 1/n]$. Determine

a. $\bigcup_{i=1}^{10} A_i$

b. $\bigcup_{i=1}^{10} B_i$

c. $\bigcap_{i=1}^{10} A_i$.

d. $\bigcap_{i=1}^{10} B_i$.

Solutions:

a. $\bigcup_{i=1}^{10} A_i = A_1 \cup A_2 \cup \cdots \cup A_{10} = [0, 1]$

b. $\bigcup_{i=1}^{10} B_i = B_1 \cup B_2 \cup \cdots \cup B_{10} = [0, 2]$.

c. $\bigcap_{i=1}^{10} A_i = A_1 \cap A_2 \cap \cdots \cap A_{10} = [0, 0.1]$.

d. $\bigcap_{i=1}^{10} B_i = B_1 \cap B_2 \cap \cdots \cap B_{10} = [0.9, 1.1]$.

More general versions of DeMorgan's laws for unions and intersections are given in Theorem 6.1.11. Also, for pedagogical reasons part (i) of Theorem 6.1.11 is proved using mathematical induction and part (ii) is proved using an element chasing proof. However, it is important to note that both parts of this theorem can easily be proved with either method.

Theorem 6.1.11 (Generalized DeMorgan's Laws) Let $A_i \subset \Omega$, $\forall\, i \in \mathbb{N}$. Then

(i) $\left(\bigcup_{i=1}^{n+1} A_i\right)^c = \bigcap_{i=1}^{n+1} A_i^c, \ \forall\, n \in \mathbb{N}$

(ii) $\left(\bigcap_{i=1}^{n+1} A_i\right)^c = \bigcup_{i=1}^{n+1} A_i^c, \ \forall\, n \in \mathbb{N}$

Proof: Let $A_1, A_2, \ldots, A_n \subset \Omega$, $\forall\, n \in \mathbb{N}$.

Proof of part (i) (by Induction): Let

$$\mathcal{P}_n := \left(\bigcup_{i=1}^{n+1} A_i\right)^c = \bigcap_{i=1}^{n+1} A_i^c$$

For $n = 1$,

$$(A_1 \cup A_2)^c = \underbrace{A_1^c \cap A_2^c}_{\text{By Theorem 6.1.9(i)}}$$

and therefore $\mathcal{P}_1$ is true.

Now, suppose that $\mathcal{P}_k$ is true for some ABF $k \in \mathbb{N}$. This means that

$$\left(\bigcup_{i=1}^{k+1} A_i\right)^c = \bigcap_{i=1}^{k+1} A_i^c$$

and if $\mathcal{P}_{k+1}$ is true, then

$$\left(\bigcup_{i=1}^{k+2} A_i\right)^c = \bigcap_{i=1}^{k+2} A_i^c$$

Now, consider $\left(\bigcup_{i=1}^{k+2} A_i\right)^c$:

$$\left(\bigcup_{i=1}^{k+2} A_i\right)^c = \left(\left[\bigcup_{i=1}^{k+1} A_i\right] \cup A_{k+2}\right)^c = \underbrace{\left(\bigcup_{i=1}^{k+1} A_i\right)^c \cap A_{k+2}^c}_{\text{By Theorem 6.1.9(i)}}$$

$$= \underbrace{\left(\bigcap_{i=1}^{k+1} A_i^c\right)}_{\text{By } \mathcal{P}_k} \cap A_{k+2}^c = \bigcap_{i=1}^{k+2} A_i^c$$

Therefore, $\mathcal{P}_{k+1}$ is true whenever $\mathcal{P}_k$ is true, and hence

$$\left(\bigcup_{i=1}^{n+1} A_i\right)^c = \bigcap_{i=1}^{n+1} A_i^c, \quad \forall\, n \in \mathbb{N}$$

Proof of part (ii) (Element Chasing Proof): Let $n \in \mathbb{N}$ be ABF, and suppose that $x \in \left(\bigcap_{i=1}^{n+1} A_i\right)^c$. Then, $x \in \left(\bigcap_{i=1}^{n+1} A_i\right)^c$ means that $x \notin \left(\bigcap_{i=1}^{n+1} A_i\right)$. Now, since $x \notin \left(\bigcap_{i=1}^{n+1} A_i\right)$, it follows that there exists $i^* \in \{1, \ldots, n\}$, such that $x \notin A_{i^*}$, and hence $x \in A_{i^*}^c$. However, since $x \in A_{i^*}^c$, it follows that $x \in \bigcup_{i=1}^{n+1} A_i^c$, and hence

$$\left(\bigcap_{i=1}^{n+1} A_i\right)^c \subset \bigcup_{i=1}^{n+1} A_i^c$$

Conversely, suppose that $x \in \bigcup_{i=1}^{n+1} A_i^c$. Then, since $x \in \bigcup_{i=1}^{n+1} A_i^c$, it follows that $x \in A_{i^*}^c$ for some $i^* \in \{1, \ldots, n\}$; and since $x \in A_{i^*}^c$ for some $i^* \in \{1, \ldots, n\}$, it follows that $x \notin A_{i^*}$. Hence $x \notin \bigcap_{i=1}^{n+1} A_i$.

Now, since $x \notin \bigcap_{i=1}^{n+1} A_i$, it follows that $x \in \left(\bigcap_{i=1}^{n+1} A_i\right)^c$ and thus

$$\bigcup_{i=1}^{n+1} A_i^c \subset \left(\bigcap_{i=1}^{n+1} A_i\right)^c$$

Therefore, by dual containment $\bigcup_{i=1}^{n+1} A_i^c = \left(\bigcap_{i=1}^{n+1} A_i\right)^c$.

■

Note that the versions of DeMorgan's laws given in Theorem 6.1.9 and Example 6.1.12 are simply corollaries of Theorem 6.1.11. Moreover, the generalized version of DeMorgan's laws states that the complement of the union (intersection) of any finite number of sets is simply the intersection (union) of their complements. For example, the generalized version of DeMorgan's laws can be used to show that

$$(A \cup B \cup C \cup D \cup E)^c = A^c \cap B^c \cap C^c \cap D^c \cap E^c$$

and

$$(A \cap A_2 \cap \cdots \cap A_{100})^c = A_1^c \cup A_2^c \cup C^c \cdots \cup A_{100}^c$$

The distributive properties for distributing union over intersection and intersection over union will now be generalized. In particular, Theorem 6.1.12 states that union distributes over the intersection of a finite number of sets and that intersection distributes over the union of a finite number of sets.

Theorem 6.1.12 (Generalized Distributive Properties) Let $A_i \subset \Omega$ $\forall\, i \in \mathbb{N}$, and let $B \in \Omega$. Then

(i) $B \cup \left(\bigcap_{i=1}^{n+1} A_i\right) = \bigcap_{i=1}^{n+1} (B \cup A_i)$, $\forall\, n \in \mathbb{N}$

(ii) $B \cap \left(\bigcup_{i=1}^{n+1} A_i\right) = \bigcup_{i=1}^{n+1} (B \cap A_i)$, $\forall\, n \in \mathbb{N}$

Proof: Let $B \subset \Omega$, and let $A_1, A_2, \ldots, A_n \subset \Omega$, $\forall\, n \in \mathbb{N}$.

Proof of part (i) (by Induction): Let

$$\mathcal{P}_n := B \cup \left(\bigcap_{i=1}^{n+1} A_i \right) = \bigcap_{i=1}^{n+1} (B \cup A_i)$$

For $n = 1$, it follows that

$$B \cup (A_1 \cap A_2) = \underbrace{(B \cup A_1) \cap (B \cup A_2)}_{\text{By Theorem 6.1.10(i)}}$$

Thus, $\mathcal{P}_1$ is true.

Now, suppose that $\mathcal{P}_k$ is true for some ABF $k \in \mathbb{N}$. This means that

$$B \cup \left(\bigcap_{i=1}^{k+1} A_i \right) = \bigcap_{i=1}^{k+1} (B \cup A_i)$$

and if $\mathcal{P}_{k+1}$ is true, then

$$B \cup \left(\bigcap_{i=1}^{k+2} A_i \right) = \bigcap_{i=1}^{k+2} (B \cup A_i)$$

Now, consider $B \cup \left(\bigcap_{i=1}^{k+2} A_i \right)$:

$$B \cup \left(\bigcap_{i=1}^{k+2} A_i \right) = B \cup \left(\left[\bigcap_{i=1}^{k+1} A_i \right] \cap A_{k+2} \right)$$

$$= \underbrace{\left(B \cup \left[\bigcap_{i=1}^{k+1} A_i \right] \right) \cap (B \cup A_{k+2})}_{\text{By Theorem 6.1.10(i)}}$$

$$= \underbrace{\left[\bigcap_{i=1}^{k+1} (B \cup A_i) \right]}_{\text{by } \mathcal{P}_k} \cap (B \cup A_{k+1})$$

$$= \bigcap_{i=1}^{k+2} (B \cup A_i)$$

Therefore, $\mathcal{P}_{k+1}$ is true whenever $\mathcal{P}_k$ is true, and hence

$$B \cup \left(\bigcap_{i=1}^{n+1} A_i \right) = \bigcap_{i=1}^{n+1} (B \cup A_i), \quad \forall\, n \in \mathbb{N}$$

Proof of part (ii): The proof of part (ii) is left as an exercise.

■

Example 6.1.14: Using the generalized versions of DeMorgan's laws and the distributive property, determine equivalent forms for the following sets:

a. $A \cap (B \cup C \cup D \cup E)$
b. $A \cup (B \cap C \cap D \cap E)$
c. $[A \cap (B \cup C \cup D \cup E)]^c$

Solutions:

a. $A \cap (B \cup C \cup D \cup E) = (A \cap B) \cup (A \cap C) \cup (A \cap D) \cup (A \cap E)$
b. $A \cup (B \cap C \cap D \cap E) = (A \cup B) \cap (A \cup C) \cap (A \cup D) \cap (A \cup E)$
c. $[A \cap (B \cup C \cup D \cup E)]^c$

$$= [(A \cap B) \cup (A \cap C) \cup (A \cap D) \cup (A \cap E)]^c$$

$$= (A \cap B)^c \cap (A \cap C)^c \cap (A \cap D)^c \cap (A \cap E)^c$$

$$= (A^c \cup B^c) \cap (A^c \cup C^c) \cap (A^c \cup D^c) \cap (A^c \cup E^c)$$

In many set problems, the set of interest, say, A, is relatively complicated and can be hard to work with. A convenient way of working with the set A is to break it into nonoverlapping subsets in a fashion such that the union of these nonoverlapping subsets is the set A. This is known as *partitioning* the set A. The general version of the definition of a *partition* of Ω is given below followed by Theorem 6.1.13, which shows how a set A can be partitioned using any partition of Ω.

Definition 6.1.16: A collection of sets $B_1, B_2, \ldots, B_n$ is said to form a *partition* of Ω if and only if

(i) $B_i \cap B_j = \emptyset$ when $i \neq j$.

(ii) $\bigcup_{i=1}^{n} B_i = \Omega$.

The collection of sets $\{B_1, \ldots, B_n\}$ is also called a *partition* of Ω, and the following theorem shows how to use a partition of Ω to partition any set subset A of Ω.

Theorem 6.1.13: Let $A \subset \Omega$. If $B_1, B_2, \ldots, B_n$ is a partition of Ω, then $A = \bigcup_{i=1}^{n} (A \cap B_i)$.

Proof: Let $A \subset \Omega$, and suppose that $B_1, B_2, \ldots, B_n$ form a partition of Ω. Then, $\bigcup_{i=1}^{n} B_i = \Omega$, and by Theorem 6.1.12(ii) it follows that

$$A = A \cap \Omega = A \cap \left(\bigcup_{i=1}^{n} B_i \right) = \underbrace{\bigcup_{i=1}^{n} (A \cap B_i)}_{\text{By Theorem 6.1.12(ii)}}$$

■

Note that a partition of a set A is simply a division of A into nonoverlapping sets such that the union of these nonoverlapping sets is A. For example, if A is the set of all prime numbers less than 40, then one way of determining the elements in A would be to first find the prime numbers between 1 and 10, then find the primes between 11 and 20, followed by finding the primes between 21 and 30, then find the primes between 31 and 40, and finally create A by listing all the prime numbers found at each step of this process. In this case, A has been partitioned by the sets $B_1 = \{1, \ldots, 10\}$, $B_2 = \{11, \ldots, 20\}$, $B_3 = \{21, \ldots, 30\}$, and $B_4 = \{31, \ldots, 40\}$.

Example 6.1.15: Let A to be the collection of natural numbers that are perfect squares and are less than or equal to 100, and let B_i be the natural numbers in the interval $[10(i-1)+1, 10i]$, for $i \in \{1, 2, \ldots, 10\}$. Then

$$A = \bigcup_{i=1}^{10} (A \cap B_i) = (A \cap B_1) \cup (A \cap B_2) \cup \cdots \cup (A \cap B_{10})$$

$$= \{1, 4, 9\} \cup \{16\} \cup \{25\} \cup \{36\} \cup \{49\} \cup \emptyset \cup \{64\}$$

$$\cup\, \emptyset \cup \{81\} \cup \{\} \cup \{100\}$$

$$= \{1, 4, 9, 16, 25, 36, 64, 81, 100\}$$

Finally, it should be noted that each of the generalizations given in this section can easily be extended to results dealing with an infinite collection of sets, say, $\{A_i : i \in \mathbb{N}\}$. For example, the generalized version of DeMorgan's laws in Theorem 6.1.11 states that

$$\left(\bigcup_{i=1}^{n+1} A_i\right)^c = \bigcap_{i=1}^{n+1} A_i^c$$

and

$$\left(\bigcap_{i=1}^{n+1} A_i\right)^c = \bigcup_{i=1}^{n+1} A_i^c, \ \forall\, n \in \mathbb{N},$$

which can be extended even further by considering the complements of the union and intersections of the sets $\{A_i : i \in \mathbb{N}\}$. An even more general version of DeMorgan's laws dealing with an infinite collection of sets is

$$\left(\bigcup_{i=1}^{\infty} A_i\right)^c = \bigcap_{i=1}^{\infty} A_i^c$$

and

$$\left(\bigcap_{i=1}^{\infty} A_i\right)^c = \bigcup_{i=1}^{\infty} A_i^c$$

Similarly, the distributive properties can be extended to the infinite versions as follows:

$$B \cup \left(\bigcap_{i=1}^{\infty} A_i\right) = \bigcap_{i=1}^{\infty} (B \cup A_i)$$

and

$$B \cap \left(\bigcup_{i=1}^{\infty} A_i \right) = \bigcup_{i=1}^{\infty} (B \cap A_i)$$

However, it also must be noted that the infinite versions of both DeMorgan's laws and the distributive properties can no longer be proved using mathematical induction; an element chasing proof can be used to prove each of these results.

6.1.4 Countable and Uncountable Sets

In this final section on set theory, a closer look at the infinite sets will be taken. Now, it usually comes as a surprise to most students that it is possible for one infinite set to be larger than another. For example, it will be shown in this section that the set of rational numbers is smaller than the set of real numbers. In fact, it was Georg Cantor who first proved that the set of real numbers is considerably larger than the set of rational numbers. In particular, the two types of infinite sets that will be studied in this section are the *countably infinite* sets and the *uncountably infinite* sets. The definitions of countably infinite and uncountably infinite sets are given below.

Definition 6.1.17: An infinite set is said to be *countably infinite* or *denumerable* if and only if there is a one-to-one correspondence between the elements of the set and the natural numbers, and a set is said to be a *countable* set if it is a finite set or a countably infinite set.

Definition 6.1.18: Any infinite set that is not countable is said to be *uncountably infinite* and is called an *uncountable* set.

Note that a set may be empty, finite, countably infinite, or uncountably infinite. For example, the set $\{1, e, \pi, 11\}$ is finite and the set of whole numbers is countably infinite, while the interval $(0, 1)$ is uncountably infinite. Moreover, it can be deduced from the definitions above that a countably infinite set is considerably smaller than an uncountably infinite set; however, two countably infinite sets are always the same size, and likewise two uncountably infinite sets will also be the same size.

Now, to show that an infinite set is countable simply amounts to showing that there is a one-to-one correspondence between the elements of the set and the natural numbers. One method of showing that there is a one-to-one correspondence between a set A and the natural numbers is to create a function $f(n)$ that (1) maps each natural number to one and only one member of the set A and (2) covers the entire set A. For example, the set of even natural numbers, say, $\mathbb{N}_E$, is countable since $f(n) = 2n$ is a function that maps the natural numbers to $\mathbb{N}_E$ and completely covers $\mathbb{N}_E$.

Recall that if an infinite set is not countable, then it is an uncountable set. Thus, showing that a set is uncountable requires showing that the set is not countable.

Example 6.1.16: Create a function $f(n)$ that shows that each of the following sets is a countably infinite set:

a. $\mathbb{W} = \{0, 1, 2, \ldots\}$
b. $\mathbb{N}_O = \{n \in \mathbb{N} : n \text{ is odd}\}$
c. $\mathcal{S} = \{w^2 : w \in \mathbb{W}\}$
d. $\mathbb{Z}$

Solutions:

a. Let $f(n) = n - 1$. Then f is a one-to-one map from $\mathbb{N}$ to $\mathbb{W}$ that completely covers $\mathbb{W}$.
b. Let $f(n) = 2n - 1$. Then f is a one-to-one map from $\mathbb{N}$ to $\mathbb{N}_O$ that completely covers $\mathbb{N}_O$.
c. Let $f(n) = (n - 1)^2$. Then f is a one-to-one map from $\mathbb{N}$ to $\mathcal{S}$ that completely covers $\mathcal{S}$.
d. Let $f(1) = 0$, and let $f(n) = f(n-1) + (-1)^{n-1}(n-1)$ for $n \geq 2$. Then f is a one-to-one map from $\mathbb{N}$ to $\mathbb{Z}$ that completely covers $\mathbb{Z}$.

The following theorem, given without proof, shows that all the subsets of a countably infinite set are countable; a proof of Theorem 6.1.14 can be found in *Real Analysis* by H. L. Royden (1968).

Theorem 6.1.14: Let $A \subset B$. If B is a countably infinite set, then A is at most a countable set.

Thus, by Theorem 6.1.14 and the result of Example 6.1.16(d), it follows that every subset of the integers is a countable set. Moreover, since the set of integers is countable, Theorem 6.1.14 shows that the set of prime numbers is a countably infinite set, as are the sets of even and odd integers, and likewise, so are the sets of negative and nonnegative integers.

Galileo (1564–1642) assumed that each infinite set had the same number of elements; however, in 1874 Cantor proved that this was not the case. In fact, Cantor proved that the set of real numbers is an uncountable set, and hence, larger than the set of rational numbers. The key to proving that the set of real numbers is uncountable is the following corollary to Theorem 6.1.14.

Corollary to Theorem 6.1.14: Let $A \subset B$. If A is an uncountable set, then B is also an uncountable set.

Proof: This corollary follows directly from Theorem 6.1.14 since it is the contrapositive theorem associated with Theorem 6.1.14.

■

Thus, from the corollary to Theorem 6.1.14 it can be deduced that if any subset of $\mathbb{R}$ is uncountable, then it follows that $\mathbb{R}$ will also be uncountable. Theorem 6.1.15 shows that the interval $[0, 1]$ is uncountable, and hence, by the Theorem 6.1.14 corollary, it follows that $\mathbb{R}$ is also an uncountable set.

Theorem 6.1.15: The interval $[0, 1]$ is an uncountable set.

Proof (by Contradiction): Suppose that $[0, 1]$ is a countably infinite set. Now, since $[0, 1]$ is a countable set, it follows that $[0, 1] = \{a_1, a_2, a_3, \ldots\}$. Furthermore, for every $i \in \mathbb{N}$, a_i has a decimal representation, say, $a_i = 0.\alpha_{i1}\alpha_{i2}\alpha_{i3}\cdots$. Suppose that the complete list of elements of $[0, 1]$ is given below.

$$a_1 = 0.\alpha_{11}\alpha_{12}\alpha_{13}\cdots$$
$$a_2 = 0.\alpha_{21}\alpha_{22}\alpha_{23}\cdots$$
$$a_3 = 0.\alpha_{31}\alpha_{32}\alpha_{33}\cdots$$
$$a_4 = 0.\alpha_{41}\alpha_{42}\alpha_{43}\cdots$$
$$a_5 = 0.\alpha_{51}\alpha_{52}\alpha_{53}\cdots$$
$$\vdots$$

Now, consider the number $b = 0.b_1b_2b_3\cdots$, which is formed by taking b_i as follows. For $i \in \mathbb{N}$, let

$$b_i = \begin{cases} 3 & \text{if } \alpha_{ii} = 5 \\ 5 & \text{if } \alpha_{ii} \neq 5 \end{cases}$$

Then, $b \neq a_i$ for any $i \in \mathbb{N}$ since $b_n \neq \alpha_{nn}$, $\forall\, n \in \mathbb{N}$. Thus, b is clearly a number in $[0, 1]$, but b is not a member of the complete list of elements of $[0, 1]$ given above. This contradicts the supposition that the list above is a complete listing of the elements in $[0, 1]$.

Hence, $[0, 1]$ is an uncountable set.

■

The proof used above to show that $[0, 1]$ is an uncountable set is known as *Cantor's diagonalization proof* (Cantor 1891). Now, since $[0, 1] \subset \mathbb{R}$, it follows by the corollary to Theorem 6.1.14 that $\mathbb{R}$ is an uncountable set.

Corollary to Theorem 6.1.15: The set of real numbers is an uncountable set.

Proof: Since $[0,1] \subset \mathbb{R}$, it follows by the corollary to Theorem 6.1.14 that $\mathbb{R}$ is an uncountable set.

■

To this point in this section, it has been shown that $\mathbb{Z}$ is countable and $\mathbb{R}$ is uncountable; however, the question of the countability of the set of rational numbers or even the irrational numbers has not yet been addressed. It turns out that the set of rational numbers is countable, but the set of irrational numbers is uncountable. The following three theorems build the foundation for the proof that the set of rational numbers is countable.

Theorem 6.1.16: Let A and B be countably infinite sets. Then $A \cup B$ is a countably infinite set.

Proof: Let $A = \{a_1, a_2, a_3, \ldots\}$ and let $B = \{b_1, b_2, b_3, \ldots\}$. Now, let $f(n)$ be defined by

$$f(n) = \begin{cases} a_{\frac{n+1}{2}} & \text{if } n \in \mathbb{N} \text{ is odd} \\ b_{\frac{n}{2}} & \text{if } n \in \mathbb{N} \text{ is even} \end{cases}$$

Then, $f(n)$ is a one-to-one mapping of the natural numbers that completely covers $A \cup B$, and therefore, $A \cup B$ is a countably infinite set.

■

Theorem 6.1.17: If $A_1, A_2, \ldots, A_{n+1}$ are countable sets, then $\bigcup_{i=1}^{n+1} A_i$ is a countable set, $\forall\, n \in \mathbb{N}$.

Proof (by Induction): Let $\mathcal{P}_n :=$ "If $A_1, A_2, \ldots, A_{n+1}$ are countable sets, then $\bigcup_{i=1}^{n+1} A_i$ is a countable set."

For $n = 1$, $\mathcal{P}_1$ is true by Theorem 6.1.16

Now, suppose that $\mathcal{P}_k$ is true for some ABF $k \in \mathbb{N}$. This means that when $A_1, A_2, \ldots, A_{k+1}$ are countable sets, then $\bigcup_{i=1}^{k+1} A_i$ is a countable set, also. If $\mathcal{P}_{k+1}$ is true, then it will be the case that $\bigcup_{i=1}^{k+2} A_i$ is a countable set whenever $A_1, A_2, \ldots, A_{k+1}, A_{k+2}$ are countable sets.

Consider $\bigcup_{i=1}^{k+2} A_i$

$$\bigcup_{i=1}^{k+2} A_i = \left(\bigcup_{i=1}^{k+1} A_i\right) \cup A_{k+2} = B \cup A_{k+2}$$

where $B = \bigcup_{i=1}^{k+1} A_i$. Since A_{k+2} is a countable set by hypothesis and B is countable set by $\mathcal{P}_k$, it follows that $\bigcup_{i=1}^{k+2} A_i = B \cup A_{k+2}$ is the union of two countable sets.

Hence, $\bigcup_{i=1}^{k+2} A_i$ is countable by Theorem 6.1.16, and therefore $\bigcup_{i=1}^{n+1} A_i$ is a countable set whenever $A_1, A_2, \ldots, A_{n+1}$ are countable sets, $\forall\, n \in \mathbb{N}$.

■

Now, Theorems 6.1.16 and 6.1.17 can be even further generalized to the following theorem showing that the infinite union of a collection of countable sets $\{A_i : i \in \mathbb{N}\}$ is also a countable set. This result is given below without proof and will be used as the basis for proving that the set of rational numbers is countable; a proof of Theorem 6.1.18 can be found in *Real Analysis* by H. L. Royden (1968).

Theorem 6.1.18: Let $\{A_i : i \in \mathbb{N}\}$ be a collection of countable sets. Then $\bigcup_{i=1}^{\infty} A_i$ is a countable set.

The basis for proving that $\mathbb{Q}$ is a countable set is Theorem 6.1.18 with the sets A_i defined to be

$$A_i = \left\{r : r = \pm\frac{z}{i},\ z \in \mathbb{Z}\right\}$$

for $i \in \mathbb{N}$. For example, the sets A_1 and A_2 are given below:

$$A_1 = \left\{r : r = \pm\frac{z}{1},\ z \in \mathbb{Z}\right\} = \{0, -1, 1, -2, 2, \ldots\}$$

$$A_2 = \left\{r : r = \pm\frac{z}{2},\ z \in \mathbb{Z}\right\} = \{0, -\frac{1}{2}, \frac{1}{2}, -\frac{2}{2}, \frac{2}{2}, \ldots\}$$

Since the set of rational numbers consists of all possible ratios of integers, except 0 as the denominator, it follows that $\mathbb{Q} = \bigcup_{i=1}^{\infty} A_i$, and hence the countability of the rational numbers follows.

Theorem 6.1.19: The set of rational numbers is a countably infinite set.

Proof: For $i \in \mathbb{N}$, let $A_i = \left\{r : r = \pm\frac{z}{i},\ z \in \mathbb{Z}\right\}$. Then, $\forall\, i \in \mathbb{N}$, A_i is countable. Furthermore,

$$\mathbb{Q} = \bigcup_{i=1}^{\infty} A_i$$

and thus, by Theorem 6.1.18, it follows that $\mathbb{Q}$ is also countable.

Therefore, the set of rational numbers is a countably infinite set.

∎

Now, since the set of rational numbers is a countable set, it follows from Theorem 6.1.14 that every subset of the rational numbers is also a countable set. In particular, the set $\left\{\frac{1}{2^n} : n \in \mathbb{N}\right\}$ is a countable set as is $\left\{\frac{F_n}{L_n} : n \in \mathbb{N}\right\}$, where F_n and L_n are the nth Fibonacci and Lucas numbers, respectively. Furthermore, since $\mathbb{R}$ is an uncountable set and $\mathbb{Q}$ is a countable set, it can easily be deduced that $\mathbb{I}$ is also an uncountable set.

Theorem 6.1.20: The set of irrational numbers is uncountable.

Proof (by Contradiction): Suppose that $\mathbb{I}$ is a countable set. Then, $\mathbb{R} = \mathbb{Q} \cup \mathbb{I}$ is a countable set by Theorem 6.1.16. However, this contradicts the fact that $\mathbb{R}$ is not a countable set, and therefore the set of irrational numbers is an uncountable set.

∎

Finally, an interesting conjecture that has perplexed mathematicians since it was first posed and studied by Cantor is the *Continuum Hypothesis*. A simple version of the Continuum Hypothesis is given below.

The Continuum Hypothesis: There is no set whose size is strictly between that of the integers and that of the real numbers.

The Continuum Hypothesis was the first of 23 important unsolved problems posed by David Hilbert in an address to the Second International Congress of Mathematicians held in Paris in 1900. The Continuum Hypothesis has been studied by many mathematicians, and these studies led to many important results in the areas of mathematics whose foundation is based on set theory; a list of Hilbert's 23 famous problems and their impact on mathematics can be found in *Hilbert* by Constance Reid (1996).

6.2 An Introduction to Group Theory

Group theory forms the foundation on which modern algebra is built and plays an important role in the theory associated with solving polynomial equations. The early roots of modern algebra and group theory came from attempts at solving algebraic equations and also the study of number theory and geometry. Today, group theory can also be shown to have ties to many other areas of mathematics, including topology, differential equations, combinatorics, and design of experiments. Group theory has even been shown to have applications in chemistry and physics with the study of crystals and symmetries of molecules and quantum theory, respectively.

Early mathematicians laying the foundation for group theory include Euler, with his study of modular arithmetic, and Gauss, who proved that every polynomial has a root of the form $a + bi$ which is known as the *Fundamental Theorem of Algebra*, as well as Joseph-Louis Lagrange (1736–1813), Niels Abel, Évariste Galois (1811–1832), Arthur Cayley (1821–1895), Felix Klein (1849–1925), and Augustin Cauchy, among others. The next generation of mathematicians to make important contributions to group theory includes the mathematicians Sophus Lie (1842–1899), William Burnside (1852–1927), Emmy Noether (1882–1935), and Ludwig Sylow (1832–1918).

6.2.1 Groups

Group theory deals with the study of a mathematical structure called a *group*, which consists of a set of elements and a binary operator. Provided that certain properties are satisfied, the set of elements and the binary operator are said to form a group. Galois, Cauchy, and Cayley all came up with definitions of a group concept; however, the modern version for the definition of a group is due to Walther von Dyck (1856–1934) and is given below.

Definition 6.2.1: Let $\mathcal{G}$ be a set and $\circ$ a binary operator. The pair $(\mathcal{G}, \circ)$ is called a *group* if

(i) $\mathcal{G}$ is closed under the operation $\circ$.

(ii) $(a \circ b) \circ c = a \circ (b \circ c), \ \forall\, a, b, c \in \mathcal{G}$.

(iii) $\exists$ an element $e \in \mathcal{G}$ such that $a \circ e = e \circ a = a, \ \forall\, a \in \mathcal{G}$.

(iv) $\forall\, a \in \mathcal{G}, \ \exists\, x \in \mathcal{G}$ such that $a \circ x = x \circ a = e$.

Thus, a group is simply a set $\mathcal{G}$ and a binary operator $\circ$ for which these four properties are satisfied. Furthermore, the group structure is an algebraic structure that forms the basis for solving everyday equations. Also, it is very important to note that the definition of the group structure states that (1) the identity e is constant with respect to each element in $\mathcal{G}$ and (2) it is not guaranteed that $a \circ b = b \circ a$. The algorithm for showing that the pair $(\mathcal{G}, \circ)$ forms a group consists of four key steps and is given below.

Algorithm for Showing that $(\mathcal{G}, \circ)$ is a Group: Let $\mathcal{G}$ be a set and $\circ$ a binary operator. To prove that $(\mathcal{G}, \circ)$ forms a group

1. Prove that $\mathcal{G}$ is closed under $\circ$.
2. Prove that $\circ$ is an associative operator on $\mathcal{G}$.
3. Prove that there exists an identity element in $\mathcal{G}$ such that $a \circ e = e \circ a$ for every $a \in \mathcal{G}$.
4. Prove that for each $a \in \mathcal{G}$, there exists an inverse element, $a^{-1} \in \mathcal{G}$.
5. Conclude that the pair $(\mathcal{G}, \circ)$ is a group.

The use of this algorithm is illustrated in the following two examples.

Example 6.2.1: Let $\mathcal{G} = \mathbb{Z}$ and $a \circ b = a + b$ (ordinary addition). Prove that $(\mathbb{Z}, +)$ is a group.

Solution: Let $\mathcal{G} = \mathbb{Z}$ and $a \circ b = a + b$.

1. **Closure:** $\mathbb{Z}$ is closed under ordinary addition (Axiom A8 of Chapter 4).
2. **Associativity:** Let $a, b, c \in \mathbb{Z}$. Then, $(a + b) + c = a + (b + c)$ since ordinary addition is associative (Axiom A2 of Chapter 4).
3. **Identity:** $0 \in \mathbb{Z}$ and 0 is the ordinary addition identity element.
4. **Inverses:** For every integer a, $-a \in \mathbb{Z}$ and $-a$ is the ordinary additive inverse element. Thus, $a^{-1} = -a$ with respect to ordinary addition.
5. **Group:** Thus, $\mathbb{Z}$ with ordinary addition forms a group.

Example 6.2.2: Show that the pair $(\mathbb{Z}, \times)$ is not a group.

Solution: Let $\mathcal{G} = \mathbb{Z}$ and $\circ = \times$.

1. **Closure:** $\mathbb{Z}$ is closed under ordinary multiplication (Axiom A9 of Chapter 4).
2. **Associativity:** Let $a, b, c \in \mathbb{Z}$. Then, $(a \times b) \times c = a \times (b \times c)$ since ordinary multiplication is associative (Axiom A6 of Chapter 4).
3. **Identity:** $1 \in \mathbb{Z}$ and 1 is the ordinary multiplication identity element.
4. **Inverses:** For every integer $a \neq \pm 1$, $1/a \notin \mathbb{Z}$, and thus $\mathbb{Z}$ does not contain all the ordinary multiplicative inverse elements.
5. **Group:** Thus, $\mathbb{Z}$ with ordinary multiplication does not form a group.

Whether the pair $(\mathcal{G}, \circ)$ in fact forms a group will depend on both the set and the binary operator. Examples 6.2.1 and 6.2.2 illustrate how the same set may form a group with one operator but not another. An example where an operator with one set forms a group, but, the same operator with a different set does not form a group is given in Example 6.2.3.

Example 6.2.3: Show that $(\mathbb{R}^+, \times)$ forms a group but $(\mathbb{R}, \times)$ does not.

Solution: The solution to Example 6.2.3 is left as an exercise.

Example 6.2.4: Let $\mathcal{G} = \mathbb{R}^+$ and let $a \circ b = a^b$ (exponentiation). Does $(\mathcal{G}, \circ)$ form a group?

Solution: Let $\mathcal{G} = \mathbb{R}^+$, and let $a \circ b = a^b$.

1. **Closure:** $\mathbb{R}^+$ is closed under exponentiation by positive powers.
2. **Associativity:** Let $a, b, c \in \mathbb{R}^+$ be ABF. Then

   $$a \circ (b \circ c) = a \circ (b^c) = a^{b^c}$$

   but

   $$(a \circ b) \circ c = (a^b) \circ c = (a^b)^c = a^{bc} \neq a^{b^c}$$

   Since $\circ$ is not associative on $\mathbb{R}^+$, $(\mathbb{R}^+, \circ)$ does not form a group.

Example 6.2.5: Let $\mathcal{G} = \mathbb{R}$, and let $a \circ b = a + b - 2$. Does $(\mathcal{G}, \circ)$ form a group?

Solution: Let $\mathcal{G} = \mathbb{R}$, and let $a \circ b = a + b - 2$.

1. **Closure:** $\mathbb{R}$ is closed under addition and subtraction, and hence $\mathbb{R}$ is closed under $\circ$.

2. **Associativity:** Let $a, b, c \in \mathbb{R}$ be ABF. Then

$$a \circ (b \circ c) = a \circ (b + c - 2) = a + b + c - 2 - 2 = a + b + c - 4$$

and

$$(a \circ b) \circ c = (a + b - 2) \circ c = a + b - 2 + c - 2 = a + b + c - 4$$

Thus, $a \circ (b \circ c) = (a \circ b) \circ c$, $\forall\, a, b, c \in \mathbb{R}$, and therefore $\circ$ is associative on $\mathbb{R}$.

3. **Identity:** To determine whether $\mathbb{R}$ has an identity under $\circ$, let $a \in \mathbb{R}$ be ABF and consider the equations $a \circ e = a$ and $e \circ a = a$. First, note that

$$a \circ e = a + e - 2 = e + a - 2 = e \circ a$$

Now, solving $a \circ e = a$ for e yields $e = 2$. Since $2 \in \mathbb{R}$, it follows that $\mathbb{R}$ does contain an identity element under the operator $\circ$.

4. **Inverses:** To determine whether $a \in \mathbb{R}$ has an inverse under $\circ$, let $a \in \mathbb{R}$ be ABF and consider the equations $a \circ x = e$ and $x \circ a = e$. Note that

$$a \circ x = a + x - 2 = x + a - 2 = x \circ a$$

Thus, solving $a \circ x = 2$ for x yields $x = 4 - a$. Since $4 - a \in \mathbb{R}$, it follows that $a^{-1} = 4 - a \in \mathbb{R}$ and $\mathbb{R}$ does contain the inverse elements under the operator $\circ$.

5. **Group:** Thus, $(\mathbb{R}, \circ)$ does form a group.

The following theorem shows that the identity element and the inverses in a group are unique. Note that the proofs are similar to the proofs that identities and inverses are unique given in Chapter 3.

Theorem 6.2.1: Let $(\mathcal{G}, \circ)$ be a group. Then

(i) The identity element e is unique.

(ii) $\forall\, a \in \mathcal{G}$, a^{-1} is unique.

Proof (Uniqueness Proof): Let $(\mathcal{G}, \circ)$ be a group.

Proof of part (i): Let e be an identity element in $\mathcal{G}$ and suppose that e is not the unique identity element. Let e_2 be any other identity element in $\mathcal{G}$ (i.e., $e \neq e_2$). Then

$$\underbrace{e \circ e_2 = e_2}_{\text{Since } e \text{ is an identity}}$$

and

$$\underbrace{e \circ e_2 = e}_{\text{Since } e_2 \text{ is an identity}}$$

Hence, $e = e \circ e_2 = e_2$, contradicting $e \neq e_2$.

Therefore, the identity element is unique.

Proof of part (ii): Let $a \in \mathcal{G}$ be ABF and suppose that a^{-1} is an inverse of a and a^{-1} is not unique. Let a_2 be any other inverse of a that is in $\mathcal{G}$ (i.e., $a^{-1} \neq a_2$). Then

$$\underbrace{a \circ a^{-1} = e = a^{-1} \circ a}_{a^{-1} \text{ is an inverse of } a}$$

and

$$\underbrace{a \circ a_2 = e = a_2 \circ a}_{a_2 \text{ is an inverse of } a}$$

Now

$$a^{-1} = a^{-1} \circ e = a^{-1} \circ \underbrace{(a \circ a_2)}_{e} = \underbrace{(a^{-1} \circ a) \circ a_2}_{\text{Since } \circ \text{ is associative}} = e \circ a_2 = a_2$$

Thus, $a^{-1} = a_2$ contradicting $a^{-1} \neq a_2$, and therefore a^{-1} is the unique inverses of a.

■

The identity element and the inverse elements are not only unique but are also the only elements in $\mathcal{G}$ that are guaranteed to commute with all the other elements of $\mathcal{G}$. When every element in a group commutes with every other element of the group, the group is called an *Abelian group*; that is, when $(\mathcal{G}, \circ)$ is a group and $\circ$ is an Abelian operator, then $(\mathcal{G}, \circ)$ is said to be an Abelian group.

Definition 6.2.2: A group $(\mathcal{G}, \circ)$is said to be an *Abelian group* if and only if $a \circ b = b \circ a$, $\forall\ a, b \in \mathcal{G}$.

As noted above, the group structure does not guarantee that $a \circ b = b \circ a$. However, in an Abelian group $a \circ b = b \circ a$, $\forall\ a, b \in \mathcal{G}$. An example of an Abelian group is $(\mathbb{Z}, +)$, and an example of a non-Abelian group is given in the following example.

Example 6.2.6: Let $\mathcal{G}$ the collection of 2×2 invertible matrices and $\circ = \times$ (the matrix multiplication operator). Then, $(\mathcal{G}, \times)$ is a group but not an Abelian group. To see that $(\mathcal{G}, \times)$ is not an Abelian group, let

$$A = \begin{bmatrix} 2 & 0 \\ 0 & 1 \end{bmatrix} \quad \text{and} \quad B = \begin{bmatrix} 1 & 2 \\ 0 & 3 \end{bmatrix}$$

Then

$$AB = \begin{bmatrix} 2 & 4 \\ 0 & 3 \end{bmatrix} \neq BA = \begin{bmatrix} 2 & 2 \\ 0 & 3 \end{bmatrix}$$

Thus, $(\mathcal{G}, \circ)$ is not an Abelian group.

Now, it is well known that if $a, b, c \in \mathbb{R}$ and $a + b = a + c$, then $b = c$, and similarly, when $ab = ac$, then it follows that $b = c$ whenever $a \neq 0$. Note that these results are easily verified using subtraction and division, even though the original equations were based on addition and multiplication. The following theorem shows that in the group structure it is also true that when $a \circ b = a \circ c$, then it also follows that $b = c$; however, the proof of this result must be based on arguments that use only the operator $\circ$. In particular, Theorem 6.2.2 shows that there are left and right cancellation laws for a group.

Theorem 6.2.2 (Cancellation Laws): Let $(\mathcal{G}, \circ)$ be a group. If $a, b, c \in \mathcal{G}$ and

(i) $a \circ b = a \circ c$, then $b = c$ (left cancellation).

(ii) $b \circ a = c \circ a$, then $b = c$ (right cancellation).

Proof: Let $(\mathcal{G}, \circ)$ be a group, and let $a, b, c \in \mathcal{G}$ be ABF.

Proof of part (i): Suppose that $a \circ b = a \circ c = d$. Then

$$a^{-1} \circ d = \underbrace{a^{-1} \circ (a \circ b) = (a^{-1} \circ a) \circ b}_{\text{Since } \circ \text{ is associative}} = b$$

and

$$a^{-1} \circ d = \underbrace{a^{-1} \circ (a \circ c) = (a^{-1} \circ a) \circ c}_{\text{Since } \circ \text{ is associative}} = c$$

Thus, $b = a^{-1} \circ d = c$, and hence, $b = c$.

Proof of part (ii): The proof of part (ii) is similar to the proof of part (i) and is left as an exercise.

■

Example 6.2.7: Let $3\mathbb{Z} = \{x : x = 3z \text{ for some integer } z\}$, and let $a \circ b = a + b$ (ordinary addition). Show that $(3\mathbb{Z}, +)$ forms a group.

Solution: Let $3\mathbb{Z} = \{x : x = 3z \text{ for some integer } z \in \mathbb{Z}\}$, and let $a \circ b = a + b$.

Let $a, b \in 3\mathbb{Z}$. Then, $a = 3z_1$ and $b = 3z_2$ for some integers z_1 and z_2. Now, $a + b = 3z_1 + 3z_2 = 3(z_1 + z_2) \in 3\mathbb{Z}$. Thus, $3\mathbb{Z}$ is closed under addition. Ordinary addition is associative on $\mathbb{Z}$, and hence $\circ$ is associative on $3\mathbb{Z}$.

$0 \in 3\mathbb{Z}$ and 0 is the additive identity.

Finally, let $a \in 3\mathbb{Z}$. Then, $a = 3z$ for some integer z and $-a = 3(-z) \in 3\mathbb{Z}$. Thus, the additive inverses are in $3\mathbb{Z}$, $\forall\, a \in 3\mathbb{Z}$, and hence, $(3\mathbb{Z}, +)$ is a group.

Note that for a group $(\mathcal{G}, \circ)$, the fact that $\mathcal{G}$ is closed under the operation $\circ$ means that $a \circ b$ is always in $\mathcal{G}$ and hence, since all the inverses are also in $\mathcal{G}$ it follows that $(a \circ b)^{-1}$ is in $\mathcal{G}$, also. In Theorem 6.2.3 $(a \circ b)^{-1}$ is shown to be equal to $b^{-1} \circ a^{-1}$ and $(a^{-1})^{-1}$ is shown to be equal to a.

Theorem 6.2.3: Let $(\mathcal{G}, \circ)$ be a group, and let $a, b \in \mathcal{G}$. Then

(i) $(a \circ b)^{-1} = b^{-1} \circ a^{-1}$.

(ii) $\left(a^{-1}\right)^{-1} = a$.

Proof: Let $(\mathcal{G}, \circ)$ be a group, and let $a, b \in \mathcal{G}$ be ABF.

Proof of part (i): Note that since inverses are unique, by showing that $b^{-1} \circ a^{-1}$ is an inverse of $a \circ b$, part (i) follows. Now

$$(a \circ b) \circ (b^{-1} \circ a^{-1}) = a \circ (b \circ b^{-1}) \circ a^{-1} = a \circ e \circ a^{-1} = a \circ a^{-1} = e$$

$$(b^{-1} \circ a^{-1}) \circ (a \circ b) = b^{-1} \circ (a^{-1} \circ a) \circ b = b^{-1} \circ e \circ b = b^{-1} \circ b = e$$

Thus, $b^{-1} \circ a^{-1}$ is an inverse of $a \circ b$, and therefore, since inverses are unique, it follows that $(a \circ b)^{-1} = b^{-1} \circ a^{-1}$.

Proof of part (ii): The proof of part (ii) is left as an exercise.

■

Note that unless a group is Abelian, it will not be the case that

$$(a \circ b)^{-1} = a^{-1} \circ b^{-1}, \ \forall\, a, b \in \mathcal{G}$$

However, if $(\mathcal{G}, \circ)$ is an Abelian group, then it does follow that

$$(a \circ b)^{-1} = a^{-1} \circ b^{-1}, \ \forall\, a, b \in \mathcal{G}$$

For instance, in the groups $(\mathbb{R}, +)$ and $(\mathbb{R}^{+}, \times)$, which are Abelian, it does follow that

$$(a \circ b)^{-1} = a^{-1} \circ b^{-1}$$

In particular, $(a + b)^{-1} = -a + -b$ for $(\mathbb{R}, +)$ and $(a \times b)^{-1} = 1/a \times 1/b$ for $(\mathbb{R}^{+}, \times)$.

Corollary to Theorem 6.2.3: If $(\mathcal{G}, \circ)$ is an Abelian group and $a, b \in \mathcal{G}$, then $(a \circ b)^{-1} = a^{-1} \circ b^{-1}$.

> ***Proof:*** Since $(\mathcal{G}, \circ)$ is an Abelian group, the Theorem 6.2.3 corollary follows directly from Theorem 6.2.3.

■

Example 6.2.8: Let $\mathcal{G}$ be the collection of 2×2 invertible matrices. Then, $(AB)^{-1} = B^{-1}A^{-1}$. Furthermore, since matrix multiplication is not an Abelian operator, it is generally the not the case that

$$(AB)^{-1} = B^{-1}A^{-1} = A^{-1}B^{-1}$$

For example, let

$$A = \begin{bmatrix} 1 & -1 \\ 0 & 1 \end{bmatrix}, \quad B = \begin{bmatrix} 1 & 2 \\ 1 & 3 \end{bmatrix}$$

Then

$$AB = \begin{bmatrix} 0 & -1 \\ 1 & 3 \end{bmatrix} \text{ and } (AB)^{-1} = \begin{bmatrix} 3 & 1 \\ -1 & 0 \end{bmatrix}$$

However

$$A^{-1} = \begin{bmatrix} 1 & 1 \\ 0 & 1 \end{bmatrix}, \quad B^{-1} = \begin{bmatrix} 3 & -2 \\ -1 & -1 \end{bmatrix}$$

and

$$A^{-1}B^{-1} = \begin{bmatrix} 2 & -1 \\ -1 & 3 \end{bmatrix} \neq (AB)^{-1} = \begin{bmatrix} 3 & 1 \\ -1 & 0 \end{bmatrix}$$

The next two theorems are generalizations of Theorem 6.2.3. In particular, Theorem 6.2.4 shows how to compute $(a \circ b \circ c)^{-1}$, and Theorem 6.2.5 shows how to compute $(a_1 \circ a_2 \circ a_3 \circ \cdots \circ a_n)^{-1}$ for any value of $n \in \mathbb{N}$.

Theorem 6.2.4: Let $(\mathcal{G}, \circ)$ be a group and let $a, b, c \in \mathcal{G}$. Then,

$$(a \circ b \circ c)^{-1} = c^{-1} \circ b^{-1} \circ a^{-1}$$

Proof: The proof of Theorem 6.2.4 is left as an exercise.

■

Theorem 6.2.5: Let $(\mathcal{G}, \circ)$ be a group, and let $a_i \in \mathcal{G}$ for all $i \in \mathbb{N}$. Then,

$$(a_1 \circ a_2 \circ a_3 \circ \cdots \circ a_n)^{-1} = a_n^{-1} \circ a_{n-1}^{-1} \circ \cdots \circ a_2^{-1} \circ a_1^{-1}, \ \forall\, n \in \mathbb{N}$$

Proof: Mathematical induction can be used to prove Theorem 6.2.5. The details of the proof are left as an exercise.

■

Much of the early work leading to the formalization of the area of group theory dealt with the problems associated with solving algebraic equations. Furthermore, the utility of any algebraic structure is based on the ability to solve equations within the structure. The following theorem shows that the group structure is an algebraic structure that allows for the solvability of equations such as $a \circ x = b$ and $x \circ a = b$.

Theorem 6.2.6: Let $(\mathcal{G}, \circ)$ be a group. If $a, b \in \mathcal{G}$, then there exists an element $x \in \mathcal{G}$ such that

(i) $x \circ a = b$.

(ii) $a \circ x = b$.

Proof (Existence Proof): Let $(\mathcal{G}, \circ)$ be a group, and let $a, b \in \mathcal{G}$ be ABF.

Proof of part (i): Since $a, b \in \mathcal{G}$, it follows that $a^{-1} \in \mathcal{G}$, and thus so is $b \circ a^{-1}$. Furthermore

$$(b \circ a^{-1}) \circ a = b \circ (a^{-1} \circ a) = b \circ e = b$$

Thus, $x = b \circ a^{-1} \in \mathcal{G}$ is a solution to the equation $x \circ a = b$.

Proof of part (ii): The proof of part (ii) is left as an exercise.

■

Note that the solutions to the equations $a \circ x = b$ and $x \circ a = b$ had to be constructed in the scratchwork leading up to the proof of Theorem 6.2.6. Similar constructions can be created to solve many equations encountered when working with a group structure. For example, the solution to the equation $x \circ a \circ b \circ c = d$ would be $x = d \circ (a \circ b \circ c)^{-1} = d \circ c^{-1} \circ b^{-1} \circ a^{-1}$, and the solution to the equation $a \circ x \circ b \circ c = d$ would be $x = a^{-1} \circ d \circ (b \circ c)^{-1} = a^{-1} \circ d \circ c^{-1} \circ b^{-1}$. Also, the properties of a group that depend only on the binary operator $\circ$, and not the individual elements of the group, are called the *algebraic properties* of the group. The results in Theorems 6.2.3–6.2.6 are all algebraic properties of a group.

Example 6.2.9: Recall that $(\mathbb{R}^+, \times)$ is a group.

a. Let $a, b \in \mathbb{R}^+$. Solve $a \circ x = b$ for x.
b. Solve $2 \circ x = 0.4$ for x.
c. Solve $(x \circ 5) \circ x = 20$ for x.

Solutions:

a. The solution to $a \circ x = b$ is found by solving the equation $ax = b$, which yields $x = \dfrac{b}{a}$.
b. The solution to $2 \circ x = 0.4$ is found by solving the equation $2x = 0.4$, which yields $x = 0.2$.
c. The solution to $(x \circ 5) \circ x = 20$ is found by solving the equation $5x^2 = 20$, which yields $x = 2$.

Now, recall that the definition of a group requires that for every element $a \in \mathcal{G}$, the identity element e and the inverse element a^{-1} must both commute with a. Thus, to show that $(\mathcal{G}, \circ)$ forms a group, it must be shown that for every $a \in \mathcal{G}$ that $a \circ e = e \circ a = a$ and $a \circ a^{-1} = a^{-1} \circ a = e$. However, the next theorem states that $(\mathcal{G}, \circ)$ is a group if $\mathcal{G}$ is closed under and associative operator $\circ$ and for every $a \in \mathcal{G}$ there is a left identity and a left inverse.

Theorem 6.2.7: Let $\mathcal{G}$ be a set and $\circ$ a binary operation. Then, $(\mathcal{G}, \circ)$ forms a group provided that the following conditions are satisfied:

(i) $\mathcal{G}$ is closed under $\circ$.

(ii) $\circ$ is associative on $\mathcal{G}$.

(iii) $\exists\ e \in \mathcal{G}$ such that $e \circ a = a$, $\forall\ a \in \mathcal{G}$ (left identity).

(iv) $\forall\ a \in \mathcal{G}$, $\exists\ \ x \in \mathcal{G}$ such that $x \circ a = e$ (left inverse).

Note that if $\mathcal{G}$ is a set, $\circ$ is a binary operator, and assuming that conditions (i)–(iv) hold, then, since $\mathcal{G}$ is closed under $\circ$ and $\circ$ is associative on $\mathcal{G}$, it is necessary only to show that the following two conditions are true in order to prove that $(\mathcal{G}, \circ)$ is a group.

1. For every element $a \in \mathcal{G}$, $a \circ e = a$.
2. For every element $a \in \mathcal{G}$, $a \circ x = e$

Before proving conditions 1 and 2, it will first be shown, in Lemma 6.2.1, that left cancellation holds on $\mathcal{G}$ with $\circ$ provided that conditions (i)–(iv) are true. This lemma will then be used in the proof of Theorem 6.2.7.

Lemma 6.2.1: Let $a, b, c \in \mathcal{G}$. If conditions (i)–(iv) of Theorem 6.2.7 are satisfied and $a \circ b = a \circ c$, then $b = c$ (left cancellation).

> ***Proof:*** Let $a, b, c \in \mathcal{G}$ be ABF. Suppose that conditions (i)–(iv) are true and that $a \circ b = a \circ c$.
>
> Since $a \in \mathcal{G}$, by condition (iv), it follows that there exists an element $x \in \mathcal{G}$ such that $x \circ a = e$. Moreover, since $a \circ b = a \circ c$, it follows that $x \circ (a \circ b) = x \circ (a \circ c)$. Now, since $\circ$ is associative, it follows that
> $$x \circ (a \circ b) = (x \circ a) \circ b = e \circ b = b$$
> and
> $$x \circ (a \circ c) = (x \circ a) \circ c = e \circ c = c$$
> Thus, $b = c$ and left cancellation holds on $\mathcal{G}$ whenever conditions (i)–(iv) are true.
>
> ∎

The proof of Theorem 6.2.7 is given below.

> ***Proof:*** Let $\mathcal{G}$ be a set, $\circ$ a binary operator, and suppose that conditions (i)–(iv) hold.
>
> **Condition 1:** Let $a \in \mathcal{G}$ be ABF. To prove condition 1, it must be shown that $a \circ e = e \circ a = a$, $\forall\ a \in \mathcal{G}$. Since $a \in \mathcal{G}$ by condition (iv) there exists $x \in \mathcal{G}$ such that $e = x \circ a$.

Consider e:

$$e = \underbrace{e \circ e}_{\text{Since } e \text{ is a left identity}} = \underbrace{(x \circ a) \circ e}_{\text{Since } e = x \circ a}$$

$$= \underbrace{x \circ (a \circ e)}_{\text{Since } \circ \text{ is associative}}$$

Also, since $e = x \circ a$, it follows that $e = x \circ a = x \circ (a \circ e)$ and hence, $x \circ a = x \circ (a \circ e)$. Now, applying Lemma 6.2.7 (i.e., left cancellation), it follows that $a = a \circ e$, $\forall\, a \in \mathcal{G}$. Hence, e is an identity element in $\mathcal{G}$.

Condition 2: Let $a \in \mathcal{G}$ be ABF. Then, $\exists\, x \in \mathcal{G}$ such that $x \circ a = e$. Consider $x \circ e$.

$$x \circ e = \underbrace{e \circ x}_{\text{By condition 1}} = \underbrace{(x \circ a)}_{e} \circ x = \underbrace{x \circ (a \circ x)}_{\text{By associativity}}$$

Thus, $x \circ e = x \circ (a \circ x)$ and by left cancellation it follows that $e = a \circ x$. Hence x is an inverse of a, and therefore $(\mathcal{G}, \circ)$ is a group.

■

A similar result holds for a set $\mathcal{G}$ that is closed under an associative binary operator $\circ$ and with the properties for the left identity and the left inverses replaced by a right identity and right inverses.

Theorem 6.2.8: Let $\mathcal{G}$ be a set and $\circ$ a binary operation. Then, $(\mathcal{G}, \circ)$ forms a group provided that the following conditions are satisfied:

(i) $\mathcal{G}$ is closed under $\circ$.

(ii) $\circ$ is associative on $\mathcal{G}$.

(iii) $\exists\, e \in \mathcal{G}$ such that $a \circ e = a$, $\forall\, a \in \mathcal{G}$ (right identity).

(iv) $\forall\, a \in \mathcal{G}$, $\exists\, x \in \mathcal{G}$ such that $a \circ x = e$ (right inverse).

Proof: The proof of Theorem 6.2.8 is left as an exercise.

■

The result of Theorems 6.2.7 and 6.2.8 provide shortcuts in verification that the pair $(\mathcal{G}, \circ)$ is a group. Thus, to show that $(\mathcal{G}, \circ)$ is a group, it is necessary only to show that

1. $\mathcal{G}$ is closed under $\circ$.
2. $\circ$ is associative on $\mathcal{G}$.
3. $\exists\ e \in \mathcal{G}$ such that $a \circ e = a$ (or $e \circ a$), $\forall\ a \in \mathcal{G}$.
4. $\forall\ a \in \mathcal{G},\ \exists\ \ x \in \mathcal{G}$ such that $a \circ x = e$ (or $x \circ a = e$).

Another important characteristic of a group is its *order*. The definition of the order of a group $(\mathcal{G}, \circ)$ is given below.

Definition 6.2.3: The *order* of a group $(\mathcal{G}, \circ)$ is the number of elements in the set $\mathcal{G}$. If $\mathcal{G}$ is infinite then $(\mathcal{G}, \circ)$ is said to have *infinite order*. The order of a group $\mathcal{G}$ is denoted by $|\mathcal{G}|$.

Note that the order of a group is simply the number of elements in the group. For example, if $\mathcal{E} = \{e\}$ and $\circ$ defined by $e \circ e = e$, then $(\mathcal{E}, \circ)$ is a group with order $|\mathcal{E}| = 1$. A group of the form $(\mathcal{E}, \circ)$ is called a *trivial group*. Two examples of trivial groups are $(\{0\}, +)$ and $(\{1\}, \times)$. Examples of groups having infinite order are $(\mathbb{R}, +)$ and $(\mathbb{R}^+, \times)$.

Moreover, when two groups have the same order, it is possible, but not guaranteed, that the two groups are structurally identical. Specifically, when two groups are structurally similar, it follows that except for the names of the elements, the algebraic properties of the two groups are identical. For example, every trivial group has the same algebraic structure; however, $(\mathbb{Z}, +)$ and $(\mathbb{R}, +)$ are not structurally the same even though the order of both of these groups is infinite. Two groups that are structurally similar are said to be *isomorphic*; isomorphic groups are discussed in most introductory texts on modern algebra and group theory (e.g., see *Classic Algebra* by P. M. Cohn (2000)).

Now, it can be shown that there are groups of order n for any natural number n. An example of a group of order n is $(\mathbb{Z}_n, \oplus)$, where $\oplus$ is *modular arithmetic* and $\mathbb{Z}_n := \{0, 1, 2, \ldots, n-1\}$. Modular arithmetic base n, also called *clock arithmetic* base n, is a system of arithmetic that can be used on the natural numbers by defining $a \oplus b = r$, where r is the Division Algorithm remainder from the unique representation $a + b = qn + r$. For example, $5 \oplus 3 = 2$ in modular base 6 because $5 + 3 = 8 = 1 \cdot 6 + 2$, and $11 \oplus 8 = 1$ in modular base 6 because $11 + 8 = 19 = 3 \cdot 6 + 1$.

Theorem 6.2.9: If n is a natural number, then $(\mathbb{Z}_n, \oplus)$ forms a group of order n.

Proof: Let $a, b \in \mathbb{Z}_n$ be ABF and consider $a \oplus b$. By definition, $a \oplus b = r$ where $a + b = qn + r$ and $0 \leq r < n$. Thus, since $0 \leq r < n$, it follows that $a + b \in \mathbb{Z}_n$ and therefore, $\mathbb{Z}_n$ is closed under $\oplus$.

Now, $\oplus$ is associative; however, the details showing that $\oplus$ is an associative operator are left as an exercise.

Since $0 \in \mathbb{Z}_n$ and since $0 \oplus a = a$ and $a \oplus 0 = a$, it follows that 0 is an identity element in $\mathbb{Z}_n$.

Suppose that $a \in \mathbb{Z}_n$; then so is $n - a$ since $0 \leq n - a < n$. Furthermore, $a \oplus (n - a) = 0$ since $a + (n - a) = n$, which has remainder 0. Therefore, $a^{-1} = n - a \in \mathbb{Z}_n$ whenever a is.

Thus, $(\mathbb{Z}_n, \oplus)$ forms a group. Furthermore, since $\mathbb{Z}_n$ consists of n elements, it follows that it is a group of order n.

■

Example 6.2.10: Let $(\mathcal{G}, \circ) = (\mathbb{Z}_9, \oplus)$. Determine

a. $3 \oplus 8$.

b. $5 \oplus 8 \oplus 6$.

c. 3^{-1}.

d. x such that $x \oplus 2 \oplus 8 \oplus 6 = 1$.

Solutions:

a. $3 \oplus 8 = 2$ since $3 + 8 = 11 = 1 \cdot 9 + 2$.

b. $5 \oplus 8 \oplus 6 = 1$ since $5 + 8 + 6 = 19 = 2 \cdot 9 + 1$.

c. $3^{-1} = 6$ since $3 + 6 = 9 = 1 \cdot 9 + 0$ and 0 is the identity element in $\mathbb{Z}_9$.

d. $x = 3$ since $3 + 2 + 8 + 6 = 19 = 2 \cdot 9 + 1$, and hence $3 \oplus 2 \oplus 8 \oplus 6 = 1$.

Theorem 6.2.10: $(\mathbb{Z}_n, \oplus)$ is an Abelian group.

Proof: The proof of Theorem 6.2.10 is left as an exercise.

■

Example 6.2.11: Let $\mathcal{G} = \{1, 2, 3, 4, 5\}$ and $\circ = \otimes$, where $a \otimes b = r$ where r is the Division Algorithm remainder in the equation $ab = qn + r$.

a. Show that $(\mathcal{G}, \otimes)$ forms a group.

b. Determine 4^{-1}.

c. Solve $2 \otimes 5 = 3$.

Solutions: The solutions to Example 6.2.11 are left as exercises.

Note that $(\mathbb{Z}_n, \otimes)$ cannot be a group because there will be no inverse for 0. However, a theorem analogous to Theorem 6.2.10 stating that the pair $(\{1, 2, 3, \ldots, n-1\}, \otimes)$ forms a group is possible.

6.2.2 Subgroups

Recall that $\mathbb{Z} \subset \mathbb{R}$ and $(\mathbb{R}, +)$ and $(\mathbb{Z}, +)$ are both groups. An important question concerning a group $(\mathcal{G}, \circ)$ is "When does a subset $\mathcal{H}$ of $\mathcal{G}$, with the operator $\circ$, form a group?" When $(\mathcal{G}, \circ)$ is a group and a subset $\mathcal{H}$ of $\mathcal{G}$ forms a group with $\circ$, then $(\mathcal{H}, \circ)$ is called a *subgroup* of $(\mathcal{G}, \circ)$.

Definition 6.2.4: Let $(\mathcal{G}, \circ)$ be a group, and let $\mathcal{H}$ be a subset of $\mathcal{G}$. The pair $(\mathcal{H}, \circ)$ is said to be a *subgroup* of $(\mathcal{G}, \circ)$ if and only if $(\mathcal{H}, \circ)$ is a group.

For example, $(\mathbb{Z}, +)$ and $(\mathbb{Q}, +)$ are subgroups of $(\mathbb{R}, +)$ and $(\mathbb{Q}^+, \times)$ is a subgroup of $(\mathbb{R}^+, \times)$. However, $(\mathbb{I}^+, \times)$ is not a subgroup of $(\mathbb{R}^+, \times)$ since $\mathbb{I}$ is not closed under multiplication.

Example 6.2.12: Let $(\mathcal{G}, \circ) = (\mathbb{Z}, +)$, and let $k \in \mathbb{N}$. Define $k\mathbb{Z}$ as follows:

$$k\mathbb{Z} = \{kz : z \in \mathbb{Z}\}$$

Then, $k\mathbb{Z}$ is simply the set consisting of the integer multiples of k. For example

$$2\mathbb{Z} = \{2z : z \in \mathbb{Z}\} = \{0, \pm 2, \pm 4, \ldots\}$$

and

$$3\mathbb{Z} = \{3z : z \in \mathbb{Z}\} = \{0, \pm 3, \pm 6, \ldots\}$$

Show that $(k\mathbb{Z}, +)$ is a subgroup of $(\mathbb{Z}, +)$, $\forall\ k \in \mathbb{N}$.

Solution: Let $k \in \mathbb{N}$ be ABF. Clearly, it follows that $k\mathbb{Z}$ is a subset of $\mathbb{Z}$. Now, to show that $(k\mathbb{Z}, +)$ is a subgroup of $(\mathbb{Z}, +)$, it must be shown that $(k\mathbb{Z}, +)$ is also a group.

Let $a, b \in k\mathbb{Z}$. Then, there exist integers i and j such that $a = ki$ and $b = kj$. Consider $a + b$

$$a + b = ki + kj = k(i + j) = kl$$

where $l = i + j \in \mathbb{Z}$, and hence $a + b \in k\mathbb{Z}$ whenever $a, b \in k\mathbb{Z}$. Therefore, $k\mathbb{Z}$ is closed under ordinary addition. Note that the associative property holds on $k\mathbb{Z}$ since ordinary addition is associative on $\mathbb{Z}$.

0 is the additive identity and $0 \in k\mathbb{Z}$ since $0 = k \cdot 0$. Thus, $k\mathbb{Z}$ does contain an identity element. Let $a \in k\mathbb{Z}$ be ABF. Then, $a = kz$ for some integer z. Now, $-a = k(-z)$, and therefore $-a$ is in $k\mathbb{Z}$ whenever a is. Since $-a$ is the additive inverse of a, it follows that $k\mathbb{Z}$ contains all the additive inverses.

Therefore, $(k\mathbb{Z}, +)$ is a group, $\forall\ k \in \mathbb{N}$.

Note that the even integers with ordinary addition (i.e., $(2\mathbb{Z}, +)$) forms a group, as does $(3\mathbb{Z}, +)$, the multiples of 3 with addition. Furthermore, since $2\mathbb{Z} \cap 3\mathbb{Z} = 6\mathbb{Z}$, it follows that $(6\mathbb{Z}, +)$ is a subgroup of both $(2\mathbb{Z}, +)$ and $(3\mathbb{Z}, +)$.

Example 6.2.13: Show that $2\mathbb{Z} \cap 3\mathbb{Z} = 6\mathbb{Z}$.

Solution: The solution to Example 6.2.13 is left as an exercise.

Theorem 6.2.11: Let $(\mathcal{G}, \circ)$ be a group. If $(\mathcal{H}_1, \circ)$ and $(\mathcal{H}_2, \circ)$ are subgroups of $(\mathcal{G}, \circ)$, then $(\mathcal{H}_1 \cap \mathcal{H}_2, \circ)$ is also a subgroup of $(\mathcal{G}, \circ)$.

> ***Proof:*** Let $(\mathcal{G}, \circ)$ be a group, and suppose that $(\mathcal{H}_1, \circ)$ and $(\mathcal{H}_2, \circ)$ are subgroups of $(\mathcal{G}, \circ)$.
>
> Let $a, b \in \mathcal{H}_1 \cap \mathcal{H}_2$. Then, it follows that $a, b \in \mathcal{H}_1$ and $a, b \in \mathcal{H}_2$. Now, since $(\mathcal{H}_1, \circ)$ and $(\mathcal{H}_2, \circ)$ are groups, $a \circ b \in \mathcal{H}_1$ and $a \circ b \in \mathcal{H}_2$ since both sets are closed with respect to the operator $\circ$. Thus, $a \circ b \in \mathcal{H}_1 \cap \mathcal{H}_2$. Therefore, $\mathcal{H}_1 \cap \mathcal{H}_2$ is closed under $\circ$.
>
> Note that $\circ$ is associative on $\mathcal{G}$, and therefore $\circ$ is associative on $\mathcal{H}_1 \cap \mathcal{H}_2$.
>
> Now, since $(\mathcal{H}_1, \circ)$ and $(\mathcal{H}_2, \circ)$ are groups, it follows that $e \in \mathcal{H}_1$ and $e \in \mathcal{H}_2$. Thus, $e \in \mathcal{H}_1 \cap \mathcal{H}_2$, and therefore $\mathcal{H}_1, \cap \mathcal{H}_2$ contains the identity element.
>
> Finally, let $a \in \mathcal{H}_1 \cap \mathcal{H}_2$ be ABF. Since $(\mathcal{H}_1, \circ)$ and $(\mathcal{H}_2, \circ)$ are groups, it follows that $a^{-1} \in \mathcal{H}_1$ and $a^{-1} \in \mathcal{H}_2$. Hence, $a^{-1} \in \mathcal{H}_1 \cap \mathcal{H}_2$, and therefore $\mathcal{H}_1 \cap \mathcal{H}_2$ contains a^{-1} whenever $a \in \mathcal{H}_1 \cap \mathcal{H}_2$.
>
> Thus, $(\mathcal{H}_1, \cap \mathcal{H}_2, \circ)$ forms a subgroup of $(\mathcal{G}, \circ)$.

■

Theorem 6.2.11 shows that when $(\mathcal{G}, \circ)$ is a group, so is the intersection of any two subgroups of $(\mathcal{G}, \circ)$. For instance, in Examples 6.2.12 and 6.2.13 it was shown that $(2\mathbb{Z}, +)$ and $(3\mathbb{Z}, +)$ are subgroups of $(\mathbb{Z}, +)$ and that $6\mathbb{Z} = 2\mathbb{Z} \cap 3\mathbb{Z}$; therefore, from these facts it follows that $(6\mathbb{Z}, +)$ is also a subgroup of $(\mathbb{Z}, +)$.

On the other hand, the union of two subgroups of $(\mathcal{G}, \circ)$ is not necessarily a subgroup of $(\mathcal{G}, \circ)$. A counterexample showing that the union of two subgroups of $(\mathcal{G}, \circ)$ is not always a subgroup is given in Example 6.2.14.

Example 6.2.14: Note that $(2\mathbb{Z}, +)$ and $(3\mathbb{Z}, +)$ are subgroups of $(\mathbb{Z}, +)$ and

$$2\mathbb{Z} \cup 3\mathbb{Z} = \{x : x = 2z \text{ or } x = 3z, \text{ where } z \in \mathbb{Z}\}$$

Now, while $2 \in 2\mathbb{Z}$ and $3 \in 3\mathbb{Z}$, $2 + 3 = 5 \notin 2\mathbb{Z} \cup 3\mathbb{Z}$ since there is no integer z such that $2z = 5$ or $3z = 5$. Hence, $2\mathbb{Z} \cup 3\mathbb{Z}$ is not closed under $\circ$, and thus $(2\mathbb{Z} \cup 3\mathbb{Z}, +)$ is not a subgroup of $(\mathbb{Z}, +)$.

For the ordinary multiplication operator an expression such as $a \times a \times a \times a$ will often written in shorthand as a^4. Analogously, with a group $(\mathcal{G}, \circ)$ and an element $a \in \mathcal{G}$, an expression such as $a \circ a \circ a \circ a$ will commonly be written as a^4, also. The *integral powers* for an element a of a group are defined below.

Definition 6.2.5: Let $(\mathcal{G}, \circ)$ be a group and $a \in \mathcal{G}$. The *integral powers* of the element a are defined as follows:

(i) $a^0 = e$

(ii) $a^1 = a$

(iii) $a^n = a^{n-1} \circ a, \ n \in \mathbb{N}$

(iv) $a^{-n} = (a^{-1})^n, \ n \in \mathbb{N}$

Note that this definition does not explicitly state that

$$a^n = \underbrace{a \circ a \circ \cdots \circ a}_{n \text{ times}}$$

for $n \in \mathbb{N}$. However, since $a^2 = a \circ a$, it follows that $a^3 = a^2 \circ a = a \circ a \circ a$, and generalizing from here on, it is clear that

$$a^n = \underbrace{a \circ a \circ \cdots \circ a}_{n \text{ times}}$$

and

$$a^{-n} = \underbrace{a^{-1} \circ a^{-1} \circ \cdots \circ a^{-1}}_{n \text{ times}}$$

Furthermore, from Definition 6.2.5 it can be deduced that $a^n \circ a^m = a^{n+m}$ and $(a^n)^m = a^{nm}$ for $n, m \in \mathbb{N}$.

Theorem 6.2.12: Let $(\mathcal{G}, \circ)$ be a group. If $a \in \mathcal{G}$, then

(i) $a^n \circ a^m = a^{n+m}$ for $n, m \in \mathbb{N}$.

(ii) $(a^n)^m = a^{nm}$ for $n, m \in \mathbb{N}$.

Proof: Let $(\mathcal{G}, \circ)$ be a group, and let $a \in \mathcal{G}$ be ABF.

Proof of part (i): Let $n \in \mathbb{N}$ be ABF, and let

$$\mathcal{P}_m := a^n \circ a^m = a^{n+m}$$

For $m = 1$, $a^n \circ a^1 = a^{n+1}$ by definition. Thus, $\mathcal{P}_1$ is true.

Now, suppose that $\mathcal{P}_k$ is true for some ABF $k \in \mathbb{N}$. This means that $a^n \circ a^k = a^{n+k}$. Furthermore, if $\mathcal{P}_{k+1}$, is true it follows that $a^n \circ a^{k+1} = a^{n+k+1}$.

Consider $a^n \circ a^{k+1}$

$$a^n \circ a^{k+1} = a^n \circ (a^k \circ a^1) = a^n \circ a^k \circ a$$

$$= \underbrace{a^{n+k}}_{\mathcal{P}_k} \circ a = \underbrace{a^{n+k+1}}_{\text{By Definition 6.2.5(iii)}}$$

Thus, $\mathcal{P}_k \rightarrow \mathcal{P}_{k+1}$, and therefore it follows that $a^n \circ a^m = a^{n+m}$, $\forall\, m \in \mathbb{N}$.

Since $n \in \mathbb{N}$ was ABF, it follows that $a^n \circ a^m = a^{n+m}$, $\forall\, n, m \in \mathbb{N}$.

Proof of part (ii): The proof of part (ii) is left as an exercise.

■

Now, for $n \in \mathbb{N}$ it follows from Definition 6.2.5 that $a^n = a^{n-1} \circ a$, and since $n = 1 + (n-1)$, it also follows from Theorem 6.2.12 that $a^n = a \circ a^{n-1}$. Furthermore, it can also be deduced from Definition 6.2.5 and Theorem 6.2.12 that $a^n \circ a^m = a^m \circ a^n$, $\forall\, n, m \in \mathbb{N}$. Also, although not proved here, it is possible to generalize Theorem 6.2.11 to similar results for the integral powers of negative numbers and also for a mixture of positive and negative integral powers. However, it is important to note that it cannot be deduced from either Definition 6.2.5 or Theorem 6.2.12, that $(a \circ b)^n = a^n \circ b^n$, unless $(\mathcal{G}, \circ)$ is an Abelian group.

Example 6.2.15: The integral powers of the elements of $(\mathbb{Z}_6, \oplus)$ are shown in Table 6.2.1.

Table 6.2.1 The Integral Powers for $(\mathbb{Z}_6, \oplus)$

a	n=0	n=1	n=2	n=3	n=4	n=5	n=6
0	0^0=0	0^1=0	0^2=0	0^3=0	0^4=0	0^5=0	0^6=0
1	1^0=0	1^1=1	1^2=2	1^3=3	1^4=4	1^5=5	1^6=0
2	2^0=0	2^1=2	2^2=4	2^3=0	2^4=2	2^5=4	2^6=0
3	3^0=0	3^1=3	3^2=0	3^3=3	3^4=0	3^5=3	3^6=0
4	4^0=0	4^1=4	4^2=2	4^3=0	4^4=4	4^5=2	4^6=0
5	5^0=0	5^1=5	5^2=4	5^3=3	5^4=2	5^5=1	5^6=0

Note that in Example 6.2.15, it turns out that $a^m = e$ for some value of m less than or equal to 6 for each of the elements in $\mathbb{Z}_6$. In particular, $1^6 = 2^3 = 3^2 = 4^3 = 5^6 = e = 0$. If $(\mathcal{G}, \circ)$ is a group and $a \in \mathcal{G}$ and $a^m = e$ for some natural number m, then element a is said to have *order* m.

Definition 6.2.6: Let $(\mathcal{G}, \circ)$ be a group and $a \in \mathcal{G}$. The *order* of an element a is defined to be the smallest natural number m such that $a^m = e$; if $a^m \neq e$ for every natural number m, then a is said to have *infinite order*.

For example, in $(\mathbb{Z}_6, \oplus)$, the order of the element 2 is 3 and the order of the element 5 is 6, while in $(\mathbb{Z}, +)$, 0 has order 1 and every other element has infinite order. The following theorem shows that the order of every element in a group of finite order n can be no larger than n.

Theorem 6.2.13: Let $(\mathcal{G}, \circ)$ be a group of order $n < \infty$. If $a \in \mathcal{G}$, then the order of a is less than or equal to n.

> ***Proof (by Contradiction):*** Let $(\mathcal{G}, \circ)$ be a group of order $n < \infty$, and let $a \in \mathcal{G}$ be ABF. Furthermore, suppose that the order of $a > n$. Since the order of $a > n$, it follows that $a^m \neq e$ for $m = 1, 2, 3, \ldots, n$.
>
> Consider $A = \{a^i : i = 0, 1, 2, \ldots, n\} \subset \mathcal{G}$ since $\mathcal{G}$ is closed under $\circ$. Now, since the order of $\mathcal{G}$ is n, it follows that at least two of the elements in A are equal, for otherwise A would contain $n + 1$ elements and could not be a subset of $\mathcal{G}$. Suppose that $a^j = a^k$ for some integers j and k with $0 \leq j < k \leq n$.
>
> Since $k > j$, it follows that $k = j + l$ for some natural number $l \leq n$. Now, $a^k = a^{j+l} = a^j \circ a^l$. Since $a^k = a^j$, it follows, by left

cancellation, that $a^l = e$. Hence, the order of a is less than or equal to l. Since $l \leq n$, this contradicts the assumption that the order of a is greater than n.

Therefore, the order of any element in a group of finite order n must be less than or equal to n.

■

Let $(\mathcal{G}, \circ)$ be a group. Then, since $\mathcal{G}$ is closed under $\circ$, it follows that the set of integral powers of any element $a \in \mathcal{G}$ will form a subset of $\mathcal{G}$. In particular, the set $A = \{a^k : k \in \mathbb{Z}\}$ is called the set of elements in $\mathcal{G}$ generated by the element a.

Definition 6.2.7: Let $(\mathcal{G}, \circ)$ be a group and $a \in \mathcal{G}$. The subset of elements generated by an element a is defined to be

$$\langle a \rangle = \{a^k : k \in \mathbb{N}\} = \{a^1, a^2, a^3, \ldots\}$$

For example, if $(\mathcal{G}, \circ) = (\mathbb{Z}, +)$, then $\langle 1 \rangle = \mathbb{Z}$ and $\langle -1 \rangle = \mathbb{Z}$. Also, the subset of $\mathbb{Z}$ generated by element 2 is

$$\langle 2 \rangle = \{0, \pm 2, \pm 4, \ldots\} = 2\mathbb{Z}$$

which is the set of even integers. Similarly, the subset of $\mathbb{Z}$ generated by element 4 is

$$\langle -4 \rangle = \{0, \pm 4, \pm 8, \ldots\} = 4\mathbb{Z}$$

Note that in general, for any integer k the subset of $\mathbb{Z}$ that is generated by k under ordinary addition is $\langle k \rangle = k\mathbb{Z}$.

Example 6.2.16: Let $(\mathcal{G}, \circ) = (\mathbb{R}^+, \times)$. Then

$$\left\langle \frac{1}{2^n} \right\rangle = \left\{ \frac{1}{2^n} : n \in \mathbb{Z} \right\} = \left\{ 1, \frac{1}{2}, 2, \frac{1}{4}, 4, \ldots \right\}$$

The following theorem shows that if $(\mathcal{G}, \circ)$ is a group and $a \in \mathcal{G}$, then $(\langle a \rangle, \circ)$ forms an Abelian subgroup of $(\mathcal{G}, \circ)$.

Theorem 6.2.14: Let $(\mathcal{G}, \circ)$ be a group, and let $a \in \mathcal{G}$. Then, $(\langle a \rangle, \circ)$ forms a subgroup of $(\mathcal{G}, \circ)$.

Proof: Let $(\mathcal{G}, \circ)$ be a group, and let $a \in \mathcal{G}$ be ABF.

Let $b, c \in \langle a \rangle$ be ABF. Then, there exist integers j and k such that $b = a^j$ and $c = a^k$. Consider $b \circ c$:

$$b \circ c = a^j \circ a^k = \underbrace{a^{j+k}}_{\text{By Theorem 6.2.12(i)}} \in \langle a \rangle$$

Hence, $\langle a \rangle$ is closed under $\circ$.

Now, suppose that $b, c, d \in \langle a \rangle$ are ABF. Then, there exist integers j,k, and l such that $b = a^j$, $c = a^k$, and $d = a^l$. Consider $(b \circ c) \circ d$ and $b \circ (c \circ d)$:

$$(b \circ c) \circ d = (a^j \circ a^k) \circ a^l = a^{j+k} \circ a^l = a^{j+k+l}$$

$$b \circ (c \circ d) = a^j \circ (a^k \circ a^l) = a^j \circ a^{k+l} = a^{j+k+l}$$

Thus, $(b \circ c) \circ d = b \circ (c \circ d)$, and hence $\circ$ is associative on $\langle a \rangle$.

Clearly, $e \in \langle a \rangle$ since $e = a^0$, and hence $\langle a \rangle$ contains the identity element.

Finally, suppose that $b \in \langle a \rangle$. Then, there exists some integer k such that $b = a^j$. Since j is an integer, so is $-j$, and thus $a^{-j} \in \langle a \rangle$. By Theorem 6.2.13, $b^{-1} = \left(a^j\right)^{-1} = a^{-j}$, and hence $b^{-1} \in \langle a \rangle$ whenever $b \in \langle a \rangle$. Thus, $\langle a \rangle$ contains all the inverses of its elements.

Therefore, $(\langle a \rangle, \circ)$ forms a group.

■

Not only does $(\langle a \rangle, \circ)$ form a group; it is also an Abelian group since

$$a^{n+m} = a^n \circ a^m = a^m \circ a^n, \ \forall\, n, m \in \mathbb{Z}$$

Furthermore, $(\langle a \rangle, \circ)$ is the smallest subgroup of $(\mathcal{G}, \circ)$ containing the element a. The subgroup $(\langle a \rangle, \circ)$ is said to be a *cyclic* subgroup of $(\mathcal{G}, \circ)$.

Definition 6.2.8: Let $(\mathcal{G}, \circ)$ be a group and a an element of $\mathcal{G}$. Then, the subgroup of $(\mathcal{G}, \circ)$ generated by a, namely, $(\langle a \rangle, \circ)$, is called the *cyclic subgroup generated by* a.

Recall that if $(\mathcal{G}, \circ)$ is a group of finite order n, then every element in $\mathcal{G}$ has order no larger than n. Furthermore, if a has order $m \leq n$, then $\langle a \rangle = \{a, a^2, a^3, \ldots, a^m\}$.

Example 6.2.17: For $(\mathbb{Z}_9, \oplus)$, determine

a. The cyclic subgroup generated by 6.
b. $\langle 5 \rangle$.
c. $\langle 8 \rangle$.

Solutions:

a. The subgroup generated by 6 is $\langle 6 \rangle = \{6, 6^2, 6^3, \ldots, 6^n\}$, where n is the order of 6. Now, the order of 6 is found by solving the equation $6^n = e$. Consider 6^n for $n = 0, 1, 2, \ldots 9$:

$$6^1 = 6, \quad 6^2 = 6 \oplus 6 = 3, \quad 6^3 = 6 \oplus 6 \oplus 6 = 0$$

Thus, the order of 6 is 3 and $(\langle 6 \rangle, \oplus) = (\{6^1, 6^2, 6^3\}, \oplus) = (\{6, 3, 0\}, \oplus)$.
b. The solution to part (b) is left as an exercise.
c. The solution to part (c) is left as an exercise.

Recall that if $|\mathcal{G}| = n < \infty$, then for any element $a \in \mathcal{G}$, $\langle a \rangle$ forms a cyclic group under the operation $\circ$. Furthermore, each of the following results holds for a group $\langle \mathcal{G}, \circ \rangle$.

Theorem 6.2.15: Let $(\mathcal{G}, \circ)$ be a group, and let $(\langle a \rangle, \circ)$ be the cyclic group generated by $a \in \mathcal{G}$. If a has order n and $b \in \langle a \rangle$, then

(i) $a^{-1} = a^{n-1}$.
(ii) $b^{-1} = a^{n-j}$ for some natural number $0 \le j \le n$.

Proof: Let $(\mathcal{G}, \circ)$ be a group, let $(\langle a \rangle, \circ)$ be the cyclic group generated by $a \in \mathcal{G}$, and suppose that a has order n.

Proof of part (i): First, since a has order n, it follows that $a^n = e$. Consider a^n:

$$e = a^n = a^{n-1} \circ a$$

Thus, $e = a^{n-1} \circ a$. Now, since $(\langle a \rangle, \circ)$ is a group, it follows that the inverses are unique, and hence $a^{-1} = a^{n-1}$.

Proof of part (ii): Let $b \in \langle a \rangle$ be ABF. Then, there exists a natural number $0 \le j \le n$ such that $b = a^j$. Now, since a has order n, it follows that $a^n = e$. Consider a^n:

$$e = a^n = a^j \circ a^{n-j} = b \circ a^{n-j}$$

Thus, $e = b \circ a^{n-j}$. Now, since $(\langle a \rangle, \circ)$ is a group, it follows that the inverses are unique, and hence $b^{-1} = a^{n-j}$.

■

Example 6.2.18: Let $(\mathcal{G}, \circ) = (\mathbb{Z}_{30}, \oplus)$. Determine

a. The order of the element 4.

b. $\langle 4 \rangle$.

c. 2^{-1}.

Solutions:

a. The order of the element 4 is 15 since $4^{15} = 0$ and 15 is the smallest natural number for which $4^n = e$.

b. The solution to part (b) is left as an exercise.

c. The solution to part (c) is left as an exercise.

EXERCISES

6.1 Let $A_i = \left[-\frac{1}{i}, 1 - \frac{1}{i}\right]$ and $B_i = \left[\frac{1}{i}, 1 + \frac{1}{i}\right]$ for $i = 1, 2, 3$. Determine

a. A_1, A_2, A_3 and B_1, B_2, B_3

b. $A_1 \cup B_1$

c. $A_3 \cap B_3$

d. $B_3 \cap A_3^c$

6.2 Let $A = 2\mathbb{Z}$ and $B = 3\mathbb{Z}$. Determine

a. $A \cap B$

b. $A \cup B$

c. A^c

d. B^c

e. $A^c \cup B^c$

f. $A^c \cap B^c$

6.3 Let $A, B \subset \Omega$. Prove that

a. If $A \not\subset B$, then $B^c \not\subset A^c$.

b. $A \cap B \subset B$.

c. $A \subset A \cup B$.

d. $B \subset A \cup B$.

e. $A \cap B \subset A \cup B$.

f. $(A \cap B)^c = A^c \cup B^c$.

g. $A \subset B$ if and only if $B^c \subset A^c$.

h. $(A \cup B) \cap (A \cap B)^c = (A \cap B^c) \cup (B \cap A^c)$.

6.4 Let A, B, C be subsets of Ω. Prove that

a. $A \cap (B \cup C) = (A \cap B) \cup (A \cap C)$.

c. $(A \cap B \cap C)^c = A^c \cup B^c \cup C^c$.

6.5 Let $A, B, C, D \subset \Omega$. Simplify each of the following expressions:

a. $(A \cup B)^c \cap B^c$

b. $(A \cup B)^c \cup B^c$

c. $A \cup (B \cap A^c) \cup (C \cap A^c \cap B^c)$

d. $(A \cap B) \cup (A \cap C) \cup (A \cap D)$
e. $(A \cup B) \cap (A \cup C) \cap (A \cup D)$
f. $(A \cup B)^c \cap (A \cup C)^c \cap (A \cup D)^c$

6.6 Let $A_i = \left[-\frac{1}{n}, \frac{1}{n}\right]$, and let $B_i = \left(0, 1 + \frac{1}{n}\right]$. Determine

a. $\bigcup_{i=1}^{\infty} A_i$

b. $\bigcup_{i=1}^{\infty} B_i$

c. $\bigcap_{i=1}^{\infty} A_i$

d. $\bigcup_{i=1}^{\infty} B_i$

6.7 Prove that if $A_i \subset \Omega$, $\forall\, i \in \mathbb{N}$, and $B \subset \Omega$, then

$$B \cap \left(\bigcup_{i=1}^{n+1} A_i\right) = \bigcup_{i=1}^{n+1} (B \cap A_i), \ \forall\, n \in \mathbb{N}.$$

6.8 Prove that if $A_i \subset \Omega$, and $B_i \subset \Omega$, $\forall\, i \in \mathbb{N}$, then

a. $\left(\bigcup_{j=1}^{m} B_j\right) \cap \left(\bigcup_{i=1}^{n} A_i\right) = \bigcup_{i=1}^{n} \bigcup_{j=1}^{m} (B_j \cap A_i).$

b. $\left(\bigcap_{j=1}^{m} B_j\right) \cup \left(\bigcap_{i=1}^{n} A_i\right) = \bigcap_{i=1}^{n} \bigcap_{j=1}^{m} (B_j \cup A_i).$

c. $\left(\bigcap_{j=1}^{m} B_j\right) \cap \left(\bigcup_{i=1}^{n} A_i\right) = \bigcup_{i=1}^{n} \bigcap_{j=1}^{m} (B_j \cap A_i).$

6.9 Let $A_1, A_2, \ldots, A_n$ be subsets of Ω with $\bigcup_{i=1}^{n} A_i = \Omega$, let $B_1 = A_1$, and let $B_i = A_i \cap \left(\bigcup_{j=1}^{i-1} A_i\right)^c$ for $i = 2, 3, \ldots, n$. Prove that $B_1, B_2, \ldots, B_n$ is a partition of Ω.

6.10 Prove that each of the following sets is a countable set:

a. $2\mathbb{Z} \cap 3\mathbb{Z}$

b. $2\mathbb{Z} \cup 3\mathbb{Z}$

c. $\{x : \sqrt{x} = b \text{ for } b \in \mathbb{Z}^+\}$

d. $\mathcal{P} = \{p \in \mathbb{Z} : p \text{ is a prime number}\}$

e. $\mathcal{C} = \{c : c = a + b\sqrt{2} \text{ for } a, b \in \mathbb{Q}\}$

6.11 Prove each of the following theorems:

a. **Theorem:** If A and B are countable sets, then $A \cup B$ is a countable set.

b. **Theorem:** If A and B are countable sets, then $A \cap B$ is a countable set.

c. **Theorem:** If A is an uncountable set and $A \subset B$, then B is an uncountable set.

d. **Theorem:** If A is an uncountable set and B is a countable set, then $A \cap B$ is a countable set.

e. **Theorem:** If A is an uncountable set and B is a countable set, then $A \cup B$ is an uncountable set.

6.12 Let $\Omega = \{(x, y) : x, y \in \mathbb{Z}\}$. Prove that Ω is a countable set.

6.13 Let $\mathcal{G} = \mathbb{R}^+$ and $a \circ b = \frac{ab}{3}$. Prove that $(\mathcal{G}, \circ)$ is a group.

6.14 Let $\mathcal{G} = \mathbb{R}^+$ and $a \circ b = \frac{a}{b}$. Prove that $(\mathcal{G}, \circ)$ does not form a group.

6.15 Let $\mathcal{G} = \mathbb{Z}$ and $a \circ b = a + b + 1$. Then

a. Prove that $a \circ -1 = -1 \circ a = a, \ \forall\, a \in \mathbb{Z}$.

b. Prove that $a \circ (-a - 2) = (-a - 2) \circ a = -1, \ \forall\, a \in \mathbb{Z}$.

c. Determine the identity element in $\mathbb{Z}$ under $\circ$.

d. Determine $a \circ (b \circ c)$.

e. Determine whether $(\mathcal{G}, \circ)$ forms a group.

6.16 Prove each of the following theorems:

a. **Theorem:** If $(\mathcal{G}, \circ)$ is a group and $a, b \in \mathcal{G}$, then $\exists\, y \in \mathcal{G}$ such that $a \circ y = b$.

b. **Theorem:** If $(\mathcal{G}, \circ)$ is a group and $a \in \mathcal{G}$, then $(a^{-1})^{-1} = a$.

c. **Theorem:** If $(\mathcal{G}, \circ)$ is a group and $(a \circ b)^{-1} = a^{-1} \circ b^{-1}, \ \forall\, a, b \in \mathcal{G}$, then $(\mathcal{G}, \circ)$ is an Abelian group.

6.17 Let $\mathcal{G}$ be a set closed under an associative binary operation $\circ$. Prove that if $\mathcal{G}$ has a right identity element under $\circ$ and a right inverse for each element under $\circ$, then $(\mathcal{G}, \circ)$ is a group.

6.18 Prove each of the following theorems:

a. **Theorem:** If $(\mathcal{H}, \circ)$ is a subgroup of $(\mathcal{G}, \circ)$, then the identity element in $\mathcal{G}$ is the identity element in $\mathcal{H}$.

b. **Theorem:** If $(\mathcal{H}, \circ)$ is a subgroup of $(\mathcal{G}, \circ)$ and $a \in \mathcal{H}$, then the inverse element of a in $\mathcal{G}$ is the inverse element of a in $\mathcal{H}$.

c. **Theorem:** Let $(\mathcal{G}, \circ)$ be a group and $\mathcal{H} \subset \mathcal{G}$. Then, $(\mathcal{H}, \circ)$ is a subgroup of $(\mathcal{G}, \circ)$ if and only if $\mathcal{H}$ is closed under $\circ$ and $a^{-1} \in \mathcal{H}$ whenever $a \in \mathcal{H}$.

6.19 Prove that if $(\mathcal{G}, \circ)$ is a group and $\mathcal{H} = \{a \in \mathcal{G} : a \circ a = e\}$, then $(\mathcal{H}, \circ)$ is a subgroup of $(\mathcal{G}, \circ)$.

6.20 Let $\mathcal{G} = \mathbb{Z}_{12}$ and $\circ = \oplus$.

a. Find the inverses of elements 3, 4, 7, and 8.

b. Find the order of elements 3, 4, 7, and 8.

c. Find $\langle 3 \rangle$ and $\langle 5 \rangle$.

d. Solve $x^2 = 1$.

e. Solve $2 \oplus x \oplus 6 = 4$.

6.21 Let $\mathcal{G} = \mathbb{Z}_{18}$ and $\circ = \oplus$.

a. Find the inverses of elements 3, 4, 7, and 8.

b. Find the order of elements 3, 4, 7, and 8.

c. Find $\langle 3 \rangle$ and $\langle 5 \rangle$.

d. Solve $x^2 = 1$.

e. Solve $12 \oplus x \oplus 16 = 4$.

6.22 Prove each of the following theorems:

a. **Theorem:** Let $(\mathcal{G}, \circ)$ be a group. If $a \in \mathcal{G}$, then $(a^n)^m = a^{nm}$, $\forall\, n, m \in \mathbb{N}$.

b. **Theorem:** If $(\mathcal{G}, \circ)$ is a cyclic group, then $(\mathcal{G}, \circ)$ is an Abelian group.

c. **Theorem:** If $(\mathcal{G}, \circ)$ is a cyclic group and $(\mathcal{H}, \circ)$ is a subgroup of $(\mathcal{G}, \circ)$, then $(\mathcal{H}, \circ)$ is a cyclic group.

6.23 Let $(\mathcal{G}, \circ)$ be a cyclic group and $a, b \in \mathcal{G}$ elements of order 2. Prove that

a. $a \circ b = b \circ a$.

b. $\{e, a, b, a \circ b\}$ is a subgroup of $(\mathcal{G}, \circ)$.

6.24 Let $(\mathcal{G}, \circ)$ be a group. Prove that if $(a \circ b)^2 = a^2 \circ b^2$, $\forall\ a, b \in \mathcal{G}$, then $(\mathcal{G}, \circ)$ is an Abelian group.

6.25 Prove that if $(\mathcal{G}, \circ)$ is a group of finite order, then there exists $n \in \mathbb{N}$ such that $a^n = e$, $\forall\ a \in \mathcal{G}$.

6.26 Let $(\mathcal{G}, \circ)$ be a cyclic group of order n. Prove that if $m \in \mathbb{N}$ and $n|m$, then there exists a subgroup of order m.

6.27 Prove that if $(\mathcal{G}, \circ)$ is group of order p and p is a prime number, then $(\mathcal{G}, \circ)$ is a cyclic group.

References

Boole, G. 1847. *The Mathematical Analysis of Logic, Being an Essay towards a Calculus of Deductive Reasoning.* Cambridge-London.

Cantor, G. 1895. "Beiträge zur Begrundung der transfiniten Mengenlehre," *Mathematische Annalen,* **46** 481–512.

Cantor, G. 1891. "Über eine elementare Frage der Mannigfaltigkeitslehre," *Jahresbericht der Deutschen Mathematiker-Vereinigung,* **1**" 75–78.

Cohn, P. M. 2000. *Classic Algebra.* Chichester, UK: Wiley.

Dauben, J. W. 1979. *GEORG CANTOR: His Mathematics and Philosophy of the Infinite.* Princeton, NJ: Princeton University Press.

Derbyshire, J. 2003. *Prime Obsession: Bernhard Riemann and the Greatest Unsolved Problem in Mathematics.* Washington, DC: Joseph Henry Press.

Dunham, W. 1997 *The Mathematical Universe: An Alphabetical Journey through the Great Proofs, Problems, and Personalities.* New York: Wiley.

Dunham, W. 2005 *The Calculus Gallery: Masterpieces from Newton to Lebesgue.* Princeton, NJ: Princeton University Press.

Gilbert, J., and Gilbert, L. 1996. *Modern Linear Algebra,* 4th edition. Boston: PWS Publishing Company.

Gödel, K. 1931. "Über Formal Unentscheidbare Sätze der Principia Mathematica und Verwandter Systeme, I," *Monatshefte für Math. u. Physik,* **38** 173–198.

Kolmogorov, A. N., and Yushkevich, A. P. 2001. *Mathematics of the 19th Century: Mathematical Logic, Algebra, Number, Theory, Probability Theory,* 2nd edition. Basel: Birkhäuser.

Koshy, T. 2001. *Fibonacci and Lucas Numbers with Applications.* New York: Wiley.

Long, C. 1972. *Elementary Introduction to Number Theory,* 2nd edition. Lexington, MA: D. C. Heath and Company.

Miller, J. 2006. "Earliest Uses of Some of the Words of Mathematics" (available online at http://members.aol.com/jeff570/mathword.html).

Nelson, D. 2003. *The Penguin Dictionary of Mathematics.* 3rd edition. London: Penguin Books.

Reid, C. 1996. *Hilbert.* New York: Springer-Verlag.

Ross, K. A. 2003. *Elementary Analysis: The Theory of Calculus.* New York: Springer-Verlag.

Royden, H. L. 1968. *Real Analysis,* 2nd edition. New York: Macmillan.

Seife, C. 2000. *ZERO: The Biography of a Dangerous Idea.* New York: Penguin Books.

Singh, S. 1997. *Fermat's Enigma: The Epic Quest to Solve the World's Greatest Mathematical Problem.* New York: Anchor Books.

Index

A

B

C

Q

R

S

Y

Z

PURE AND APPLIED MATHEMATICS

A Wiley-Interscience Series of Texts, Monographs, and Tracts

ADÁMEK, HERRLICH, and STRECKER—Abstract and Concrete Catetories
ADAMOWICZ and ZBIERSKI—Logic of Mathematics
AINSWORTH and ODEN—A Posteriori Error Estimation in Finite Element Analysis
AKIVIS and GOLDBERG—Conformal Differential Geometry and Its Generalizations
ALLEN and ISAACSON—Numerical Analysis for Applied Science
*ARTIN—Geometric Algebra
AUBIN—Applied Functional Analysis, Second Edition
AZIZOV and IOKHVIDOV—Linear Operators in Spaces with an Indefinite Metric
BERG—The Fourier-Analytic Proof of Quadratic Reciprocity
BERMAN, NEUMANN, and STERN—Nonnegative Matrices in Dynamic Systems
BERKOVITZ—Convexity and Optimization in $\mathbb{R}^n$
BOYARINTSEV—Methods of Solving Singular Systems of Ordinary Differential Equations
BURK—Lebesgue Measure and Integration: An Introduction
*CARTER—Finite Groups of Lie Type
CASTILLO, COBO, JUBETE, and PRUNEDA—Orthogonal Sets and Polar Methods in Linear Algebra: Applications to Matrix Calculations, Systems of Equations, Inequalities, and Linear Programming
CASTILLO, CONEJO, PEDREGAL, GARCIÁ, and ALGUACIL—Building and Solving Mathematical Programming Models in Engineering and Science
CHATELIN—Eigenvalues of Matrices
CLARK—Mathematical Bioeconomics: The Optimal Management of Renewable Resources, Second Edition
COX—Galois Theory
†COX—Primes of the Form $x^2 + ny^2$: Fermat, Class Field Theory, and Complex Multiplication
*CURTIS and REINER—Representation Theory of Finite Groups and Associative Algebras
*CURTIS and REINER—Methods of Representation Theory: With Applications to Finite Groups and Orders, Volume I
CURTIS and REINER—Methods of Representation Theory: With Applications to Finite Groups and Orders, Volume II
DINCULEANU—Vector Integration and Stochastic Integration in Banach Spaces
*DUNFORD and SCHWARTZ—Linear Operators
Part 1—General Theory
Part 2—Spectral Theory, Self Adjoint Operators in Hilbert Space
Part 3—Spectral Operators
FARINA and RINALDI—Positive Linear Systems: Theory and Applications
FOLLAND—Real Analysis: Modern Techniques and Their Applications
FRÖLICHER and KRIEGL—Linear Spaces and Differentiation Theory
GARDINER—Teichmüller Theory and Quadratic Differentials

*Now available in a lower priced paperback edition in the Wiley Classics Library.
†Now available in paperback.

GILBERT and NICHOLSON—Modern Algebra with Applications, Second Edition
*GRIFFITHS and HARRIS—Principles of Algebraic Geometry
GRILLET—Algebra
GROVE—Groups and Characters
GUSTAFSSON, KREISS and OLIGER—Time Dependent Problems and Difference Methods
HANNA and ROWLAND—Fourier Series, Transforms, and Boundary Value Problems, Second Edition
*HENRICI—Applied and Computational Complex Analysis
Volume 1, Power Series—Integration—Conformal Mapping—Location of Zeros
Volume 2, Special Functions—Integral Transforms—Asymptotics—Continued Fractions
Volume 3, Discrete Fourier Analysis, Cauchy Integrals, Construction of Conformal Maps, Univalent Functions
*HILTON and WU—A Course in Modern Algebra
*HOCHSTADT—Integral Equations
JOST—Two-Dimensional Geometric Variational Procedures
KHAMSI and KIRK—An Introduction to Metric Spaces and Fixed Point Theory
*KOBAYASHI and NOMIZU—Foundations of Differential Geometry, Volume I
*KOBAYASHI and NOMIZU—Foundations of Differential Geometry, Volume II
KOSHY—Fibonacci and Lucas Numbers with Applications
LAX—Functional Analysis
LAX—Linear Algebra
LOGAN—An Introduction to Nonlinear Partial Differential Equations
MARKLEY—Principles of Differential Equations
MORRISON—Functional Analysis: An Introduction to Banach Space Theory
NAYFEH—Perturbation Methods
NAYFEH and MOOK—Nonlinear Oscillations
PANDEY—The Hilbert Transform of Schwartz Distributions and Applications
PETKOV—Geometry of Reflecting Rays and Inverse Spectral Problems
*PRENTER—Splines and Variational Methods
RAO—Measure Theory and Integration
RASSIAS and SIMSA—Finite Sums Decompositions in Mathematical Analysis
RENELT—Elliptic Systems and Quasiconformal Mappings
RIVLIN—Chebyshev Polynomials: From Approximation Theory to Algebra and Number Theory, Second Edition
ROCKAFELLAR—Network Flows and Monotropic Optimization
ROITMAN—Introduction to Modern Set Theory
ROSSI—Theorems, Corollaries, Lemmas, and Methods of Proof
*RUDIN—Fourier Analysis on Groups
SENDOV—The Averaged Moduli of Smoothness: Applications in Numerical Methods and Approximations
SENDOV and POPOV—The Averaged Moduli of Smoothness
SEWELL—The Numerical Solution of Ordinary and Partial Differential Equations, Second Edition
SEWELL—Computational Methods of Linear Algebra, Second Edition
*SIEGEL—Topics in Complex Function Theory
Volume 1—Elliptic Functions and Uniformization Theory
Volume 2—Automorphic Functions and Abelian Integrals
Volume 3—Abelian Functions and Modular Functions of Several Variables
SMITH and ROMANOWSKA—Post-Modern Algebra
ŠOLÍN–Partial Differential Equations and the Finite Element Method

*Now available in a lower priced paperback edition in the Wiley Classics Library.
†Now available in paperback.

STADE—Fourier Analysis
STAKGOLD—Green's Functions and Boundary Value Problems, Second Editon
STAHL—Introduction to Topology and Geometry
STANOYEVITCH—Introduction to Numerical Ordinary and Partial Differential Equations Using MATLAB®
*STOKER—Differential Geometry
*STOKER—Nonlinear Vibrations in Mechanical and Electrical Systems
*STOKER—Water Waves: The Mathematical Theory with Applications
WATKINS—Fundamentals of Matrix Computations, Second Edition
WESSELING—An Introduction to Multigrid Methods
†WHITHAM—Linear and Nonlinear Waves
†ZAUDERER—Partial Differential Equations of Applied Mathematics, Second Edition

*Now available in a lower priced paperback edition in the Wiley Classics Library.
†Now available in paperback.

CPSIA information can be obtained at www.ICGtesting.com
Printed in the USA
267996BV00006B/51/P

9 780470 042953